T5-CVD-948

NATURAL SCIENCES IN AMERICA

AMERICAN GEOLOGY

BY EBENEZER EMMONS.

Volumes I and II

ARNO PRESS
A New York Times Company
New York, N. Y. • 1974

Reprint Edition 1974 by Arno Press Inc.

Reprinted from a copy in the University
of Illinois Library

NATURAL SCIENCES IN AMERICA
ISBN for complete set: 0-405-05700-8
See last pages of this volume for titles.

Manufactured in the United States of America

Library of Congress Cataloging in Publication Data

Emmons, Ebenezer, 1799-1863.
American geology.

(Natural sciences in America)
Reprint of the 1855 ed. published by Sprague, Albany.
1. Geology. 2. Geology--United States. I. Title.
II. Series.
QE26.2.E55 1974 557.3 73-17818
ISBN 0-405-05734-2

AMERICAN GEOLOGY

AMERICAN GEOLOGY,

CONTAINING A

Statement of the Principles of the Science,

WITH FULL ILLUSTRATIONS OF

THE CHARACTERISTIC AMERICAN FOSSILS.

WITH

AN ATLAS AND A GEOLOGICAL MAP OF THE UNITED STATES.

BY EBENEZER EMMONS.

VOL. I.

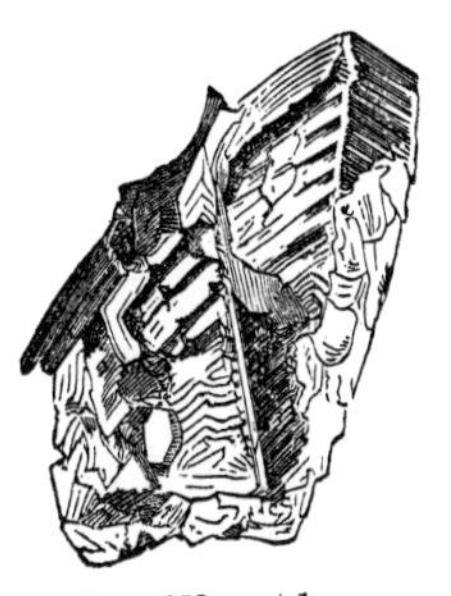

Page 158, part 1.

ALBANY:

SPRAGUE & CO., 51 STATE STREET.

1855.

J. MUNSELL, 78 STATE STREET.

TO THE

REV. CHESTER DEWEY, D. D., LL. D.,

LATE PROFESSOR OF CHEMISTRY AND NATURAL PHILOSOPHY IN WILLIAMS COLLEGE.

SIR:

As one of the earliest Teachers of Geology and Natural History in this country, as a lover and successful cultivator of these departments of learning, as one of high scientific attainments and moral worth, and especially as a friend, the author feels that you deserve a higher testimonal of his regards than the dedication of this work; still, he hopes that it may be received as a mark of esteem, and an expression of his appreciation of your kindness and friendship continued through many years.

Albany, Sept. 20th, 1855.

PREFACE.

The object of this work is to furnish the American student with a guide to the geology of this country. To accomplish this requires a statement of the principles of Geology, derived from phenomena which the student will observe in his field exercises, a statement of the arrangement and composition of the rocks composing the earth's crust, and so much relating to palæontology as shall embrace the characteristic fossils of the sediments. I have therefore pursued a plan which accords with these views.

In the first part I have stated, as it seems to me, the most important principles of geology, and have attempted to apply them in an explanation of phenomena which belong to the primitive crust, composed as it is of the pyrocrystalline formations, and also to the business of mining.

In the second part, its principles are virtually applied to the formation of the sediments, and their succession; and I have proceeded so far as to state their physical characteristics, and to describe and locate the fossils of the oldest sediments known to geologists. There may be a difference of opinion among geologists, as to what extent a work of this kind should be illustrated, in order to carry out its plan. On this question, it has appeared to the author very desirable that all classes of animals should be well represented, not only for the purpose of serving its direct objects, but also to convey information respecting the forms under which life has been clothed in the earliest periods of the earth's history.

It is proper to state, in this connection, that as it respects its palæontological illustrations, the work is more fully stocked than was intended when prepared for publication. This occurred in consequence of the expressed wishes of friends to use the wood cuts which had been made for the illustration of the palæontology of New York, and which had been mostly abandoned, notwithstanding the great expense which had been incurred in their preparation. On application, therefore, to the proper persons, for the use of these cuts, they were readily put in my possession, and have, accordingly, been used in the plates accompanying this volume. There were, it is true, some objections to their use; but it was urged, that though it would not directly increase the stock of geological information, still, it would serve to disseminate or diffuse knowledge, which is one of the great objects of the publication of books. But it may be said that I have exposed myself to the charge of plagiarism, or of servile imitation. It must, however, be well known, that I have labored in the same field, that I have aided in the development of the palæozoic riches of the Silurian rocks, and especially of the Lower Silurian, and that a large proportion of those given in this book were brought to light by my own labors, or by my friends, in the district assigned me in the Geological Survey of New York. But in making the foregoing claim as one founded in right, it is still proper to acknowledge the use of several plates, made up by the state palæontologist,

Mr. Hall, who, I am cheerful to acknowledge, has acquired an enviable reputation in his department.

In regard to the Taconic system, I do not know that I am indebted to any one for favors, or for suggestions. Indeed, nothing very flattering has ever been said, or published, respecting the views I have maintained upon this subject. But I have the pleasure of knowing that a few of my friends, and those, too, who have the best opportunities for judging of facts, if they do not agree with me in every particular, still agree with me in the main, respecting the relation of the Taconic system to the Lower Silurian. It is not for me to express an opinion how this question will be received by geologists now, or whether I have stated my views in the form best calculated to establish my positions. Geologists of higher abilities than myself would probably have succeeded better in placing the evidence on record; still, the most important facts will be found in the work; and, ultimately, they must exert an influence in favor of the doctrines I have stated; for I have full faith in them, and believe that they are founded upon established principles, and can not be set aside.

Having been familiar with the persons who have been employed in the New York Geological Survey, and well acquainted with its history, I feel that some acknowledgment is due to one who retired from the survey at a comparatively early stage. I allude to T. A. Conrad, the first palæontologist of the state, who, I believe, laid the foundation for a correct knowledge of its palæontology; and who brought to the department a profound knowledge of the lower forms of organic remains. His services, though unacknowledged, still deserve to be remembered and held in high estimation, and this reminiscence is only a partial acknowledgment of what he accomplished while connected with the survey.

Probably no one has furnished more important material for illustrating the palæontology of the Silurian system, than John Gebhard, Esq., who devoted many years almost exclusively to this subject, in one of the most interesting fields in the state; and who is now deservedly placed over the state collection, as its Curator. It is probably due to this gentleman's labors, that palæontology acquired so much interest and favor.

In the arrangement of matter, and also for a corrected list of fossils belonging to the Lower Silurian, I am, in a great measure, indebted to M. d'Orbiny, who is regarded as one of the ablest palæontologists in Europe. It may happen, however, that when this distinguished savant has depended solely on figures, for the determination of fossils, he may have been led into mistakes. Certainly, this is the fact in one instance, viz: that of his reference of the Discophyllum, described in the first volume of the Palæontology of New York, which is by no means a stony coral, belonging to the Cyathophyllidæ; it is rather a membranous polypi, which may be allied to the Graptolites. In the Brachiopoda, I have followed Davidson, who has given a most admirable monograph of this order, in the Transactions of the Palæontographical Society.

It will be observed that I have added a few new species to the list of fossils of the Lower Silurian rocks, in addition to those belonging to the Taconic system. These additions have been made only in cases which required them; or, at least, where descriptions and figures of them could not be found.

While executing this work, I did not expect that I should be able to secure perfect accuracy, but I am able to say, when the volume is finished, that I have put in requisition all the means within my reach, to secure an end so desirable. As it is, I submit the work to the student with the hope that it will prove a useful guide in the study of American Geology.

EBENEZER EMMONS.

Albany, Sept. 20th, 1855.

CONTENTS.

PART I.

PART II.

AMERICAN GEOLOGY.

PRELIMINARY REMARKS.

§ 1. The science of geology is of recent origin. The first attempts which were made towards the construction of a system, date no farther back than the middle of the last century. In its progress it has undergone many changes, as has every other science dependent upon observation and experiment.

The object of geology is to give a rational explanation of the structure of the earth. To accomplish this, it examines the phenomena presented at the surface of the earth, and its interior, where it is accessible; and it attempts to discover the causes of those phenomena, and to find the true reason for their existence, and also to fix the dates when remarkable changes occurred. The advantages resulting from the study of geology are numerous. It gratifies a laudable curiosity; it informs us where we may find the most valuable natural productions, as coal, salt, iron, gold, silver, manganese, copper, lead, marble, and many other useful substances; it enlarges our views of the field of nature; it enables us, by our knowledge of the present, to look far backward into the past; it reveals to us a vast duration whose limit we can not fix—a succession of changes in the physical condition of the earth, which exhibit a progress towards an ulterior end which seems to have had reference to the existence and well being of man. We see in the earth's changes, and its brute inhabitants, a progressive movement along an upward scale, not in a direct track, but rather in the ultimate results. It teaches us that order has prevailed in the operations of the natural elements through the lapse of ages—

that there was a plan in the divine mind which has been working with a special reference to the good of our race; and lastly, that the plan of creation, and its scheme of construction, belongs to but one system, however far we may go back into the past. All our observations respecting the past and present lead to the conclusion that the plan of creation is one—that the laws and forces which are now in operation have been the same from the beginning: therefore, the true method for an interpretation of the past, is by those laws and forces which govern the present.

§ 2. Our knowledge of the earth is confined to the earth's crust, by which we mean to include all that part which is accessible to human observation. This part is the theatre upon which geological events have been acting from remotest periods, and still it is safe to draw inferences respecting phenomena belonging to the deeper seated parts, provided they are in accordance with established principles, or with what we know.

§ 3. The earth's crust is composed of rocks, in which term geologists include not only consolidated materials, but sands, clays, soils, and fluids. Strictly speaking, the earth's crust is composed of rock and water. We might perhaps reckon also the free gaseous bodies confined in caverns, which, under favorable circumstances, escape into space, as atmospheric air, carbonic acid, nitrogen, and ammonia; or they may be regarded as things contained in the crust, and as agencies through whose force and power the solid crust has changed its phases in time and its position in space. Heat should be added to the foregoing; it operates *per se*, and gives activity and life to the liquids and gases which permeate the crust and fill its empty spaces.

§ 4. The monumental records of the past are of two kinds, the physical and the organic. To the former belong the impress of the movements of the earth's crust upon itself, or upon the different strata which were deposited in different periods; to the latter, the preservations of plants and animals. Their remains occur in groups, and represent the forms of the different periods. No two periods are represented by similar groups, and their dissimilarity is practically important, furnishing the facts by which the periods themselves may be distinguished from one another. It is also an interesting feature in a historical sense, proving by comparison a progression in development along the measures of an ascending scale. The periods as they approach the present are represented by plants and animals more akin to the living, while in the more remote their resemblances as a whole are less. Notwithstanding this, the four types of the living are represented in the successive periods, with the exception of the first.

§ 5. Prior to the creation of plants and animals there was a period very clearly marked by the reign of physical forces; it is azoic, and the rocks of this period have no parallel—they are all crystalline. Heat was the predominant and active element. Stability and form was given to the earth in this period, principally by the escape of heat into space; and condensation of aqueous vapor upon a cooling crust, gave origin to the surface waters of the globe. Seas and oceans were formed in all the basins and great depressions, and the culminating points gave origin to streams which flowed oceanward; but the oceans, ere the waters had filled their bosom, began to lose their contents by the vaporization of this new element. When it had saturated the atmosphere, it fell again to the earth. Thus began that vast machinery by which the earth is supplied with rain and dews. It has known no suspension. The vapors rise upward, and the streams flow onward in perpetual cadence. At this stage the consolidated crust begins to wear, and its debris also begins to be transported and borne onward, and its progress is only arrested by the plains and depressions, where it accumulates, giving origin to another class of rocks whose parentage is indicated by rounded particles and masses. The activity of fire diminishes—that of water increases. The two forces are antagonistic of each other. One levels the surface, the other breaks it up; the first is constant in its action, the latter is paroxysmal. The fire slumbers longer, but never expires. It has retired more inward toward the earth's core, and it is girt about by stronger and stronger bands; but it yet preserves its outward vents, and often warns us of its power in the trembling of its bands and the molten rocks which flow from the fire chasms which it opens.

The interpretation of the varied phenomena to which we have just alluded, must be in accordance with human experience and observation. Observations turned to those phenomena which belong to each of the periods in the earth's history, prove that the plan of creation—the ideas which that plan expresses—is a unity. It proves more than this, that the present is only a part of the past, and belongs to it. As the whole of a thing is made up of its parts, and is imperfect in the absence of one, so the present is imperfect without the past, and the past would be imperfect without the present. The interpretation of the present is perfected only by reference to the past, and the past would be unintelligible without the present. Each period then is a fragment; but the present is a greater fragment than all the past put together. The present does not date its beginning with man, but with the earliest species of plants and animals which now live, and whose primordial forms are not as yet extinct.

§ 6. The life of plants and animals is controlled by a single element; that is oxygen. The adaptation of organs is in accordance with its properties. It has always been so. We presume, too, that its supply has been constant—that there has been no period when its quantity was either greater or less than it is now. A different view is not sustained by the fact that at one period huge lizards predominated in its fauna, for analogy proves that this class would have perished with a less proportion than that which exists in our atmosphere at the present time.

In cases of this kind reasoning from special structural affinities to general physical conditions is not always safe. The lias and oolitic periods abound in the remains of cold-blooded animals, whose respiratory apparatus was undoubtedly imperfect, like the lizards of the present. It is a favorite inference with a certain class of the progressive geologists, that the oxygen of the atmosphere at that period was less in proportion to its mass than it is now. But then may we not inquire, if there was less oxygen in the present atmosphere than there actually is, could the membranous reptile lung supply the demands of the system; and is not the constituent proportion of oxygen the quantity required to give the creature the power to breathe at all? I say it is not always safe to reason from structural affinity to physical conditions. If we take any other organ, as the eye, and draw from its structure and condition analogous inferences concerning the quantity of light, we may see where it will lead us. A class of progressive geologists, maintaining that in the early periods of life the light of the globe was dim, and that but few rays shone through the hazy atmosphere, find in support of this doctrine the fossil remains of a fish or a lizard with enormous eye sockets. He believes that the large eye was adapted to a dim state of the atmosphere. Another person finds a fossil with very small bony sockets. In this case, too, it may be said the eye was very small, and hence it was adapted to an exceeding intense light—to the sun when it shone fiercely from its throne in the heavens. But again, a fossil is found entirely destitute of an eye socket, and not a vestige of an organ of vision can be found; hence there was a time when the earth was shrouded in darkness, for in darkness animals have no need of eyes, and light would be useless to animals destitute of the visual organ. But then we find all these states of the eye in the present arrangements for supplying the world with light. The Pomatomus telescopium, a fish of the Mediterranean, which lives in very deep water, has a remarkably large eye. It is the position which it occupies that requires the large eye, and that large eye is adapted to its abode; and if only one-half of the light of the sun was extinguished, it probably would be unable to see at all. And just so with the reptile, if one-half of the oxygen of the atmosphere was withdrawn from it, all reptiles would die. The mole has a very small eye, but that small eye would be of no use if the ground was lighted by fewer rays. The blind fish and the blind animals of the mammoth cave live in a period when the earth is lighted up most gloriously: their abodes are dark, but yet the sun shines without in all its strength. We find, then, all conditions of the eye and the lungs at the present time. But it does not follow, that because the structure of the fossils of a given period may be found whose organs belong to a certain type, that the physical conditions of the earth were materially different from what they are now. This view of the subject does not conflict with the doctrine of adaptation, but rather sustains it. The physical conditions are first established; the organic kingdoms afterwards come in with their separate adaptations. The organisms are perfect in their adaption to the conditions in which they are to live, as well as to the position in space, and the mediums in which they are to be placed. The doctrine of progressive development, as usually represented, seems to be untrue. It proceeds on the ground that the earliest beings were the least perfect, and that progression consisted in the creation of those animals which were more perfect in their structures than their predecessors. But who can not see that the world is full of the same imperfections in animals now as in the beginning. Progression has no reference to perfection of structure, but to rank. Structures have been always perfect, but rank has been progressive.

§ 7. Water acts upon the earth's surface in many ways. It is a solvent. Temperature and pressure modify this property. Pressure and temperature combined increase it to an indefinite amount. It also acts mechanically. It permeates the solids and penetrates the fissures of rocks, and in the cold regions divides asunder the particles and masses in freezing. It is a carrier or transporter of the divided matter. Flowing in streams over the surface, it carries along from the higher to the lower levels the broken-down matter. When moving in masses, in the form of tides and waves of seas and oceans, it bears away and moves to distant points the mud and sand committed to it; as a body moving in great ocean rivers, as the Gulf stream, it also bears these materials forward to certain and well determined regions. It thus arranges and moulds the ocean's bottom from the matter committed to it by the terrestrial rivers. By mechanical force the waves break down the strong rocky barriers of coasts, as well as those shores which are girded with sand. The ocean then is a moulder and distributer of all the plastic matter committed to its bosom. In fine, moving water, under whatever name it has received, whether rivers, waves, tides, or ocean currents, is both destructive and constructive, according to the conditions and circumstances of the moving mass.

The changes of temperature which a country undergoes, and the amount of rain which it receives, produce important changes in the physical condition of its surface. The observations which have been hitherto recorded are however too few to become the basis of important geological reasoning; yet they are sufficiently so to require some notice in this place. The following facts are recorded in the periodicals of the day, and are among the most important of this class:

Latitude.	Place.	Mean an. temp.	Rain in inches.
	Huntsville (9 years),		51·13
	Natchez (8 years),	64·76	
	Columbia, S. C.,	56·8	49·90
	Washington, D. C.,	56·57	
39° 56′	Philadelphia,	53·42	
39° 06′	Cincinnati, 510 ft. above tide water,	53·78	
41° 14′	Hudson, Ohio (3 years,) 9 A. M.,	48·7	37·63
	(The amount of rain varies some nine or ten inches.)		
	The average at 3 P. M.,	55·6	
	Trenton, N. J.,	48·22	
	For six summer months,	60·60	
	Winter months,	35·79	
	Prevailing winds S. W.		
	State of New York—		
40° 37′	Flatbush; L. I.,	51·25	
41° 30′	Newburgh (17 years),	48·96	35·54
	Albany, do	48·27	39·91

Latitude.	Place.	Mean an. temp.	Rain in inches.
	Rochester, 620 ft. above tide water (11 years), ...	47·48	27·16
	Barometrical mean,. 29·56 in.		
	Range,............ 1·34		
	Malone, ...	52·66	
42° 42′	Penn Yan (16 years),................	46·87	27·8
43° 6′	Utica 437 ft. above tide water (5 years),	46·35	44·12
	Barometrical mean,. 29·64 in.		
	Range,............ 2·03		
	Syracuse—barometer (400 feet), ... 29·33		
	Range, ... 2·26		
	Salem, Mass. (33 years), ...	48·65	
	Williamstown, 595 ft. above tide water (23 years), ...	45·59	37·48
			Snow in inches.
	Montreal (3 years), ...	41·07	65·85
			Rain in inches.
	do		18·60
	Warmest day, ... +90		
	Coldest, ... —13		
46° 47′	Quebec, 340 ft. above tide water, ...	37·19	
	Detroit, ...	52·66	32·79
	Brunswick, Me. (11 years),..........	.....	37·15
	Mean temperature for latitude of 41° 43′,	49·75	
	do do do 42° 43′,	48·15	

Decrease of temperature by elevation, one degree for every 325 feet, for New York.

Mean temperature for the state of New York, reduced to the standard of Albany and level of the sea, 48·95.

Mean quantity of rain, in inches, 39·55

Mean quantity of rain for places near the lakes, and western courses, 24·52 inches. Thus Lewiston has only 20·40 inches of rain, Ogdensburgh 24·61 inches, and Rochester only 28·69 inches.

The area of the state of New York is 48,000 square miles. I have no observations which show how much of the rain evaporates, and how much passes off in drainage. The tributaries of the Ohio river* rise over an area of 24,337 square

* Smithsonian Contributions, by Charles Ellet, jr., Civil Engineer.

miles. Here the total annual fall of rain is approximately thirty-six inches. Forty per cent of this quantity passes off in the drainage of the branches above Wheeling; sixty per cent is evaporated, or is employed by the vegetable kingdom. The average discharge of the Ohio at Wheeling for six consecutive years, was found to be 835,323,000,000 cubic feet. The quantity of solid matter which this quantity of water holds suspended in a cubic foot, is approximately $\frac{1}{14600}$ part.

The quantity of rain which a country annually receives is connected with the amount of degradation which that country is undergoing; and the amount of matter dissolved out of the exposed surfaces of limestone and other rocks, is also related to the quantity of rain which flows over its surface. The quantity of carbonic acid and ammonia which is required to confer fertility upon a country, stands connected with the number of inches of rain with which that country is supplied. When observations have been made upon the quantity of rain which a given area receives annually, together with the amount of sediment which drainage carries away, it will be possible to form an approximate calculation of the rate the degradation is going on, as well as the rate at which the valleys are filling up. There are also many other problems which will receive a solution when the related facts or data shall have been obtained.

Questions of a practical kind, and which are closely related to geology, are constantly arising; and upon their answer some of the most important interests of society are involved. For example, our rivers overflow their banks, and inflict heavy damages upon private and public property. Can any practical scheme be devised, by which these injuries shall be avoided or prevented? Can those streams be controlled so far as to render their swellings harmless. We have a natural illustration how nature sometimes counteracts her own evils: the Androscoggin, when in flood, flows through a short cut into Umbagog lake. Hence, its waters being partially diverted for a time, finally pass down to the ocean harmlessly. In a valuable treatise on the physical geography of the Mississippi valley, by Mr.

2

Charles Ellet, he proposes to put in execution an analogous plan, to distribute the waters of the Ohio in a more equable manner by means of reservoirs, so as to preserve a given quantity on its bar at Wheeling. It must be noted, that in the execution of all similar projects certain geological results will follow. The detritus will be arrested in artificial basins; the stream, in its onward course, will be freed in part from sediment; the accumulations which have been collecting at the mouths of rivers and in shallow ground, will be diminished in quantity, and their nature somewhat changed. The changes which improvements in navigation by dams, by diversion of streams by canals, are and quite important. These changes are not confined to the sediments. Certain species of fish become more widely distributed by means of channels of communication being opened between the lakes in the interior of the country and its coasts. The proteus of lake Erie has found its way to the Hudson river by the Erie canal, and the different species of limneas and unios now occupy its bed throughout its whole extent.

§ 8. Notwithstanding the great extent of land, the North American continent is well watered. This is especially true of the United States. Situated between two oceans, it has a breadth of 2500 miles of land. The great western lakes are inconsiderable areas compared with the wide interval between the oceans. From the gulf of Mexico to the great lakes, or country of lakes, it is 1200 miles by the shorter route.

Notwithstanding the great area of unbroken soil, the distribution of the forces which supply water to 2,500,000 square miles is such, that the whole country can be traversed and cultivated. It is true that at the base of the Rocky mountains there is an arid country—one too dry to be inhabited. The westerly winds which are known to prevail in this country, are deprived of a large proportion of their water by the ranges intervening between this dry country and the Pacific ocean.

Taking the whole globe into view, we may learn by an inspection of its map, that land and water are unequally distributed. As it regards the area of land, it is found that the northern hemisphere contains three times as much as the southern; and as it regards the expanse of water, its superficial area in the southern exceeds that in the northern hemisphere.

§ 9. The Atlantic ocean is prolonged north and south so as to extend from pole to pole, while its breadth does not exceed 5000 miles. Its depth has been stated at about three miles; its area is 20,000,000 square miles. The Pacific ocean is prolonged from east to west. If measured on a line extending from Peru to the eastern coast of Africa, it is 16,000 miles. This great expanse of waters contains 70,000,000 of square miles, exclusive of the areas which are occupied by its islands. Its depth is four miles; but many points have not been fathomed even with lines six miles in length. The sounding in all waters, whether oceans or seas, or inland fresh and salt water lakes, demonstrates that their bottoms possess all the diversities of surface as the land, sinking in many places to profound and unfathomed depths; in others, banks and terraces spread out far and wide. These banks or terraces are probably made by the joint operation of the waves and of submarine and superficial currents, which are common to all great bodies of water.

The area of dry land does not exceed 35,500,000 of square miles. Its mean elevation is about 1000 feet; hence it follows that the entire surface of dry land may be covered with water. The great disproportion of dry land to water is a provision which is necessary to the well being of plants and animals.

§ 10. If water covers four-fifths of the earth's surface, it is evident that its influence, as a geological cause, should not be overlooked. The North American continent being skirted by two great oceans, and being supplied also with large inland lakes, and the largest and longest water courses in the world, we may expect to find those phenomena which are due to aqueous action upon the grandest scale; while the other element, volcanic fire, seems to be so far exhausted in its power in the United States, that it is impossible to obtain specimens for laboratory illustration. The Atlantic coast is remarkably indented. The coast of Maine, Massachusetts, and Connecticut is deeply gashed and serrated. The southern coast and shores are penetrated by deep bays. The Chesapeake and Delaware are two of the most important. The inshore sea, which has received the name of sound, is another feature of the coast which may properly claim attention. Long Island, Albemarle, and Palmico are the largest. These sounds have inlets which are liable to be closed by coast storms; or on the contrary, new inlets may be formed by the action of winds and waves upon a sandy barrier. For this reason, the sounds upon the coast of North Carolina vary much in the amount of their saline matter; for this reason, too, they undergo changes in their marine faunas. No fact, however, so conclusively proves the variable condition of the ground occupied by these sounds, as the fact that their bottoms are everywhere studded with the stumps of the common pines of the country. The grounds which may be selected for fishing, require the removal of these stumps by gun powder before a net can be drawn. At the first view it may be inferred that the bottoms of the sounds were dry land but very recently; but the stumps of pine, when immersed in water, are almost imperishable, lasting for centuries: still, geologically speaking, the pines belong exclusively to the present, as they are evidently the same species of pines as those which now live upon the coast.

The coast is protected by belts of sand, which in time support a stunted vegetation, and admit of pasturage for mules, horses, and sheep. The horses which run wild upon those semi-deserts belong to the pony breed; but they are tough and hardy. They invariably refuse corn when first taken. The sand reefs and barrens are entirely due to the action of winds and waves. The sands are constantly accumulating along the coast line. Portions of wrecks, fish-spears, coins from wrecked vessels, wash upon the beach after due time. These facts illustrate the action of the waves. This coast, by the incessant action of its waves, has traveled eastward two hundred miles since the Eocenic period. Cape Hatteras moves in advance of the general coast line. The inclination of the Eocenic plain is equal to one foot per mile. It extends inward to Raleigh, which is two hundred feet above the sea level. Upon the Atlantic coast we learn the nature of the action of water moving in wave masses. The slope of the beach is gentle to the surf line; here the bank steepens, and the crested wave rotates vertically upon itself, giving origin to the ground wave or undertow, while a portion shoots forward in thin sheets, rippling the sands over which it flows. Upon the Carolina coast this action is mainly constructive.

§ 11. The constructive action of water is equally manifest in the formation of shoals. The tide-wave, which travels northeast, transports detritus, which is deposited at any point where an obstruction lies in its way. A portion of a wreck is sufficient to form a shoal.

§ 12. It has been said already that the bottoms of oceans and seas are not spread out in level plains. They have all the diversities of dry land, rising in some places into mountains and hills; in others, sinking into deep valleys. It is in these deep valleys that sounding lines fail to reach the bottom. Extensive and comparatively level banks exist, where the water has only a moderate depth. The Atlantic's shore is skirted by extended ridges, which are formed or moulded by the joint action of tides and waves.

§ 13. Animals live upon the ocean's bottom; but different kinds inhabit it at different depths. They are rarely found living below the depth of 180 feet. Vegetables grow in the ocean. They can subsist at the depth of 300 feet. The most favorable positions for animals and plants are near the shore, where the water is comparatively shallow. In deep water they select the slopes of ridges. The summits are generally avoided on account of the disturbance by waves. Great pressure, and the absence of light in deep water, are unfavorable to life; and the more profound abyses, like the heights of the Himalaya and Andes, are dreary wastes—the one from its darkness and pressure of the superincumbent water, the other from its excessive cold and thin atmosphere. The causes which are now operative in excluding animals and vegetables from deep sea bottoms and high mountains, have also been operative in all periods of the earth's history.

§ 14. The earth's surface did not receive its present configuration at its creation. Its mountains and valleys had no existence in the original constitution of the globe. Even its highest mountains, the Himalayas, have been raised long since animals and vegetables were created; and oceans but lately rolled over lands which are now the highest points of continents. Powerful agents have therefore been operative in these changes. The most effective of these agents are water and fire. Water is operative in many ways. It is a solvent of many of the materials composing the earth's crust. The rocks are dissolved by water. It is the most effective when aided by heat and pressure. In the deep parts of the earth's crust, where there is both heat and pressure to aid it, silex is dissolved; and if it rises to the surface in a heated state, and there cools, its silex is deposited upon the soil or rocks, as at the Geysers of Iceland. This deposit is called *siliceous sinter.* Cold water readily dissolves carbonic acid, which is diffused in the atmosphere and soil; and cold water, aided by carbonic acid, dissolves carbonate of lime and other carbonates, together with the oxides of iron, manganese, and phosphate of lime. Spring water, which holds them in solution below the surface and under pressure, deposits them at the surface. Incrustations of lime and porous beds of it, are formed around these springs. These deposits are called tufa or travertine. It should be noted that travertine is a rock formed on the dry land—it is a subaerial formation. Ochrey iron ore, mixed with carbonate of iron, and manganese are also deposited around springs, and upon dry land, under similar conditions. All these are subaerial deposits, and should be distinguished from the subaqueous. Water percolating through the soil, holding in solution carbonic acid, dissolves carbonate of lime and iron, which it deposits on the coarse and fine materials, when they become cemented together. They are then less porous in structure, and consequently less pervious to water. The cemented stratum is often called hard pan. When impervious, or only partially so, water is retained too long in the soil for profitable cultivation. Beds of gravel and pebbles are also cemented under similar conditions. These are called pudding stones; they are also subaerial formations, and should be distinguished from conglomerates, which are subaqueous. Limestone, permeated in the same way, yields to the action of water. A shelving ledge, or the roof of a cave below, is often hung with pendent masses, like icicles from the eaves of a house. These are called stalactites. Their formation begins with a deposit of lime in the form of a ring, which is gradually prolonged by additions from the water. A part of the water drops to the ground or floor, and there forms another deposit of carbonate of lime. This is called stalagmite.

Numerous instances, illustrating the agency of water in the mode I have stated, are found in all parts of the Union. Thus stalactites and stalagmites occur on a large scale in all the great caves of Kentuckey, Tennessee, and Virginia; also in the smaller caves of New York in Albany and Schoharie counties. Tufa is deposited from numerous springs in Onondaga, and other western counties in New York. The hydrous peroxides of iron and manganese are derived from mineral springs formerly existing in the tertiary formations of the southern states, as well as in the more ancient rocks of the primary belts. Silica is often separated from its solution in hot water and from steam, and has furnished the siliceous sinter surrounding hot springs in almost all regions of the globe. Amethysts and coatings of chalcedony upon common quartz in Nova Scotia and the trap ranges of our country, and upon crystals of calcspar in Edwards, N. Y., were the products of hot water holding this substance in solution.

But water is more eminently a solvent for those bodies which have taste, as common salt, alum, copperas, &c. Extensive rocks are known, which consist of nearly pure salt or chloride of sodium. The elements of common salt exist in many of the stony matters upon the earth's surface. Chemical action has separated these elements, and united them again so as to form this compound. It is dissolved by the streams, and transported to the ocean, which has become salt by small additions from time to time. The water of the ocean which is carried off by evaporation is fresh; and hence, while water continually flows in charged with a little salt imperceptible to the taste, and none is carried off, it will finally become salt. In shallow seas or bays in warm climates, where evaporation is rapid, salt is rapidly formed; and though it may be mixed with mud, yet when it crystalizes, as it always will, it becomes pure. Layers of crystals are successively formed under favorable conditions, and these united form thick beds of salt. The quantity of salt in the oceans and seas is enormous, and long periods must have passed before it became perceptible to the taste. The small quantity of salt in the water in the early periods may have modified the forms both of animals and vegetables. Brine springs may originate from beds of rock salt, or water in the deep parts of the earth's crust, and by the aid of chemical affinity may form common salt from its elements. These elements are known to exist there; and the water being charged with salt, rises to the surface as brine springs, or else the brine may remain below in reservoirs until they are reached by the industry of man. The brines or salines of New York seem to be formed in this way; that is, from the elements of salt which exist in the rocks in other combinations.

Water, then, acting as a solvent, produces many important changes upon the earth's surface. It dissolves and consolidates rocks when aided by carbonic acid, heat, and pressure. It dissolves and transports the saline bodies to the common reservoirs, where they are concentrated by one of the natural processes—that of solar evaporation. It is a result of the utmost importance to the well being of man.

Bodies of salt water are never formed so long as there is a supply of rain to maintain a drainage to the ocean. If, however, the rains only supply sufficient water to fill a basin, and there is no discharge or surplus water, the constant evaporation of fresh water concentrates the saline matter of the lake, and in time it becomes salt. The great salt lake of Utah became saline by evaporation of its waters, the evaporation and drainage of the valley being sufficient to equalize each other.

There is a high probability that the saltness of the seas and oceans had attained a large amount of saline matter prior to the palæozoic period. The silurian system furnishes brine springs, and the fossils of this early period indicate that they were the inhabitants of the ocean. The vast quantity of salt in the oceans of the globe indicates also the lapse of long periods during which the saline matter was accumulating.

Water separates the parts of rocks from each other. This is the result of congelation; this, often repeated, ends in comminution or disintegration. Soils are comminuted or pulverized rocks. The process of comminution is more rapid on the tops of mountains. Slaty, schistose, and jointed rocks favor this result by admitting water between their lamina and joints. The broken and comminuted rocks are thus prepared for a removal to a lower level. The small but rapid streams first take upon themselves this office. A part of the disintegrated rock remains upon the tops and sides of mountains. It bears a scanty herbage, whose roots confine it more securely. The streams lose a portion of their burthen at all the levels, where small meadows are formed, and where grass springs up. The streams unite in plains below, where a rich vegetation is nourished, and where climate favors the organic kingdoms. The united mountain streams form rivers, which flow oceanward; but before they reach the great reservoirs, the tides check their currents. Here deposits are again made. Shoals are the uniform results of the meeting of river currents with the tidal wave. But the return tide favors the river current, and its detrital matter, which it has borne along, is delivered up to the ocean wave. The quantity of earthy matter which is thus transported, varies with the season. The Missouri is always muddy, and a thick deposit subsides in vessels in which it stands.

From the foregoing facts it follows that rocks are now forming. The muddy sediments, or the sand and gravel, will in time consolidate. There are, however, no additions made to the earth's crust: the gain of one place is at the expense of another. The contribution which the mountains make to the plains, and to the ocean's bottom, is from their tops and sides; and though the plains are raised, the mountains are lowered.

§ 16. Water moving in masses, as in waves, modifies the character of the earth's surface. It is upon and near the shore that waves produce the greatest modifying effects. The long swell of waves, as they break successively upon a shore, raises the mud and fine sand, and bears it onwards to the land. The material, whatever it may be, will accumulate in ridges, which have a steeper slope on the land than upon the ocean's side. But ridges are formed also upon the shores of lakes on a scale commensurate with their size. As examples of ridges upon a large scale, I may cite Long island and its parallel outer oceanic ridge, and the ridge upon the south side of lake Ontario. These ridges are formed by the waves which bear some of the sand to its crest, when it falls over it to the land side. This is a permanent addition to the ridge. The ancient shores of lake Champlain, in Clinton county, may be traced by similar ridges of sand and gravel. There are no less than four nearly parallel ridges, which mark the former positions of the lake. These show that the land for long periods occupied positions at lower levels than at the present time, and that these changes of level have occurred since the drift period. There is generally towards the land a sheet of shoal water, which becomes a marsh, and which in time may be filled up.

Water moving in mass, in the form of tides, modifies also the earth's surface. The tide entering the mouth of a river flows up its channel, but meeting the down current of the river, it checks its flow. Some of its burthen of mud falls to the bottom, and forms a bar or a shoal. The outward tide, however, receives the rest of the burthen, and bears it oceanward, some of which may be carried to the ocean rivers. The tide of our own coast flows northward, and carries its burthen onward; and at every opening mouth of a river it receives accessions, which, as a common carrier, it bears along, or gives it a new distribution. The conjoint actions of waves and tides are not only to carry, but break down, the bars and ridges which they have formed across inlets and bays. Bays and inlets may have been shut off from the ocean for centuries, or until their waters become fresh, and peopled with fresh water tenants, both from the animal and vegetable kingdoms. Again a high tide, accompanied with high waves and winds, breaks down the bar, when it is at once filled with salt water. The animals and vegetables of fresh water die, and are replaced by the marine. Such alternations are well established facts. These views are illustrated by the changes in Albemarle and other sounds upon our southern coast.

Water moving in masses, as in ocean rivers, modifies the present surface of the earth. These great rivers are only carriers; they take what is committed to them, but they do not furnish the matter they carry. The Gulf stream is an ocean river. The great terrestrial rivers, as the Amazon, La Plata, Orinoco, Mississippi, Ganges, and Brahmapootra, are the contributors to the ocean rivers. The density of sea water aids in the wider distribution of matter committed to it. The great river, called the Gulf stream, originates in the Atlantic. Beginning near cape Horn, it is there divided: its main current flows down the western coast of America, but it turns suddenly to the west, and is lost in the great equatorial current of the Pacific; it then crosses the ocean in the parallels of 26 and 24 N. It is 350 miles broad when it impinges upon the coast of China, the eastern peninsula, and islands of the Indian archipelago, where it is again divided. A portion is deflected to join the great equatorial current of the Indian ocean, when it is impelled by the south-east trade wind: it however maintains a westerly course between 10° and 20° parallels of south latitude. It is divided a third time by the island of Madagascar. One part bends round its northern end, and flows through the

Mozambique channel, and before doubling the cape of Good Hope is joined by the other stream, when it flows outside the Aguillas bank. It there takes the name of the South-Atlantic current. It now flows up the west coast of Africa to the parallel of St. Helena, where it is deflected by the coast of Guinea. It now forms the great Atlantic equatorial current, and flowing westward, it splits upon cape St. Roque. One stream flowing along the eastern coast of South America, its force is finally spent in southern latitudes before it reaches the straits of Magellan, except a single branch, which is deflected to the cape of Good Hope. The other branch, or the great equatorial current, flows northerly along the coast of Brazil with great force. It encounters the river currents of the Orinoco and Amazon, and yet it speeds its way to the Carribean sea and gulf of Mexico. It here gets great accessions of heat, its temperature rising to 88° 52′. It now flows up the coast of North America under the name of the Gulf stream. It is deflected to the east by Newfoundland. It is again divided into several streams: one stream flows towards Britain and Norway, and being still farther divided, a branch flows onward to Spitzbergen, where it mitigates the severity of its climate. From Newfoundland a branch strikes off for Baffin's bay. Another branch also takes a sweep southwards to the Azores, and being aided in its march by the north trade winds, it rejoins the great equatorial current. In consequence of this great bend, a vast expanse of water is left nearly motionless and stagnant, in which sea weed, trees, bodies of drowned animals, float for a time, when they are finally cast upon the shores of the Azores. The greatest velocity which the Gulf stream acquires, that of seventy-eight miles in twenty-four hours, is when it leaves the Florida straits. The rapidity of its current is variable.

Currents also flow from the antarctic and arctic circles. These bring with them icebergs. The arctic brings down icebergs to the latitude of Newfoundland, and even to the Azores. In these latitudes they are melted in the Gulf stream. In consequence of the meeting of the floating masses of ice with the warm stream and warm atmosphere, fogs are generated, which obscure the air for great distances around. Another cold current from the Arctic sea flows inside of the Gulf stream down the coast of North America. It lowers the temperature of the coast. This effect is unfavorable upon New England, but favorable to the southern states situated upon the seaboard. These great ocean carriers, wherever they impinge upon the bottom of the sea, or upon a shore, deposit a part of their burthen. The banks of Newfoundland, St. George's bank, Sable island, have been made by contributions from the Gulf stream. While it distributes these burthens of matter, it is also a great distributer of heat. The climate of Europe and of the northern seas is mitigated by this great current carrying a warmer water than the surrounding ocean. As westerly winds prevail upon the Atlantic, the coast of New England receives only a small amount of the heat of the Gulf stream.

§ 17. *The Atlantic tidal wave.* The ocean, under the influence of astronomical forces, is acted upon in mass, and its waters rise in advancing or retreating waves, according to the position of the sun and moon, which are the positive agents in effecting these movements. By the rise and fall of the wave the detrital matter is borne forward in the direction of the advancing undulation. The great tidal wave of the Atlantic moves from south to north along the eastern coast of the United States, impinging upon its irregular shore, by which the inward terminus of the wave is retarded. It flows up the bays and harbors, encounters the river currents, which it retards, and in consequence of which the heavier portions of their burdens are deposited. In its onward progress to the north it is deflected to the east by the subaqueous banks. The easterly direction is continued till it passes the eastern soundings of Newfoundland; it then resumes its northerly course. To the interposition of an irregular coast, causing a retardation of the free forward movement of the advancing wave, great offshore and bank deposits are due; the latter of which are well known as the George's and Newfoundland banks, Sable island, &c.

The line of soundings and the sand deposits lie in the general tide wave which washes the shore of the continent, but which acquires a maximum accumulation in the banks which have just been spoken of. These great deposits are due mainly to conflicting currents, which are here met with, particularly the confluent tide wave of the Atlantic with the divergent wave of the American coast.

§ 18. Local and specific deposits belong to the same agency as that producing the great cumulative masses of George's, Newfoundland banks, and Sable island, &c. The tides, as they are usually understood, or as they are known to the common observer, exert a constructive influence upon the materials conveyed to them by river currents. The flood tide, advancing into a bay, sound, or estuary, bears onward its burden as has been already described. It deposits it along the shore in consequence of its conflict with the bottom and irregular sides of the projecting land, forming thereby a sandy, ridgy border of greater or less extent, according to the amount of the detrital matter it has received. The outward flow, or ebb tide, distributes the materials in a more central track, but is less effective than the flood tide: it renders the deeper parts of the bay more shoal, as it partially gathers the detritus from the sloping shores under the convex crest of the retiring wave in the deeper channels. Aided by the river current, detritus and floating bodies are moved far out to sea. If the cumulative process were confined to the bay or estuary, the bottoms would be raised much more rapidly than at present. One of the effects of the tide wave is to drive across the mouths of the deep coast indentations and bars of sand. In process of time these bars rise to the surface, and inclose a bay, producing thereby a lagoon, which becomes first brackish, and afterwards fresh water. This fact has an important bearing in geological reasoning, as it is in these reclaimed areas that we find oceanic or pelagic shells, estuary, and fresh water, and land remains superimposed upon one another. The breaking down of the bar by powerful waves may convert an inland lake to a bay of fresh or salt water, which in time will be peopled with marine inhabitants, forming a complication of deposits of the most interesting character.

§ 19. *Wind Wave.* The sea, as well as fresh water lakes, are raised into waves by winds. The movement of such waves will exert a constructive action upon the coast close inshore. The obvious effect of wind waves is to raise the sandy deposits in a ridge, sloping more rapidly upon the land than upon the sea side. This process often cuts off a bay from the sea, the ridge running parallel to the main shore from two projecting points. On the land side there may be a lagoon and a marsh, which in due time will be filled up by aquative plants intermixed with soil or sand. The parallel roads and ridges traversing a country in the axis of its valleys, are often wind ridges, which mark the former shores of a lake or of an arm of the sea. These may be hundreds of feet above the sea level. They do not indicate a depression of the ocean level, but that the land has been raised in successive stages. Under favorable circumstances the wind wave pushes onward the sand detritus upon the land. The march inland devastates the soil, and spreads over it barrenness and sterility. The inland sands form the dunes. They may be arrested by a vegetation which delights in an arenaceous soil, such as the beach grass (Calamagrostis arenaria), or some of the species of pine.

A gentle breath of wind produces a ridged surface upon the sandy bottom. Waves of great strength destroy this rippled surface. The ripple mark will be formed in a vessel of water charged with earthy matter. The slightest agitation, as walking across the floor, will be sufficient to arrange the sediment like ripple marks upon a beach. Raised beaches are often the joint effect of the tidal and wind wave, especially where the direction of the wind is constant. The transporting power of flood tide is not obliterated by strong contrary winds, as the north-east winds upon our coast and the advancing northerly tide wave fully show.

A constructive action is not the uniform result of the tide wave; its action is often destructive. Sea bluffs are broken down, and the soil is washed away, and carried to other points upon the coast. Portions of Long island, Martha's Vineyard, are undergoing changes of this kind. Points of land where fortifications and lighthouses have been built, have often been undermined by sea action. The sea encroaches on the land. But the constructive action has greatly exceeded the destructive on our own coast.

§ 20. Ice and semifluid ice is an instrument of change which should be noticed in this connection. Water congeals, and remains so the whole of the year, upon mountains which rise above 15,000 feet above the sea under the equator, and at still lower levels at points north or south of it. In consequence of this, there are vast accumulations of snow and ice upon the tops and in the high valleys of many mountains. These beds of snow and ice are called glaciers. This ice, as it approaches the lower parts of the mountain, softens, and becomes movable upon the inclined plane upon which it rests. The glaciers freeze, however, during the night, and become stationary; but becoming softened, or partially thawed, during the day, they again move on at a certain rate. The middle moves faster than the sides. The glaciers are now regarded as instruments of change upon the rocks over which they move. The position of these glaciers is such that they receive all the rocks and debris of the mountains where they are formed. They exist more or less throughout their icy beds, and hence it often happens that rocks stand out from their inferior surfaces. Owing to this circumstance, the glacier, as it moves, forces these rocks over the rocks in place beneath: they are therefore abraded and loosened, and the melted ice, as it flows away, is charged with mud, which is merely the matter worn off from the rocks. The glaciers then work mechanically, and with great power; and thus they aid in the process of leveling the mountains and filling up the valleys at their base. Glaciers, when they reach the sea in high latitudes, carry directly the abraded matter to the ocean. Glaciers in high latitudes jut over the sea, and the sea edge being left unsupported, break off from the main mass, and fall into the ocean. These are icebergs. They then drift away, and are carried by currents out to sea, and often reach, in the northern hemisphere, the latitude of the Azores. On their march, they melt and distribute their burthens of earth and rock over the ocean's bottom. It is maintained by eminent geologists, that glaciers were formerly far more extensive than now, and hence there was a period which deserved the name of glacial period.

§ 21. The general tendency of the operation of water, as described in the foregoing paragraphs, is to form accumulations on the lines and planes of flow, or a little outside of them; which, beginning upon the mountain tops, terminate in the broad ocean rivers. The coarsest matter is left near the summits, and the finest is carried to the bottoms of the ocean rivers. Another belt of coarse materials is along the ocean shore and margins of lakes. These materials are mostly hard and siliceous, and resist change a long time; but they become pebbles and sand by attrition. These form the conglomerates.* Farther out from land there is a mixture of fine silex and clay; and still farther, in the flow of ocean rivers and currents we find suspended the finest particles of carbonate of lime and alumine still mixed with the finest sands. In this distribution of materials, we learn that three contemporaneous deposits may go on, the siliceous, the conglomerate and sandstone, the clay slates and shales, and the limestone rocks. As marine animals and plants occupy different stations, some living in shallow and others in deep water, it is plain that these rocks which are being deposited, will very likely contain the remains of animals of different species, though of contemporaneous formation. The nature of the bottom, too, influences the law of distribution of animals, as well as plants. Hence, at the same depth, and under other circumstances which are equal, a sandy shore is

* Conglomerates are cemented beds of pebbles formed in water. Pudding stones are cemented pebbles formed above water upon dry land, by percolation of water holding carbonate of lime and silex in solution. So the travertin and tufa are formed above water.

peopled with different species than the soft muddy ones. All these facts are important in geological reasoning.

From the foregoing we may also learn the influence of water as a formative agent. It moulds and forms, as it were, all the coarse and fine materials which are abraded from preexisting rocks. It spreads them out upon bottoms, and deposits layer after layer. We may have a glimpse also of the time required to form rocks. Only a few thin layers, like paper, are deposited annually; and it is probably a rare occurrence, that streams are so loaded with mud that living shell fish are buried beneath it.

§ 22. The foregoing details of the distributive, as well as constructive agencies of river and oceanic waves and currents, will be more complete by a relation of the distribution of an animal life upon the ocean bed. The profound depths of the ocean are tenantless wastes, except for the dead, who have here found their resting places, where no wind or wave can move them, or bring up their sacred relics to light, and cast them once more upon a troubled shore. Along shore, in the reach of soundings, the waves distribute their burdens in ridges. These ridges are also in the main desert lands; but the valleys being protected by the ridges, teem with activity and life. Upon the slopes the rounded and worn materials are cast together with the remains of the exuvia of organic life, which have been cast off. The deeper valleys are suited to one class, while the shallower portions are sought by others. From shallow water, or the high water mark, to a depth of thirty fathoms, forms the main range in depth of marine animal life. Vegetable life, however, rises up from greater depths. Thus the gigantic sea weed of the Falkland islands rises from a depth of 300 feet. The sea bottom, therefore, like the earth's surface, presents all the variations of contour which are necessary to give life the widest exhibitions of form and character. The profound depths, like the snow-tops of the Andes and Himalaya, are dreadful wastes, devoid of life; the one awful from its profound solitude, the other fearful from the howling blasts which sweep their towering tops.

§ 23. The coast of the United States is flanked within soundings by an arenaceous deposit, arranged in the manner already described. The shore deposits are more thoroughly arenaceous than the more distant depths occupied by the valleys. Here the formation is more muddy, and partakes of an argillaceous character; while it is highly probable that farther from land the calcareous matter will be found. The West Indian archipelago may well be regarded as the repository of the calcareous formations. Here, aided by the incessant toil of the coralline animals, a limestone bed or rock is in the progress of formation, equal in extent to the Onondaga limestone of New York, and like that abounding in branched corals and massive madrepores, fragments of shells, together with the perfect animal forms which abound in the Carribean sea. This formation becomes the conservatory of the constructive works of man, as well as the burial place of his remains. Guadaloupe has furnished an instance verifying this assertion, by the discovery of a human skeleton nearly perfect in its parts. No sea is so richly freighted with the remains of animal life from the highest to the lowest—from man to the polyp. An entire record of human life, since the day the Carribee set his foot upon these islands, is treasured up in the archives of its deposits. Every layer is a leaf bearing the impress of the past; and the medals strewed profusely upon its bottom tell of the strange vicissitudes of a lost continent, whose existence is proclaimed by the crests of Cuba and the volcanic peaks of the lesser Antilles.

§ 24. The complexity which is created in geological researches by the contemporaneous formations, may well create a hesitancy in pronouncing upon the age of a given deposit. Looking upon the present as a type of the past, we see in the arenaceous shores of America the exterior muddy deposits, which may be regarded as argillaceous, and the coralline formations far from its coasts, whose rocky nature is completed by the cementing agency of calcareous matter in solution and suspension. Three contemporaneous formations of vast extent and importance, and which, being judged of by their lithological and fossiliferous characters, would be regarded as formations of different eras. Taking it then as a type of the past, we may well doubt the correctness of many of the geological data which have formed the foundation of our reasoning. Tidal waves, normal oceanic currents, and river currents, with their burdens of detritus, have ever exerted their powerful agency in distributing the waste materials of continents, and in constructing the fossiliferous mountains of the globe. Conglomerates, coarse and fine sandstones, if we regard the foregoing facts, may always be considered as shore deposits, and argillaceous and calcareous rocks as pelagic formations; especially may we recognize in most of the calcareous rocks formations similar to the recent beds in the Pacific and West Indian islands.

§ 25. The systems of relief of the North American continent are not as yet well determined. The mountain ranges, however, pursue a northerly and north-easterly directions, by which it appears that the force which raised the continent acted in those directions, or that this force preponderated over all the forces acting in other directions. On the Atlantic coast it is north-easterly, or parallel with the coast line; on the Pacific coast it is northerly, and parallel with the Pacific coast line.

Considered as water sheds, the ranges of the United States may be reduced to five: 1. The Appalachian, which is parallel to the Atlantic coast; 2. The Green mountain range, which runs north, or six or seven degrees east of north; 3. The Rocky Mountain range, which, according to the best maps, is also directed to the north; 4. The Pacific coast ranges, which are north; 5. The Lawrentine range, which pursues an easterly and westerly course. These chains of mountains are often flanked by parallel ones, which attain at many points greater heights; yet they are evidently subordinate to them, since they are broken through by the water courses, and are not coextensive with them in length. Thus the Taconic range flanks the Green mountain on the west in western Massachusetts and Vermont; yet Graylock of the former rises nearly 3600 feet above tide, and the latter only 2500. Black mountain flanks the Blue ridge in North Carolina: the former attains a height of 6200 feet, while the pinnacles of the latter are about 5000 feet only.

The slopes of the water shed of the Appalachian range is to the southeast, or to the Atlantic, on the east side; but to the northwest on the west side, or towards the main trunk of drainage, the Ohio river. But it has also a southern slope, by which it furnishes a drainage into the gulf of Mexico.

§ 26. The Appalachian range begins in the northern part of Alabama, and terminates with the valley of the Mohawk. The culminating point is the Black mountain in North Carolina; and the culminating ridges extend north from Black mountain to Grandfather mountain, by Table rock in Burke county, North Carolina. Here are five close-pressed parallel ridges, of which the Blue ridge is a subordinate to them all; but when traced in either direction it becomes the main and principal range of upheaval, and forms withal the crest which divides the Gulf system of waters from the Atlantic system. The Green mountain range runs north six or seven degrees east. It is regarded by President Hitchcock as the Meridional system of Massachusetts and Vermont. It begins upon the Sound, and extends into Canada East. The two principal rivers which drain their slopes are the Hudson and Connecticut, whose courses are parallel with each other. The Taconic range is separated by a well defined valley from the Hoosic range. The latter is the eastern rim of the rocks of the Taconic system, and which have suffered many dislocations since their deposition. The fractures are parallel to the main range. Igneous injections are almost unknown in that part of the Taconic system which lies north of the Highlands and south of Rutland, Vermont. The Hudson river runs upon a line of fracture which extends from New York to Montmorenci in Canada East, lake Champlain being a wider and deeper fissure than that along which the river flows.

The Atlantic slope bordering the ocean is exceedingly gentle, indeed: the country is nearly flat until we encounter the first low granitic ridge, which creates a line of falls in all the southern rivers, viz: the Rappahannock at Fredericksburg, James river at Richmond, the Roanoke at Weldon, the Tau at Rocky Mount, the Neuse six miles east of Raleigh, the Cape Fear near Haywood in Chatham county, or at the falls of the Buckhorn. Above the falls the rivers are more rapid, but their ascent to the base of the Blue ridge is still gentle, though rapid at many places. At the base of the crest of the Blue ridge their height above the ocean rarely exceeds 500 feet. From this point up the ascent on the east side is exceeding rapid for five miles. On the west side the slope is again gentle. It appears, therefore, that the slope is on the west side towards the Mississippi valley, while the counter slope is on the east, and contrary to that which prevails in Pennsylvania and New York.

A remarkable feature in the Atlantic plain, forming part of the eastern slope of the Appalachian range, is the country of the Pine barrens. These are sandy plains, undulating like a sea bottom, and clothed with the long-leaved pine. Although of considerable extent, and traversed by rivers and streams, the sand, though a marine formation, furnishes no fossils, except silicified wood, which is derived from the triassic beds of Deep and Dan rivers. These monotonous barrens are analogous to the prairie lands of the west. Towards the north the Atlantic slope becomes less sandy, and its vegetation is in accordance with a gradual change in climate and soil. To the south this plain extends westward and southward, and connects itself with the Mississippian, by turning around the southern points of the Appalachian chain in the north parts of Georgia and Alabama.

§ 27. The Mississippi flows upon a low ridge or anticlinal axis. The country westward swells and rises gently, and finally attains, at the base of the Rocky mountains, an elevation of about 5000 feet at the South pass. A pass lower by 2000 feet has been discovered by Governor Stephens. The culminating points of the Rocky mountains are near Fremont's peak and the three Tetons, as here the Colorado, the Missouri, and the Columbia take their origin. The Rocky mountains are belted by a sandy desert some 400 or 500 miles wide, which is prolonged northward to the mouth of the McKenzie river, a distance of 1500 miles.

The coast range of the Pacific, and the Sierra Nevada, are parallel chains, and separated by the valley of San Joachim and San Francisco. These ranges, prolonged into Oregon, are succeeded by the Cascade mountains. These three ranges frequently rise above the line of perpetual snow. The outer range is only 380 miles from the Pacific ocean. Considered as continental ranges, their slope is towards the Atlantic, and the counter slope to the Pacific. They are in the ratio of two to one. The great valley and its slopes drained by the Mississippi, and its thirty-four navigable rivers, contain an area of 3,245,000 square miles. The Mississippi trunk is navigable to the falls of St. Anthony, and the Missouri high up the waters of the Yellow Stone.

§ 28. The Lawrentine chain is comparatively low, not exceeding 2500 feet. It is but little known. It divides the waters of the St. Lawrence, the Mississippi, and the rivers of the British territories. The chain varies but a few degrees from east to west.

A small mountainous tract lies north of the Mohawk, between the St. Lawrence and Lake Champlain. The four distinct ranges by which this tract is traversed are parallel with each other. Their axes are directed to the north-east. The main chain rises at Little Falls, and pursuing a north-east course, terminates abruptly at Trembleau point on lake Champlain. The culminating point of this range is mount Marcy. This mountain is the center of the Adirondack group, and rises to the height of 5467 feet. From this group the drainage is composed of the Ausable, Saranac, Racket, Black, and the branches of the Hudson river. The lakes situated upon the table land, and from which these rivers rise, are from 1500 to 1800 feet. This level is about the same as that of Connecticut lake, and not greatly inferior to the lakes which give origin to the Mississippi.

How do we know that the present valleys and mountains were not coeval with the foundations of the earth, or that many of them have been formed since it was inhabited? This fact is determined, like all other facts, by observation. Though we do not witness their formation, still our observations are not the less certain and true. We learn first what rocks compose the mountain and its valley—the arrangements of their strata, and the relative position of the principal and subordinate masses. We examine its cliffs, its fractures, and veins. With equal care we examine the valley, and compare its formations with those of the mountain. We find they agree. If a traveler should find by the roadside the parts of a broken walking stick, how would he know that they were parts of one stick? He would find that the pieces were the same kind of wood, that they were colored and polished alike, and that the ends of the fractured parts fitted each other. So, in the same way, the strata of the mountain are the same as the valley, and the fractured ends, if brought together, would fit each other. But the cliffs are a thousand feet above the valley, and their present position is incompatible with the mode of formation of all sedimentary rocks. They are not in the position required by sediments, hence a part has been broken from the mass and elevated, and now forms the mountain mass, while parts of it still form the valley below; and the rocks themselves still retain the marks of the operating force in their curved and contorted beds. They too are the repositories of fossil remains of the same kinds. Long since the time they formed the ocean bed, they were raised from the depth of the sea, and their fractures and dislocations attest the action of forces which elevated them to the positions they now occupy.

§ 29. *River systems.* The machinery by which the earth is watered is extremely simple. The atmosphere, set in motion by heat, is the carrier, and mountains and hills the condensers of moisture. The rivers receive their supply of water from an infinitude of streams flowing from the sides of mountain chains. The simple process of condensation of the moisture of clouds and winds keeps a perpetual flow from the mountain to the sea; and the wind current, in passing over the ocean, loads itself with vapor, which is ready to fall in mist and rain upon surfaces cooler than itself.

In these simple facts we find an explanation of the origin of the river systems of this country, and of all countries. The Appalachian, the Rocky Mountain, and Pacific Coast chains, with their numerous spurs and branches stretching from the Gulf to the British possessions, form an immense condensing surface, sufficient to irrigate and fertilize 3,000,000 of square miles. It is a singular fact, that the United States is watered by rivers which rise in its own borders. The crest of the great water shed dividing the river systems which flow to the north and south, formed by the Lawrentine chain, rises not far from the boundary between the British and American territories. The rivers of the Atlantic slope are short, and comparatively small. The Potomac, the Delaware, and Susquehannah spread out into wide and deep bays. This results from the porous nature of the tertiary deposits which belt the coast. The Hudson and other northern rivers are rock-bound, and hence their bays are narrow and inconsiderable. The tide flows up the Hudson 160 miles from the ocean. It cuts the primitive rocks of the Highlands and some of its branches, and flows over the lower silurian formation beyond.

The Mississippian is the great river system of the United States. It is second only to the Amazonian. The Missouri is the great trunk of this system. It rises high up in the Rocky Mountain chain. The innumerable streams draining the eastern slope, converge and form four great rivers, which, uniting in the distant plain, form the Missouri. This unites with the Mississippi twenty-five miles above St. Louis. Measuring its windings, it has already reached a point 1500 miles from its source, and its journey is only half finished. Its current is rapid, and it carries mud and sand, derived from the soft cretaceous and tertiary formation through which it has flowed. A particle of water, starting from the steep sides of the Rocky mountains, reaches St. Louis in about twenty-five days. But the most interesting fact which the Mississippian current reveals to us is contained in the sediment. The last resting place of this matter was in a cretaceous sea. Where they then came from, and how many transportations they had undergone, will never be revealed. In our geological reasoning they are destined to form the most modern deposits. They have undoubtedly passed through all the historical periods since aqueous deposits began to be formed. The particles are as old as the foundations of the earth, but the formations of which they are destined to form a part, are becoming the newest. They have been associated with the oldest organic beings, but they are now brought in contact with the most recent—with the people of the present age. The recent is made of particles derived from every known period. The water of the Mississippi is clear compared with the Missouri. Above its confluence with the latter it has a long and gentle descent. But its progress, in one respect, differs from that of the Missouri: it passes through many lakes, a fact which is unknown upon the course of the latter river.

The Mississippian system, unlike the Atlantic system of drainage, has two slopes, an eastern and western. The former, however, has more than twice the area of the latter. They unite, and form the basin of the Mississippi.

Above New Orleans, where all the great trunks of this system of waters flow in one channel, the quantity of water is immense. According to the most reliable calculations relative to the quantity which tbis river discharges into the ocean annually, it amounts to 14,883,360,636,880 cubic feet. The amount of sediment transported to the ocean by the Mississippi is 28,188,083,892 cubic feet. This sediment is sufficient to form an annual deposit one mile square, and 1000 feet thick. As the delta of this river contains 13,000 square miles, and as the sediments of the delta are at least 1056 feet thick, it is evident that the time required to accumulate so much material must have been greatly protracted. Fourteen thousand years has been stated as the result of the best observation which has hitherto been made. But the finest of the sediment probably passes over the river bar, and may be carried far from the delta by oceanic currents. The actual time, then, consumed in the formation of the delta is greater than the calculated.

What takes place by the instrumentality of the Mississippi, takes place in the same mode by all rivers. As the winds are the carriers of water, so the rivers are the carriers of sediments. By the combined machinery of wind and water all the sedimentary rocks are formed. Water, acting upon these plastic materials, spreads them evenly upon deltas and over wide areas upon the ocean's bottom. Here, subjected to a great pressure, they become consolidated into rock.

§ 30. *Winds, as the distributive agents of heat and moisture.* The geological agency of the wind is modified by its direction. The trade wind of the Gulf furnishes a supply of water for the western slope of the Appalachian chain, and the northeast wind of the Atlantic brings a supply for the New England and Middle states. The mountains of Oregon condense the moisture of the northwest winds which have passed over the Pacific. To the Eastern states the same wind is dry and cold, having been robbed of its moisture, as well as cooled, by the highlands over which it has passed. A continent is indebted to the agency of winds for the supply of water, both for its vegetation and that which is required to feed its rivers and streams. They dry the earth's surface when wet. Sixty per cent of the rain which falls in the valley of the Ohio, is restored directly to the atmosphere, or is taken up by vegetables. The southwestern winds are warm and damp from having passed over the Mexican gulf. This great body of water is of the utmost importance to the well being of the Appalachian slopes and valleys, imparting moisture and a subdued temperature where in its absence it would be dry and hot.

§ 31. The earth's surface is acted upon mainly, as I have already stated, by two distinct and diverse agencies, fire and water: the former, by its well known properties, which are manifested in the simple expansion and fusion of matter; the latter, by its transporting power, and by the aid of frosts in breaking up the strata and disintegrating the exposed surfaces. The entire operation may be summed up in two processes, one of which degrades the more elevated parts of the earth's surface, and the other fills up the depressed portions. Fire or heat operates in four ways: 1. In the elevation of areas by the application of expansive forces beneath the earth's crust, by which it is raised up in mass. 2. By the transference of fused matter from the interior to the surface, and which it overflows, and thereby makes an additional thickness to the visible strata. The addition being transferred from the interior to the exterior, may be in the form of melted matter, semifluid matter, or in the form of mud, or in pulverulent matter, in the condition of ashes and semifused mass ejected from the craters of volcanoes. 3. In consequence of the loss of matter thus thrown out from the interior, areas of subsidence are formed, and the superficial strata are engulfed suddenly, or else slowly subside and sink below their former levels. 4. Areas are elevated or depressed by the simple expansion of strata by heat and their contraction by cold.

The force generated by heat is proportionate to its intensity. It pervades, in a limited degree, the zones of rock immediately beneath the earth's surface. This is proved by its increase downward from the limit of solar influence, which is a point of no variation for the year. The ratio of increase for this country is one degree of Fahrenheit for every fifty-five or sixty feet, and for Europe one degree for every forty-five or fifty feet. These facts point to a source of heat in the earth's interior. This view is supported by the overflow of immense quantities of incandescent and melted matter from volcanic vents. Like all other bodies, rocks are expanded by heat and contracted by cold, and these changes in volume are connected both with changes of level in the earth's crust and in its disruption, or the forcible separation of continuous strata, and the formation of intervening fissures. In the simple expansion of rocks by heat, and their subsequent contraction by cold, we have an element which is competent to explain many phenomena connected with changes of the earth's crust.

Rocks are fused in the interior of the earth's crust, and in that state may rise to the surface. The fused rock often overflows the brim of craters, and flowing down the mountain sides, breaks and vitrifies the rocks over which it flows, and fills the hollows in its line of march. So, also, fused matter may rise in the fissure formed by disruption, and after reaching the surface flow like a lava current over its edge, or force itself between the layers of a sedimentary rock.

Internal heat must be regarded as an arrangement which conduces to the well being of the earth's inhabitants. It no doubt ameliorates the cold, and sustains that degree of temperature which is best fitted to the organic and structural conditions of living beings.

§ 32. *Time is an element in geological dynamics.* It is measured by forces whose operations we may witness. In the estimation of time we follow two methods, the results of which agree. The first method, we estimate the rate at which deposits accumulate in the present seas. The measuring line which we thus obtain is applied to the past. The second method is the reverse of the first. We estimate the amount of matter in the sedimentary rocks of all periods. This matter is composed of the waste of former continents and former mountain ranges. The sediments of each period are immense; and immense as our present mountain systems may be, still the sediments and wastes which have accumulated since animals and plants have lived, are sufficient to compose many such mountain systems as now exist. Life materially aids us in making our computations. It determines the slow rate of the accumulations of which we have spoken. We are not to presume that the Prime Mover, the great Efficient Cause, has hastened events because he has the power. Events are preceded by preparatory steps, and time and events develop themselves in stages and periods. Like the planets in their orbits, they may be accelerated in motion for a moment, and the forces may act more intensely as they pass disturbing bodies, still the sum of the results in long periods are the same.

In geology time is only relative. It can not be absolute, or at least absolute time can be reckoned only for those changes which have taken place in the historical period. But absolute time is diminutive compared with relative or geologic time.

§ 33. It has been shown that sediments of the same age occupy positions which are determined in part by their size and weight, or the form of their particles. The large and heavy particles are deposited early, while the fine are transported far out in the ocean. The sediments are distinguishable by the forms of their particles, their peculiar arrangement, or by the presence of fossils. Attrition rounds the salient angles, though it often happens that particles are consolidated while they are still angular. They have a foliated arrangement, being superimposed upon one another. This foliation has received the technical name of stratification. Stratification is manifested by folia of different kinds of matter, as sand, mica, and talc, or by folia of different colors. An amorphous mass of materials, as cart loads of sand, gravel, and stones, become stratified by the percolation of water when thrown down into heaps. Lamination closely resembles stratification: it is the separation of a mass into thick or thin layers by an imperfect or unfinished crystalization. Gneiss, mica slate, and talcose slate, are examples of lamination. Where the planes of separation are indistinct, the term sublaminated may be employed. It is proper to distinguish these two forms of the separation of the parts of rocks. The lamination of gneiss, mica slate, &c., can not be regarded as a true stratification, as the arrangement of their parts is not due to the same causes. There is no evidence that the mica or feldspar planes in gneiss, or the mica and quartz planes in mica slate, were the result of a sedimentary process. Where heat has been sufficiently powerful to fuse pebbles, it must also perfectly destroy the stratification, and the present so-called stratification of gneiss must be due to the heat and fusion the mass has suffered. I would therefore prefer the adoption of the term *lamination* to be applied to the rocks designated, rather than to continue and extend the use of the term *stratification*. I would restrict this to sediments, or transported matter, which have been subjected to the action of water.

§ 34. We may observe frequently a condition in massive and stratified rocks which is not due to the conditions which existed at the time of their formation. In granite, for instance, where it is exposed to disintegrating agencies, there may be observed a separation of its mass into laminæ, or into thick tabular masses, simulating a laminated rock. When this condition is examined it may be referred to a molecular force, or to a concretionary movement of its particles. The separation takes place in parallel planes, but they are usually curvilinear. Sedimentary rocks undergo changes from the operation of the same forces; the original planes of stratification are obliterated and the new planes which are formed are concentric, and arranged around a nucleus. Another change takes place in rocks whose particles are bathed in water. Clays, and clay slates, and limestones of all ages, contain rounded masses which are known as septaria, clay stones, or concretions, nodules, &c. This is a most interesting change. These bodies may be either purely siliceous, or they may be calcareo-aluminous. The siliceous concretions are abundant in the carboniferous limestones of Missouri near St. Louis: the flint nodules and layers in chalk is another example of the kind. The slates furnish the calcareo-aluminous bodies, which in clays are known as claystones. They are abundant in most clay beds or marls of all ages; and those of the slates which are known as septaria, differing from the former by their septa, are formed of crystalline limestone, barytes or strontian. We are obliged in all these instances to recognize a force, by virtue of which the molecules are really transferred to central points, where, by constant accumulation, they form a nodule, or septaria, or concretion. This force is operative at all times, and upon all rocks. Molecules are never at rest until they have acquired a symmetrical arrangement. Concretions and nodules are symmetrical bodies. The parallel planes of the rhombic forms so common in limestones and slaty rocks are due to this force. The jointed structure admits of the same explanation.

§ 35. *Composition of the earth's crust.* The mineral kingdom is composed of a large number of distinct species of substances, the knowledge of which is highly important. The composition, however, of the rocks or masses is represented by an extremely small number of simple minerals, which are repeated over and over again in the layers of the rocks. The rocks are either mechanical mixtures of a few simple minerals, or they are made up of a single simple or homogenous mineral by itself. Granite, gneiss, sandstones, and conglomerates, are examples of the former; and limestone, gypsum, serpentine, and hornblende, of the latter. The elementary bodies are extremely rare in nature, or in the mineral kingdom. Sulphur is common in volcanic districts, but is a product of decomposition. Carbon, nearly pure, exists under the form of anthracite. The metals, gold, silver, copper, and mercury, may be said to be of frequent occurrence, but can not be claimed as component parts of the masses geologically considered.

The minerals which predominate in the earth's crust are the siliceous, aluminous, and calcareous. Silica, as a constituent part of the rocks, occurs under two forms: the first and most obvious and common is quartz, as it exists in flint, white sand, or an aggregation of sand in the form and condition of sandstone; the second is an acid, and is combined with one or more bases, and forms those bodies which are called silicates: feldspar, hornblende, mica, and pyroxene are examples. Some rocks contain examples of both forms, as granite, where it is in the first form as particles of quartz, and in the second as silicates in the feldspar and mica. The aluminous minerals are represented by common clay, as it everywhere occurs, or by slates which are consolidated by pressure, or baked clay still more consolidated and changed by heat. These examples, however, are not those of pure alumine; they are mixtures of silicates and of fine and impalpable sands. Their purity or approach to alumine is indicated by their whiteness.

Limestone is composed of carbonic acid and lime. As marble, it is nearly pure. When acid is poured on limestone it effervesces or boils by the escape of carbonic acid. Lime is also found combined with sulphuric acid, when it forms gypsum. It may be distinguished from the carbonate by its softness, its fusibility, and the absence of effervescence in the presence of acids.

The sediments are mixtures of the silicates, sand or quartz, in fine or coarse grains, pebbles, clays, sandy clays, limestones, or sandy and aluminous limestones. These mixtures, however, never form chronological successions; neither do they occur in modes or ways by which their lithological properties may be used as characteristics of age or place. To say that a rock is limestone, sandstone, or slate, conveys no idea of its place. It is a mineralogical fact which has some importance.

Among the chemical and mechanical mixtures iron is rarely absent. Its presence is usually indicated by red and brown colors, which it imparts to the mixtures containing it. The red and brown sandstones are stained with it. Of the simple bodies, however, oxygen must be regarded as the most general, and most widely diffused in the mineral kingdom. In a state of purity it is aeriform. Its properties are better known to us in its mixture with nitrogen, forming the atmosphere. Very few substances are known which do not contain it. The iron which has just been referred to, is a compound of iron and oxygen, or it is an oxide of iron. Sulphuric acid, which forms a part of gypsum, is a combination of sulphur and oxygen; and carbonic acid a combination of pure carbon and oxygen; quartz or flint is silicon and oxygen; and quick lime is calcium and oxygen.

§ 36. It is a point upon which all geologists agree, that the earth's crust is composed of rocks which have been formed at different periods. Both the rocks and periods being numerous, it is important they should be arranged into groups or classes,

6

according to characteristics which belong and are common to each of the respective groups or classes. Attempts were made at an early day to construct such an arrangement of rocks as would meet the ends in view. These attempts embody the views of the prevailing systems of geology, and the nomenclature employed to express the generalizations of the authors have been found both defective, as well as expressive of fundamental errors. The more recent attempts of classifiers have been confined mainly to the use of terms which express facts. The rejection of the terms *primitive*, *transition*, and *secondary*, and perhaps *tertiary*, seems to be acceded to on the ground that they express a theory which is untenable in the light of modern discoveries; although the names might continue to be employed without endangering the interests of the science, provided those names were used simply as names, without regard to the theoretical views of the authors who first used them. It is an interesting fact, that the terms transition, secondary, and tertiary, the three periods to which they have been applied, stand forth the prominent *triads* of geologic time. It is no less certain that primitive or primary express also truly in the main the fact they were originally designed to convey. The nomenclatures of all the schemes of arrangement are objectionable, inasmuch as they are not consistent with the demands of science. The slight modifications which I have proposed in nomenclature it is hoped will not be regarded as an unwarrantable innovation, as they are the expression of admitted facts. Still there is a want of unity in the names, which may be corrected hereafter as discoveries are made. The systems into which the hydroplastic rocks are divided are arranged chronologically, but the names which have been given to these systems are by no means chronological. But by dividing these systems into three groups, we may express approximately their chronology in the terms palæozoic, mesozoic, and kainozoic. At present the most fashionable, and perhaps too the most useful names of systems, are taken from localities where those systems are well developed, of which we have an eminent example in the word *silurian.* So long as this name for a system of rocks is retained, so long will its example find imitators.

CLASSIFICATION.

Class	Section	Stratified.
III. HYDROPLASTIC, ...	c. *Kainozoic*, ..	Alluvial and drift,
		Pleistocene,
		Pliocene,
		Meiocene,
		Eocene.
	b. *Mesozoic*, ..	Cretaceous,
		Lias and Oolite,
		Trias.
	a. *Palæozoic*, ..	Permian,
		Carboniferous,
		Devonian,
		Silurian,
		Taconic.
II. PYROPLASTIC, ...	b. *Sub ærial*, ..	Lavas, Tufa, or Volcanic Products.
	a. *Sub marine*,	Greenstone, Porphyry,
		Basalt, Trap.
I. PYROCRYSTALLINE,	b. *Laminated*, .	Gneiss, Mica, Slate, Hornblende, Talcose Slate, &c.
		Laminated Limestone,
		Laminated Serpentine.
	a. *Massive*, ...	Granite Sienite,
		Hypenthene Rock, Pyrocrystal-line, Limestone,
		Serpentine, Rensselaerite,
		Octahedral Iron Ore.

In the foregoing arrangement, the rocks which have been called *metamorphic* and *azoic* by several eminent writers, have not been recognized as classes, or even as subdivisions of sections, inasmuch as they can have no special peculiarities which make them applicable for such purposes. Metamorphism occurs, or may occur, in all the series of rocks from the earliest to the latest sediments. It is true, the term metamorphic has been confined to gneiss, mica slate, hornblende, talcose slate, &c.; but its use is theoretical, and was thus applied on the hypothesis that those rocks are altered sediments, of which there is no evidence. The term *azoic* is still more objectionable: it presupposes that our observations have made certain that which from the nature of our evidence must ever remain doubtful. There is no doubt but that granite, gneiss, mica slate, &c., are azoic, but no one would think it proper to apply it to those rocks.

§ 37. *Structure of the massive pyrocrystalline rocks.* The peculiar mode in which these rocks became consolidated furnishes a clue to their essential structure. This is crystalline. They are not only composed of crystallized minerals, but they are crystalline in the mass. This statement is sustained by the fact that, in the quarry where large masses are raised, they split readily in certain directions: it is in fact a cleavage similar in form to that of a simple mineral. These directions or joints of cleavage are developed by the disintegration of the rock by atmospheric causes, the action being always more perceptible in the direction of the cleavage planes; they appear to separate spontaneously, and to extend deeply into the rock. The rock in this condition shows all the directions in which it may be split. The annexed cut (fig. 1) illustrates the

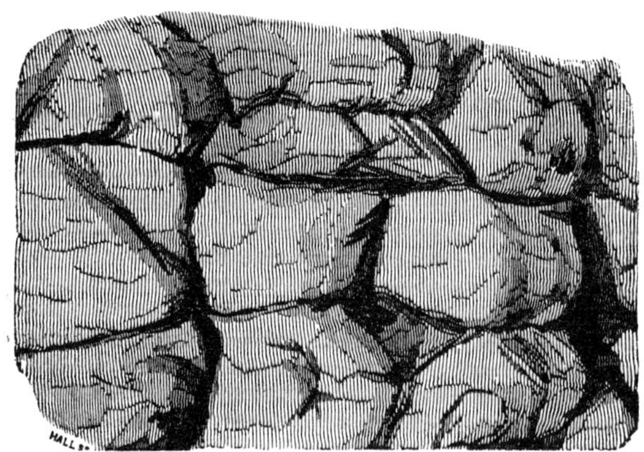

Fig. 1.

appearance of a mass undergoing the changes alluded to, and by which it is separated into angular parts. This result is not to be regarded as a lamination, inasmuch as lamination is the result of the arrangement of different minerals in parallel planes; and the ready splitting in the direction of those planes is due to the diminished cohesion between two different minerals in part, and in part, also, to diminished cohesion which always exists between the broad planes of crystals. The separation of the folia of mica or talc through their broader planes illustrates this fact.

§ 38. *Age of pyrocrystalline rocks.* The consolidation of the earth's crust resulted in the production of the pyrocrystalline rocks. If any part of the cooled pellicle thus formed remains, that would be the oldest rock. A pellicle must have been ultimately formed, and which still maintains its existence as a constituent part of it. From the manner in which the surface cools, the consolidated masses which successively form must lie in contact with the inferior surface of the first-formed pellicle. The thickness of the crust increases by additions below. This mode of consolidation differs materially from the increase of the crust by additions to the hydroplastic rocks, as these increase by new overlying deposits—a mode by which the newest or latest formed rocks are superior; while in the former the newer are beneath and the older above. When, however, the crust contracts fissures are produced, through which the still fluid matter finds its way to the surface, and may overflow the consolidated surface. The age of two rocks thus related is determined by very obvious facts. The rock intersected by fissures and filled with melted matter must be the oldest, and the intersecting mass the newest. Three, and even more masses may be thus related to each other. This mode of formation, as well as the indications of age, belong

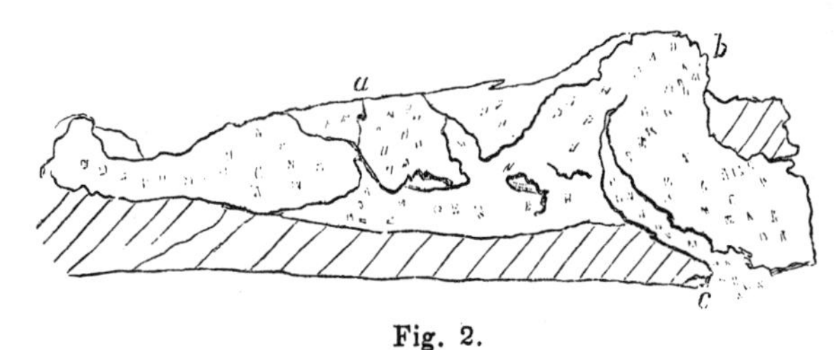

Fig. 2.

exclusively to this class of rocks, and in a series of adjacent beds we are to look for these peculiar relations, when it is desirable to determine which is the oldest and which the newest. The rule has a general application, as it is obvious that all intersecting masses of rock must be more recent than the intersected, whether the latter belongs to the oldest or newest classes. Fig. 2 illustrates the mode of formation of the pyro-crystalline rocks, and the relative age of each mass.

§ 39. *The age of rocks deduced from the perfection of their crystalline state.* Assuming the former fluid condition of the earth by heat, and its present condition by the loss of it, it may be inferred that the greater intensity of heat produced fusions far more perfectly in the earlier than in the later periods. We may conceive, then, that the products of a perfect will differ in structure from those of an imperfect fusion. The former will be longer in cooling, and the particles of the mass will be in a more favorable condition to move freely and arrange themselves according to their respective affinities. The first products arising from the cooling of the earth's crust are the granites. These rocks are preeminently crystalline, and their perfect crystalline condition resulted from the former high temperature to which their masses were subjected when the whole earth was in a molten state. The products of the subsequent periods, when the earth's crust had materially cooled, are less crystalline. Thus granites in mass, the first products of cooling, are traversed by granitic veins which constitute a second stage in granitic productions. In periods still later trap dykes appear, and we find in their structure indications of a diminished fluidity by fusion: they are black granular or compact and homogeneous, or imperfect crystals of feldspar pervade the mass. The traps proper traverse the earlier and later rocks, and cut the more ancient granitic veins, but are themselves more rarely intersected by granitic veins; and even the massive greenstones are rarely, if ever, intersected by granitic veins. Granitic veins and beds, however, are products of all periods as late at least as the chalk; but their frequency is diminished in the ratio of a thousand to one, comparing the granitic with the cretaceous periods.

The foregoing considerations seem to favor the doctrine that the imperfect crystalline structure of the later igneous products is due in part to a diminished heat in the earth's crust. We have never seen a granitic vein intersect a trap dyke or a mass of greenstone, although some granitic veins are newer than many veins and masses of trap.

The evidence of age, however, when deduced from structure alone can not be relied upon, only so far as it indicates a general diminution of temperature; traps and greenstones never forming those parts of the earth's crust which belong to the most ancient periods—the rocks of the most ancient periods being represented by granites and gneiss, whose structures are eminently crystalline.

Metallic veins—those of iron in northern New York, the auriferous quartz veins of Virginia and North Carolina—are traversed by dykes of trap or greenstone, and hence the former are older than the latter; and we have never seen the former traversing the latter, though in Derbyshire, England, metallic veins pass into them. Taking a general view of facts, however, as they are developed upon a large scale, we are inclined to adopt the opinion, that the prevalence of the intersecting dykes of trap are due to general and not to local causes, and that this cause will be found connected with the cooling of the earth's crust.

§ 40. *Special considerations respecting the elements of granite and its allied rocks.* Feldspar is the predominating element in all the massive pyrocrystalline rocks. It is a fusible compound, but the mass of rock in which it is so abundant may have been more fusible than feldspar by itself. The three principal kinds of feldspar, all of which are quite common in American granites, are composed of

	Prismatic Feldspar.	Albite.	Labradonite.
Silex,	65·40	70·7	53·70
Alumina,	18·60	19.8	29·90
Lime,	00·20	0·3	12 00
Potash,	15·70	0·0	0·90
Soda,	0·10	9·0	4·50

The presence of the alkalies and alkaline earths, while they promote as fluxes the fusion of the mass, materially contribute to its disintegration and decomposition. In consequence of this last property, those rocks which have undoubtedly contributed to the formation of subsequent rocks, have had a great influence upon the character of the soil. The most important minerals associated with feldspars are hornblende and pyroxene. They are composed of

	Hornblende.	Pyroxene.
Silex,	46·26	54·08
Lime,	13·36	23·19
Magnesia,	19·33	11·49
Prot. ox. iron,	3·43	10·02
Ox. manganese,		0·61

The fusibility of hornblende and pyroxene is greater than feldspar, and as they are both associated with feldspar, they probably increase the fusibility of the compound. Hornblende and pyroxene rocks, however, decompose more slowly than feldspathic, in consequence of the absence of the alkalies. The latter rocks, however, contain in this country sulphuret of iron, and in consequence of its presence, these rocks undergo greater and more rapid changes than they would were they free from this substance. The feldspathic rocks, however, contain a much smaller proportion of sulphuret of iron; it is not associated so frequently with it. In the first group feldspar is the most important one of its compounds. In this country granites are the principal rocks of the group, especially since it seems to be proper to include under granites all the mixtures in which the three kinds of feldspar are found. In this country we are unable to add many of the minor compounds—those, for example, which are found in the ancient volcanic districts of Europe, as clintstone, porphyry, graystone, diallage rock, &c.

In certain compounds hornblende or pyroxene predominate, though feldspar is always present.

Epidote should be noticed in this connection. It is no uncommon fact to find this mineral where the change in a rock is comparatively slight. Chloritic slates, when acted upon but feebly by heat, almost always contain epidote. It may be massive or crystalline. Its peculiar yellowish green color denotes the presence of this mineral. Its composition is represented below:

	Epidote.
Silex,	37·0
Alumina,	21·0
Lime,	15·0
Oxide of iron,	24·0
Manganese,	1·5

It is closely allied in composition to pyroxene.

GROUPS OF PYROCRYSTALLINE ROCKS.

§ 41. Three kinds of feldspar, the prismatic, albite, and labradorite, are frequently associated with hornblende and pyroxene. They form groups which belong to the later formed rocks, or to the pyroplastic rocks. Hornblende, taking the place of mica in granite, forms a compound which is called sienite, and it may be surmised that it is not the oldest kind of granite. The feldspars, when associated with hornblende and labradorite, constitute the greenstones, porphyries, basalt, and trap. These associations do not appear to have been formed at all in the earliest consolidations of the earth's crust.

§ 42. Feldspar, however, is not confined to the greenstones and basalts. We find it incorporated with many of the later formed pyroplastic rocks. Thus the claystones and clinkstones are compact rocks, in which the elements of feldspar predominate. The latter, when struck with a hard body, rings like a piece of baked earthenware. The first is often a porphyry.

The rock which is known as petrosilex is a reddish compact feldspar, spotted with crystals of white feldspar. Trachyte being composed of hornblende and glassy feldspar, belongs to this group. Diallage rock might be placed in this connection, as it is composed of diallage and feldspar; but geologically, it is associated with serpentine, and seems in this respect to be separated from the true greenstones.

The dolerites are regarded by some geologists as combinations of labradorite and augite. They may be placed therefore with the pyroxenic compositions, and also as associates of the feldspathic rocks. Graystone contains seventy-five per cent of feldspar—the rest is pyroxene. Euphotide is composed of labradorite and sausurite, according to Rose.

The feldspathic group of rocks is quite extensive, including sienite, greenstone, basalt, porphyry, trap, diallage rock, dolerite, graystone, sausurite in euphotide.

Mica, although associated with feldspar in granite, can not be regarded as one of its constant companions: it rarely accompanies it except in the granites of the older periods. The following analyses express the compositions of two kinds of mica:

	Mica.	Lepidolite.
Silex,	46·36	49·86
Alumina,	36·80	33·61
Lithia,		3·60
Potash,	9·22	4 18
Fluoric acid and water,	1·81	3·45
Oxide of manganese,		1·45
do. iron,	4·53	4·18

The composition of mica is very variable; the iron amounts to fourteen per cent. The presence of mica in a rock promotes its disintegration mechanically. Its structure favors the entrance of water into the mass of which it is a constituent part. But this is not all; it contains potash, and hence, like feldspar, it is subject to decomposition. Granites, therefore, composed of large folia of mica, and large crystals of feldspar, are less stable and more subject to crumble than rocks composed of carbonate of lime, or which consist of a simple mineral as hornblende; or if the granite is composed of fine particles, it is more permanent than the coarser varieties. The Quincy and Maine granites are of this description.

Mica is very rarely a constituent of the pyroplastic rocks. The oldest lavas of Etna contain it, and a mass of metamorphic magnesian limestone of the Onondaga salt group occurs near Syracuse, and is associated with serpentine, another product of metamorphic action.

Feldspar seems to be associated with certain vitreous minerals, as pitchstone, obsidian, &c.; or they may be regarded as fused feldspars and augite, in which the former may occur in obscure crystals. Their composition may be stated as below:

	Obsidian.	Pitchstone.
Silex,	72 00	73·00
Alumina,	12·50	14·56
Potash and soda,	10·00	
Oxide of iron and manganese,	2·00	1·10
Lime,		1·00
Water,	0·09	8·50

Pearlstone and pumice are products of volcanic action quite similar to the foregoing. They are composed of

	Pearlstone.	Pumice.
Silex,	75·25	72·52
Alumina,	12·00	17·50
Potash and soda,	4·50	3·00
Oxides of iron and manganese,	1·60	1·75
Lime,	0·50	
Water,	4·50	

There are other products of igneous action, among which siliceous minerals are the common companions, and might be regarded as real geological groups. Thus most of the greenstones, toadstones, and traps furnish varieties of uncleavable quartz, as chalcedony, cacholon, cornelian, jasper, siliceous sinter, &c. All these varieties are nearly pure silica. Their origin is due to the agency of heated water holding silica in solution. They are of course of posterior origin to the greenstones and traps containing them. If we extend this kind of grouping a little further, we shall find associated with the foregoing a family of minerals which were formerly called zeolites, consisting of analcime, laumonite, chabasie, stilbite, heulandite, thompsonite, mesotype, phrenite, &c.; calcspar is frequently associated with them. They are not confined to the greenstones and traps, as talcose and mica slate rarely furnish them. But the pyroplastic rocks are the true repositories of the zeolites, or according to systematic mineralogy, *kouphone spars.* The following tables exhibit the composition of the most important:

	Analcime.	Laumonite.
Silex,	58·00	48·30
Alumina,	18·00	22·70
Soda,	10·00	
Lime,	2·00	12·10
Water,	8·50	16·00

Laumonite is remarkable for its instability, which is due to loss of its water, when it becomes a white powder. It effloresces in the dry atmosphere.

	Stilbite.	Heulandite.
Silex,	58·08	59·00
Alumina,	16·10	7·19
Lime,	9·20	16·87
Water,	16·40	13·45
Oxide of iron,		10·00

The foregoing family of minerals seem to be due to the solvent action of water. The elements existing in the parent rock are dissolved out under favorable circumstances. The igneous rocks are more or less porous, and hence admit of its transfusion through the mass. If a cavity be opened in a rock, however deep its situation, it is almost impossible to exclude the entrance of water into it, provided the rock is not absolutely anhydrous. At the surface the pressure by which water would be forced into a cavity (which may be regarded as a vacuum) equals fifteen pounds to the square inch: under water the pressure would be greatly increased. The solvent power of water is aided by pressure; hence the transfusion of water into cavities may be accounted for, and indeed provided for, and in its progress through the rock must necessarily dissolve and carry forward the soluble matter with which it meets.

The idea of the transfusion of water into cavities and pores in rocks beneath the sea, is illustrated by Dr. Scoreby's account of a boat pulled down to a considerable depth by a whale, after which the wood became too heavy to float, the air being forced out and replaced by water. So also the filling of empty bottles at great depths in the sea. But the water when forced through the pores of a rock into cavities, becomes a powerful solvent of the earthy salts, which suffer also a transference of matter, which, on crystallizing, constitutes the regular crystals of geodes, cavities, or fissures.

The result of the action of water proves to us that the alkalies and alkaline earths are present in the rock, and that water is competent to dissolve silica. The following exhibits the composition of

	Mesotype.	Leucite.
Silex,	54·46	53·75
Alumina,	19·70	24·62
Lime,	1·61	
Soda,	15·09	
Potash,		21·35
Water,	9·83	

It is proper to remark that the foregoing minerals belong to the submarine division of the pyroplastic rocks. The condition of the submerged rocks is favorable to the development and formation of this natural family of minerals, while the subaerial divisions rarely contain minerals in their cavities, not indeed until they have been placed in favorable conditions for their production. In some instances the foregoing minerals appear to have been formed directly by heat. Those instances may be cited where a rock, as clay slate, has been altered by contact with a trap dyke. Both analcime and garnet have been formed in the slate by the heat of the trap. It is not however clear but that water in this, and most other cases of the kind, has been instrumental in the formation of the minerals under consideration.

§ 43. *Quartz and its group of associates.* Quartz, when interlaminated with mica, forms mica slate, and when associated in the same way with talc, forms the common talcose slates. These mixtures are variable. Sometimes one and sometimes the other predominates. But quartz, although it occurs in the relations I have stated, still it does not seem to hold that relation to talc or mica that feldspar holds to augite or hornblende. The proportion of quartz is very great at some localities in Maryland and North Carolina, and so large that the mass is regarded as a sandstone. Quartz, however, has its associates among the metals and oxides and sulphurets of the metals. The auriferous formations are so constantly associated with quartz, that it is scarcely ever absent. Even the fine talcose slates, which appear at first much like talc alone, when examined with a glass are found to be made up mostly of fine grains of quartz.

The oxides of iron, when in mass or in veins, are usually accompanied with quartz. Carbonate of lime, which is so common, and as the veinstone of Galena, is rarely, if ever, the veinstone of the oxides of iron. The sulphurets of iron and copper are usually accompanied with quartz, especially if auriferous. It is not well determined how the fissures now occupied with quartz have been filled. Some seem to be disposed to regard them as products of fusion. Black tourmalin is common in quartz, penetrating it in a very remarkable manner.

The talco-micaceous slates furnish the staurotides, garnet, and kyanite. The two former are frequently so abundant that they protect the rock from weathering in consequence of their hardness. Garnet and staurotide are composed of

	Garnet.	Staurotide.
Silex,	43·00	33·00
Alumina,	16·00	44·00
Lime,	20·00	3·84
Oxide of iron,	16·00	13·00
Manganese,	0·25	1·00

Garnet gives different results by analysis; in some the lime is wanting, and in others the iron is increased sufficiently to warrant its use as an iron ore. The manganese too is variable in quantity, and in consequence of the difference of composition it furnishes several varieties, as the precious garnet, the melanite, colophonite, manganesian garnet, &c.

§ 44. *Serpentine group.* Serpentine must be regarded as an igneous product, and like other rocks of this class, it has been formed at different periods. It is associated with bronzite, diallage, talc, rensselaerite, schillerspar, steatite, chromite of iron, chalcedony, and calcspar. It will be perceived that the grouping of serpentine differs from other pyrocrystalline or pyroplastic rocks. Serpentine itself is composed of

	Serpentine.
Silex,	42·50
Magnesia,	38 63
Alumina,	1·00
Oxide of iron,	1·50
do. chrome,	0·54
Lime,	0·25
Water,	15·00

The associates of serpentine contain magnesia in different proportions; thus bronzite and diallage:

	Bronzite.	Diallage.
Silex,	60·00	52·00
Magnesia,	27·50	15·91
Alumina,		3·18
Oxide of iron,	10·50	7·47
Lime,		19·59

Diallage is also associated with feldspar, and forms with it diallage rock. In this country, however, it is a rare rock. Steatite and talc contain magnesia, but less iron than the foregoing. They are found to be composed of

	Talc.	Steatite.
Silex,	62·00	48·3
Magnesia,	27·00	26·65
Oxide of iron,	3·50	2·00
Alumina,	1·50	6·18
Water,	6·00	9·05

The soapy bed of talc and steatite is characteristic of these minerals. They scarcely differ, as will be seen by the foregoing. Chlorite, a deep green mineral, which often looks and feels like a green talc, differs however in composition. Chlorite is somewhat important as a mineral species; it is usually associated or grouped with quartz, particularly with milky quartz, as in a part of the Taconic range of mountains in Berkshire, Mass. A supposed chloritic compound forms chloritic slate in combination with quartz, in which respect it is analogous to talcose slate. Its composition is as follows:

	Chlorite.
Silex,	26·0
Alumina,	18·5
Magnesia,	8·0
Soda,	2·0
Oxide of iron,	43·0

Chromate or chromite of iron occurs in serpentine, but where calcspar or carbonate of lime is associated with serpentine, it is often absent, as in northern New York and New Jersey. Octahedral iron, as well as the specular iron, is often associated with serpentine.

§ 45. *Hornblende group.* The associates of hornblende are feldspar, pyroxene, and quartz. It sometimes stands by itself, or has no intermixture of feldspar or quartz. But generally feldspar is present, and the two minerals are arranged in parallel stripes, appearing like a stratified rock. Quartz is never abundant in this compound.

Hornblende is associated with feldspar in another class of rocks—the pyroplastic, the greenstones, or as they are sometimes called, dolerites and diorites. Their structure is more or less granular, and the feldspar may be seen in white crystalline grains, sometimes by the naked eye, but in many instances the eye requires the aid of a magnifying glass. It appears that the hornblende in these compounds is greater than that of feldspar. Sometimes again the feldspar is in quite large crystals, giving the rock a spotted appearance. It has become a greenstone porphyry. This is the case in many places situated in the outburst of greenstone along the Hudson and Connecticut rivers.

Chemists have paid but little attention to the composition of rocks, and hence it is impossible to group them as perfectly as it is wished. Our trap dykes, which are usually so homogeneous and compact, afford no external clue to their composition. They may be hornblendic, or they may be pyroxenic. The absence of magnesia would lead us to place them in this group. Epidote and titanium are common associates of hornblende, especially the rutile and silico-calcareous oxide of titanium. Hypersthene is often associated with it, but commonly with labradorite. It is composed of

	Hypersthene.
Silex,	54·25
Alumina,	2·25
Magnesia,	14·50
Lime,	1·00
Oxide of iron,	24·50
Water,	1·00

§ 46. *Pyroxenic group.* In combination with feldspar, pyroxene forms basalt. This rock is black and perfectly compact, or formed of particles which are indistinguishable by the naked eye. It is the most important of the group. The melapyres are also combinations of these minerals, to which may be added obsidian, pitchstone, and peperino. These volcanic products have but little interest in the eyes of an American geologist. As a mineral, pyroxene is associated with, and hence might be grouped with, our pyrocrystalline limestone and serpentine. It is not common to those serpentines which are not associated with lime. But neither hornblende nor pyroxene enter into combination with limestone as a rock.

§ 47. *Limestone group.* The class of limestones under consideration, though they contain many minerals, yet as a rock, it is not associated with any important ones, except serpentine and its congener rensselaerite. Serpentine in this grouping is subordinate to the limestone. Specular oxide of iron occurs in beds in this rock in northern New York.

The circumstances under which this rock occurs in this country warrants its recognition as a rock quite as distinct from all others as granite. It is by no means a metamorphic mass. When this rock occurs among granites, it is massive and unlaminated; when it occurs among gneiss, mica slate, and hornblende rocks, it is laminated. It might perhaps be grouped with all these rocks, as in New York, Massachusetts, and New Jersey, where it frequently accompanies them.

The pyrocrystalline limestones abound in pyroxene, hornblende, chondrodite, brown tourmalin, spinelle, sulphurets of iron and copper, and phosphate of lime. In veins of Galena calcspar is often the gangue. The foregoing minerals never form rocks in combination with limestone, excepting serpentine. The minerals imbedded in limestone have a peculiar composition, though it may not be due to the rock in which they occur. Thus chondrodite and spinelle are composed of

	Chondrodite.	Spinelle.
Silex,	32·66	2·00
Magnesia,	51·00	12·00
Peroxide of iron,	2·33	16·00
Fluoric acid,	4 08	
Potash,	2·10	
Water,	1·00	
Alumina,		68.00

In the compositions of tourmalins boracic acid is found. It is a rare substance. It is volatile at a high temperature, but possesses active solvent powers. The dissolved matters crystallize from its solutions. Black tourmalin is composed of

	Black Tourmalin.
Silex,	36·03
Alumina,	35·82
Magnesia,	4·44
Lime,	0·28
Potash,	0·73
Soda,	1·56
Oxide of iron,	12·71
Manganese,	0·75
Boracic acid,	4·02

Tourmalins accompany the coarse granites. They are quite rare in the greenstones. A vein of green, red, and blue tourmalin occurs in the coarse granite at Chesterfield. They are more commonly disseminated in the rock, especially the indicolite, and sometimes so abundantly as to have given its name (schorl rock) to the compound. Graphite in tables, and sometimes in regular hexahedral tables, is one of the most common minerals of pyrocrystalline limestone. When it is considered that graphite is a furnace product, we can scarcely entertain a doubt respecting the agent which was instrumental in its origin. This mineral, however, is absent in those rocks which are usually regarded as the metamorphic limestones, as the marbles of Berkshire, Mass., and which are prolonged to Canada on the north, and to Georgia on the south. The green crystallized mica, the brown zircons, sphene, corundum, sulphuret of iron in fine crystals, prismatic feldspar, crystals of quartz in dodecahedrons, carbonate of iron, yellow and brown tourmalins, are among the simple minerals of this rock.

The simple minerals which we find commonly associated with certain rocks may be regarded as having originated under three conditions: the first, those which belong to the granites and pyrocrystalline limestone; these have been developed through the agency of high temperature. The second, those which belong to the mica and talcose slates; these have required for their production only a moderate amount of heat. The kyanites, garnet, staurotide, andalusite, belong to this series. Third, those which are developed in thin seams (not veins) and cavities of rocks, mainly through the instrumentality of water. As examples, stilbite, heulandite, thompsonite, chabasic, &c., may be cited.

The rocks admit of grouping to a certain extent according to the period during which they were formed, or according to the minerals which enter into their composition. The feldspathic rocks belong to different periods—the most crystalline to the earliest, the amorphous and subcrystalline to later periods. To the former belong the granites, and to the latter the greenstones. The schistose rocks, gneiss, mica and talcose slates, together with hornblende, are closely related, and were evidently formed under a diminished temperature. It was temperature, and not water, which arranged their laminæ into parallel layers, a result which is sometimes imitated in furnace operations. We find no sediments beneath, or intercalated between them.

PYROGENIC OR ERUPTIVE ROCKS.

§ 48. *General considerations respecting pyrogenic or eruptive rocks.* Pyrogenic or eruptive rocks have five phases, each of which should be described. The five phases are recognized by the structure of their masses. The first have a symmetrical arrangement of their component parts in consequence of their crystallization. Thus the feldspar of granite is crystallized; and however small its particles, it is perfectly separated from the other minerals of the mass, whatever they may be. The mica also is crystallized, and the quartz, though massive, is equally distinct. Each mineral composing the rock is clearly defined in its boundaries, and is a fact attested by the eye. This structure, which belongs to the mass and each mineral composing the mass, is the first phase among the pyrogenic rocks. The second phase preserves the isolation of particles, whose crystallization took place at the time when, and at the place where, they now remain; but the particles of minerals of the same in kind are arranged in parallel stripes or layers. The feldspar is arranged in its stripes or bands, the mica or hornblende in their bands respectively, and they may alternate with each other. The rock presents a striped aspect. Such an arrangement of parts is properly called lamination. Gneiss, mica and talcose slates, and hornblende are examples of this kind of structure. The third aspect, the separation of particles, is too indistinct to be recognized, or it is far less so than in granite. A single element of the rock may be imperfectly crystallized and isolated, while the particles of the mass remain indistinct, or it may be granular. Greenstones and porphyry are the most distinctive examples; the first is both massive and columnar, and it may be laminated, but the lamination is not distinguished by the arrangement of different minerals in parallel bands or stripes; but the laminæ are all of the same kind of matter while separated from each in thin sheets. This structure is not uncommon. The fourth aspect, the rock is vesicular. In fusion the mass became pasty, and the confinement of the air, which is expanded by heat, forms the vesicles, the sides or walls of which are stiffened before the heated air escapes. The vesicles are large and small, and these may be arranged in stripes. Sometimes the vesiculation expands the mass sufficiently to render it buoyant on water as in pumice. Crystalline structure is wanting in the vesicular rocks. The fifth aspect which rocks of igneous origin present, is that of a glass; it is a vitrification of the rock; it is sometimes homogeneous or striped. To the eye it appears like a furnace production. Under certain circumstances the vitreous mass may be converted into fine glassy spiculæ. These spiculæ often cluster together, and form a flaxen appearance. The rock glasses contain less alumina than feldspar. Their composition, though variable, may be represented as follows:

	1 Obsidian.	2 Obsidian.
Silex,	60·52	84·00
Alumina,	19·05	4·64
Oxide of iron,	4·22	5·01
do. manganese,	0·33	
Lime,	0·59	2·39
Magnesia,	0·19	
Potash,	10·63	
Soda,	3·50	3·55

The 1st is from Teneriffe, the 2d from Iceland.

The sixth aspect occurs in those rocks where heat and mechanical action is so combined as to reduce the mass to powder. Volcanic ashes are examples of this form. The particles are buoyant in the air, and are carried or transported by winds sometimes for hundreds of miles. The foregoing examples are all distinct in the extremes; indeed, except in a few cases, they may be recognized by the student without difficulty. They may, it is true, graduate into each other. It is sufficiently plain, however, that fire, acting with different degrees of intensity under different circumstances, and acting too on compounds variable in fusibility, must furnish a variety of results which are not perfectly classifiable. There will necessarily occur some intermediate results, belonging in part to one kind of structure, and in part to another; still the examples described in the foregoing paragraphs, constitute the distinctive kinds of structural arrangement, which result from the common action of heat upon rocks, and the structure which results from the different degrees of it, furnish the grounds upon which they may be separated into classes or groups.

We have already taken occasion to speak of structures, as affording indications of the age of the pyrogenic rocks, and it appears that at one extreme of time the rocks formed were all crystallized, while at the other extreme they all want it. The first belongs to the most remote period which any of the geological phenomena recorded in earth's history, have furnished; the last belongs to the present, or to the action of the present periods, and form only lavas, slags, sands, and porous products, but no granites. The pyrocrystalline, therefore, differ from other rocks of the pyrogenic kind, in structure and age, though they are not confined to one age or period. By the modification of structural arrangements under the influence of heat, varying in intensity, we may separate the pyrocrystalline rocks into two groups. In the first the massive structure prevails, in the second the laminated. These distinctions have been already illustrated.

The special characters of each group, together with its members, will be given in the proper place.

CHARACTERISTICS OF THE PYROGENIC OR ERUPTIVE ROCKS.

§ 49. *The massive pyrocrystalline rocks.* The first characteristic possessed by this class of rocks is, the perfect separation of each individual of the mass, by crystallization. The second is the indiscriminate arrangement or mixture of the minerals, without regard to lamina, bands or stripes. They are composed of feldspar, mica, quartz, limestone, hornblende and augite. But a separation of parts of the mass is effected by crystallization, which has affected the rock as a whole. They are represented by those which are referred to under the first phase, § 48.

§ 50. *The laminated pyrocrystalline rocks.* They possess the first characteristic of the preceding class. Their second characteristic consists in the arrangement of the component minerals into parallel bands or stripes. In the third characteristic they agree again with the preceding section. They are referred to under the second phase of the preceding section.

§ 51. *The pyroplastic rocks.* The first characteristic consists in their homogenity, or an approach to it. When compact, they are perfectly homogeneous; when granular, it is sometimes possible to discover the mixed nature of the mass by the occurrence of whitish particles in a granular ground, or the ground or base may furnish individuals distinguishable in size, as the basalts, the greenstones, trap, and porphyry. They belong to rocks indicated under the third phase of the section already referred to. The mass may be laminated, or rather sheeted, columnar or massive, or the mass may be vesicular, but the vesicles are not empty. The circumstances connected with their cooling have modified their structure. A part have cooled beneath water, or the sea, and a part have cooled in the atmosphere, and hence the subdivision of the class into *submarine* and *subærial.* The subærial products are numerous. They may be porous, vesicular, glassy or vitreous and compact; vitreous and fibrous, like hair, or in the condition of an ash. In the vesicular structures, the vesicles are usually empty. They are referred to under the fifth and sixth phase of § 48. They are the modern volcanic products.

OF THE MEMBERS OF THE MASSIVE PYROCRYSTALLINE CLASS.

§ 52. GRANITE. *The primary, hypogene and igneous, of different authors.* It consists of feldspar, quartz and mica, commingled together, forming a mass in which their arrangement has no order which can be discerned. Each mineral may predominate in different localities, though it is rare for the quartz to exist in excess over and above the feldspar and mica. The individual minerals have no allotted size; the mass may consist of small particles, or they may be very large. Hence, when the particles of composition are regarded, a granite is fine or coarse. Granites differ in color. The fine are gray, usually; the coarse are white, or nearly so. Granites of an intermediate texture may be either gray or flesh color. The fine and very coarse are rarely flesh color. The quartz is sometimes rose red, but usually gray, and never crystallized. The mica in the fine granite, is nearly black. In the coarse, the mica is greenish, and in some cases black or very dark green. Mica is frequently wanting. Sometimes its place is supplied by hornblende. This last commixture of minerals constitutes the sienite of authors, provided the arrangement is granitic; or if the mica is intermixed with hornblende, it is still regarded as a sienite.

The variability of granite is seen in its coarseness or fineness. The extreme of these kinds will be found in the veins, traversing gneiss, or mica slate. The mica and feldspar is in large sheets and blocks; the former occupying, very frequently, the middle of the vein, and standing with its edges to the center. In other varieties where albite is present, this occupies the center, and is arranged in imperfect stellated laminæ, which are usually hemitropic. Such is the case with the veins of coarse granite at Chester and Chesterfield, Mass. These veins are well defined at their borders, and usually contain some variety of tourmalin. It is mostly indicolite at Chester, but at Chesterfield, black, blue, green, and red occur. So at Topsham and Brunswick, Me., the coarse granites resemble those already referred to. Their width varies from one inch to forty or fifty feet. The hills of primary rocks, of which these coarse veins form a characteristic feature, are peculiar to the New England states, extending on the south to the Long Island sound, and to Maine on the north.

The coarse granitic beds occur at numerous places. Chester, Russell, South Hampton lead mine, Granville, northern New York, Pennsylvania, New Hampshire and Maine. The feldspar is white and bluish white, and predominates in the mass, while the mica is poorly represented. Feldspar in moderately large blocks, of a flesh red, occur at Granville. Another coarse granite is found in Williamsburgh, Mass., in which the mica is plumose. These coarser granites are frequently porphyritic, and the feldspar is the most prominent mineral; but these varieties pass into the finer kinds, and also into gneiss, and might be designated by the descriptive name, granitic gneiss. This variety is well exhibited in the rocks of St. Lawrence county, N. Y., where the granitic rocks are more frequently of a character which places them in intermediate positions. The fine granites of New England, fine in texture, have long been known at Quincy, Chelmsford, Fitchburg, and Sharon. In Maine the granites are both fine in texture and fine in quality. They form the caps of hills, and the mass not being remarkably thick, it has frequently been removed, exposing the gneiss or mica slate upon which it rested, and also bringing to view the dykes through which the molten matter had reached the surface. It is probable that these masses, capping the hills, have been greatly reduced in thickness by denudation. Some of the granites are fine, and have a red color of a uniform tint; others are gray, with a greenish tinge; and others still dark green, from the presence of both mica and hornblende. It is unnecessary, however, to attempt to describe all the varieties of granites of this country. Some of the best, for building stone, belong to the gray fine-grained kinds, which flowed through narrow fissures in mica slate or gneiss, and which appear to have overspread large areas, as those beds in the neighborhood of Augusta and Hallowell, in Maine; while those which are coarse occur in veins in gneiss or mica slate. These usually furnish feldspar in larger blocks, free from iron, and it is often suitable for the manufacture of porcelain. The granites whose mica is in large folia, are unsuitable for building, or works of construction. The granites of the Rocky Mountain range resemble the common gray and flesh colored granites of New England.

§ 53. *Age of granite.* Granite, as it is described in the foregoing paragraphs, may or may not be connected with the oldest masses of the globe. Its age and position is indeterminable, in consequence of concealment by the soil, or by the adjacent rocks. Fig. 5 illustrates this position. If a complete section of a hill in which a mass of granite cropped out, it might disclose granites of three periods, a, b, c (Fig. 2, p. 45), but if a portion only of the mass, a, could be seen, and a portion only of b, it would be difficult, if not impossible, to say whether the two masses were of the same age or not. Sir Charles Lyell has demonstrated that the oldest granites usually rest upon the newer, and hence the term *hypogene*, the nether formed rock. The newer rocks may be connected with the surface by dykes or veins, as at a. In this case the masses of the rock with which they are connected, have cooled against the under side of a more ancient mass. Now that it is understood that cracks and fissures are formed in the rocks by cooling, it is no longer difficult to explain how veins of granite, as well as the metalliferous veins, are occasionally found in sedimentary as well as in those of igneous origin. Those rocks which repose upon an igneous mass, are more frequently traversed by veins than others which are superimposed upon them, proving that they are nearer the source whence all the fused materials originate. The age of granite, whether in veins or in dome-shaped masses, can be determined only approximately. If a mass of granite overlies the carboniferous rocks, it is certainly as new as those rocks, but it is possible it may be more recent than the trias, or new red sandstone.

Fig. 3.

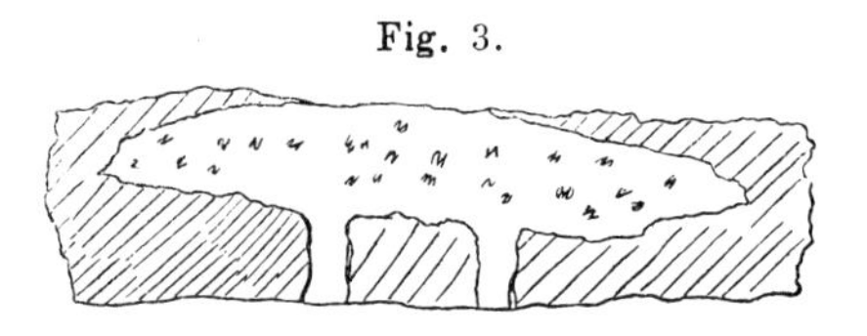

Fig. 3 illustrates a small mass of granite, in Chester, Mass., where the bed is connected with two veins which penetrate through a fine mica slate. The bed itself was formed undoubtedly by an overflow of granite which passed through the vertical veins. This illustrates, on a small scale, the formation of some of the large beds of granite in Maine, and other parts of the country. We have not as yet discovered granite as new as the cretaceous system in this country, while in the Alps this rock penetrates the chalk formation. It appears, from observation, that many of the most imposing outbursts of granite were accompanied with important changes of level of the earth's surface, and consequently affected more or less animal and vegetable life.

§ 54. *Distribution of granite in the United States.* The granites of this country furnish the usual varieties which have been described by foreign authors. Two of the most common departures from the normal kind is composed of quartz and feldspar, the mica being absent, and that which is composed of quartz, feldspar, and hornblende or pyroxene—the two replacing the mica. Whatever change the rock has undergone, it retains the massive structure of granite. A less common variety receives talc in the place of mica, and is called protogene, and forms that kind which is liable to decomposition, and which furnishes one of the porcelain clays. The soda granite at Gouverneur contains large crystalline masses of albite, though frequently in perfect hemitrope crystals, associated with carbonate of lime.

§ 55. *Granites of northern New York.* In northern New York the granites are exceedingly variable in composition. In this district they become metalliferous, and in this respect differ from those of New England. In Clinton county a flesh-colored granite is traversed by lodes of magnetic iron. The state prison, located in that county, is built upon lodes of this rock. The magnetic iron of Arnold hill, in the same county, is in the same rock. One vein of this hill is a peroxide. The granite of St. Lawrence county is in part similar to the New England granite, particularly that variety which is found at Alexandria bay. It forms the Thousand islands of the St. Lawrence, and contains schorl and imperfect garnets and epidote. The most common kind of this county is associated with limestone. It is underlaid and traversed by seams or veins of coarse crystallized limestone. The rock itself is also coarser than it usually is. It is also the repository of many minerals. The peroxide of iron, barytes, strontian, carbonate of strontian, albite, pyroxene, hornblende, fluorspar, and sulphurets of iron and copper belong to this rock. It should be stated that the rock decomposes readily, and where exposed upon the lake or river shore, becomes cavernous. The ores and minerals occur in nests and strings, which run out, and hence has ever proved an unsafe rock in mining. The most common variety of granite occurring in the low ranges of the Blue ridge is composed of feldspar and quartz. It is always in irregular veins, and is sometimes auriferous. In other respects it is barren of minerals, and in decomposition forms a porcelain clay. It is common in Guilford, Davidson, Cabarros, Mecklenburg, and Rowan counties, in North Carolina, and is associated with greenstone in dykes. A similar granite occurs in Macon and Cherokee counties. It is not uncommon in the Nantahala range, a spur of the Blue ridge.

§ 56. *Granites of the New England states.* In Vermont, granite occupies a portion of the eastern slope of the Hoosick Mountain range. It does not appear in the western part of the range, but comes in east of Montpelier.

Maine furnishes, however, some of the finest fields of architectural granite.* It is light gray, of fine texture, and works easily. Columns from thirty to fifty feet in length may sometimes be split out from the quarry. The granite of Hallowell lies in sheets or thick laminæ, which may be reduced to columns by splitting in lines parallel with the grain of the rock. These granitic beds may be said to be sheeted, in consequence of the easy and ready cleavage of the mass on a large scale. Indeed, it is a spontaneous separation into laminæ, varying in thickness from one to three feet. It is not well determined how the granite received this peculiar structure. It is probable, however, that it owes its sheeted structure to its flow at the time of its eruption, and the consequent cooling

* Jackson's Maine Reports.

of the mass. It approaches gneiss in its structure, but differs from it in wanting the arrangement of the mica planes. This example, however, proves that a sheeted structure is not due to the action of water, and was never arranged into beds like the sedimentary rocks.

The granites of the New England states lie in ranges, inclined upon the eastern slope of the Hoosick Mountain range. At the south-western corner of Vermont a field of granite forms a portion of the western side of the Hoosick mountain, upon which repose the lower members of the Taconic system. But most of the New England granite lies adjacent to the Atlantic coast. It is divisible into those granites which have been erupted from comparatively narrow fissures in gneiss, mica slate, and an older granite, or those granites which have erupted from fissures, but which seem to have overflowed wider areas, and whose structure is more or less sheeted, like the granite of Hallowell in Maine; and lastly, those granites which are still more widely spread, and more ancient than the preceding. It is impossible, however, to draw lines of distinction between the last two kinds of granite, except when the last is traversed by the preceding granites in veins.

§ 57. *Granites of the Appalachians.* The granites and sienites of the southern Highlands upon the Hudson river resemble those of New England. They pass southerly into Pennsylvania, some of which contain zircons and octahedral iron. In the county of Philadelphia the gneiss and mica slate is traversed by coarse veins of granite, the feldspar of which decomposes into kaolin. So near Manayunk, the mica slate which abounds in garnets is intersected by numerous veins of granite, in which feldspar predominatès. Chester and Lancaster counties furnish numerous localities of granite.

In Virginia, granite and sienite are not common rocks Those which occur form a part of the Blue ridge. Both occur in Halifax and Campbell counties on Staunton river and Whipping creek; also on James river four miles above Greenway. This is a formation of sienite, and is well adapted for works of construction, dams, culverts, viaducts, &c. It extends many miles in length. Granites only slightly elevated above the general level of the country form a low and rather continuous ridge, extending through Virginia, North and South Carolina. This ridge forms the first waterfall of the Rappahannock, James, Roanoke, Tau, Neuse, and Cape Fear rivers. It is often a variety which may be called a gneiseoid granite.

Granite underlies in part the counties of Guilford, Davidson, Rowan, Mecklenburg, and Lincoln. The belt extends south into South Carolina. But a compound of feldspar and quartz is one of the most common rocks at the base of the Blue ridge. Granite is by no means a common rock in the higher parts of the Hoosick mountain and the Blue ridge. The most common rocks of these high ranges are mica and talcose slates, hornblende and gneiss.

On the west side of the Blue ridge granite is equally scarce, and when it occurs it is mostly in veins, and consists of quartz and feldspar, a rock which is sometimes extensive. It decomposes, and forms a great abundance of white clay. About four miles west of Ashville, Buncombe county, a handsome granite suitable for architecture crops out and crosses the road; but most of the rocks skirting the French Broad river are gneiss and mica slate. Granite occurs four miles east of the Warm springs: it rises in a dome-shaped mass, and supports quartz and slate rocks, the lower members of the Taconic system. South-westward and westward from the Warm springs all the rocks to the Mississippi river belong to the sedimentary class. Granites of the same kind and character appear in Macon and Cherokee counties, and from thence they extend into Georgia. They never form large and important masses among the rocks. To the south-west and west of the Mississippi granite occurs, and forms in part a low range of mountains, which have been called the Ozark mountains. Their tops rise like islands in the midst of cretaceous seas. In the Thousand isles of the St. Lawrence, and so onward to the west in the Lawrentine chain, granites and sienites are far more common than in the Appalachian range. This range is flanked on all sides with the oldest sedimentary rocks. In the British provinces, Nova Scotia and New Brunswick, granites resembling those of Maine are well known. They range, with some intermission, from Canso to Halifax, bordering a low rocky coast.

§ 58. *Granites of Oregon and California.* Granite of several kinds is a constituent part of the great mountain ranges of Oregon and California. It is associated with traps, basalts, sienites, and mica and talcose slates.* According to Dr. Pickering, granite forms a part of the Cascade mountains, having met with it about twenty miles north of mount Rainier. The summit of the pass of this range is however trachyte. North of Okanagan, and east of fort Colville, granite is the prevailing rock. But it occupies, according to Dr. P., an anomalous position, the summit of the range being formed of basalt or trap, while the sides only are granitic. Farther south, according to Professor Dana, or between Oregon and San Francisco, albitic granite is a common rock of the principal ranges, especially of the Shasty mountains. This is sometimes porphyritic. The color of the granite is usually light, and fine grained, constituting a firm mass, and little subject to decomposition; though when changed it is as white as chalk. Sometimes the albite is red. The rock is barren of minerals. The Rocky Mountain range has its share of granite in its composition, though sedimentary rocks reach the principal passes of the chain on the eastern slope.

§ 59. *Granites of lake Superior.* The northern side of lakes Superior and Huron, together with the highlands of Wisconsin and Michigan, and between the Upper Mississippi and Michigan and Superior, is another extensive field of eruptive rocks, among which granitic and sienitic protrusions are very numerous. It is a region covered with drift, and hence no small parts of the rocks are hidden from view.

* Dana's Report U. S. Exploring Expedition.

SIENITE.

§ 60. Sienite is a granite in which hornblende takes the place of mica. This is more abundant than mica. The latter may be present in small quantities, and it may be absent. This rock is to be distinguished from hornblende rock, however, by the absence of lamination. Its structure is granitic in the true sense of the word. Taking granite as the type of sienite, it may, like the former, be divided into varieties, either by the absence of one of its elements, or by the form or shape which one of its elements has assumed.

§ 61. *Feldspar and hornblende.* By reference to the composition of greenstone, it will be seen that this is also composed of the same substances, but the particles in greenstone are minute, and unless it is porphyritic, it appears homogeneous. But this variety is made up of distinct particles of hornblende and feldspar. There is, however, a gradation of this variety into greenstone, or an approach to greenstone. This fact may be observed at Nahant in Massachusetts.

Fig. 4.

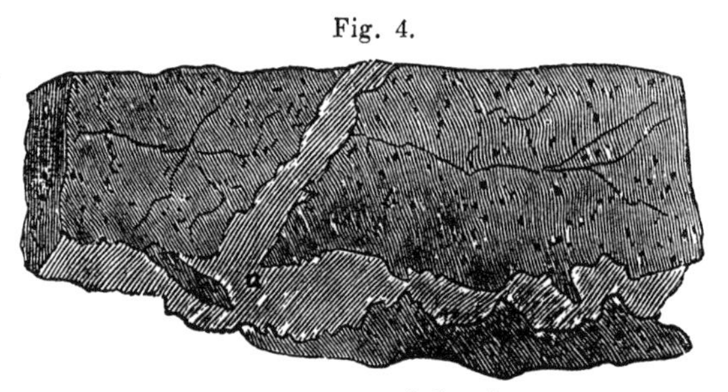

a Limestone, *b* Granite.

Sienite and granite are often associated with limestone. Orange and St. Lawrence counties furnish many instances: fig. 4 is an example of the relation of the two rocks at Fowler, St. Lawrence county. The figure shows the structure of granite.

That limestone is strictly a pyrocrystalline or eruptive rock at all the localities which I have cited, appears from the fact that where it is in contact with the Potsdam sandstone the latter rock is vitrified. It has lost its granular structure, and near its junction with limestone it has become porous, preserving at the same time its vitrified character, and its disposition to break with a conchoidal fracture. Its luster is resinous rather than vitreous. One of the most interesting localities is at Theresa on Indian river. The junction of these rocks may be observed in a gorge below the falls of the river. Many interesting points are exposed in the vicinity of this place, either in the iron mines or the various ravines connected with the creeks in the vicinity.

Fig. 5.

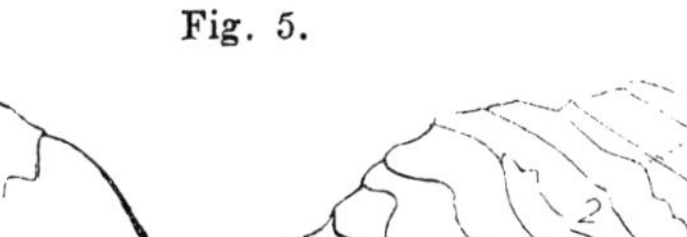

1 1 Gneiss, 2 2 Primary Limestone, 3 3 Potsdam Sandstone, changed at its junction with the limestone into vitrified quartz.

The common kind of sienite is composed of feldspar, quartz, and hornblende, arranged as the minerals are in a true granite. The quartz is usually a light smoke gray, the feldspar is also usually gray; but in some cases the latter is red. These minerals are mixed uniformly together, and the individuals are small. If mica is present, it does not materially alter the appearance or character of the rock: ususally both are black and in small particles. The whole compound will be fine-grained, and in this condition constitutes a good building stone.

§ 62. *Porphyritic sienite.* The feldspar in this variety appears in large individuals, imbedded in a finer ground. Either of the preceding varieties may pass into this by a change in the size of the particles of feldspar.

Sienite occurs in Orange, Essex, and Clinton counties in New York, and in several counties in the eastern part of Massachusetts, particularly in the vicinity of Boston. So also sienite is a common rock in Pennsylvania, Virginia, and North and South Carolinas. In Virginia it occurs in Nelson and

10

Augusta counties, forming one of the rocks of the Blue ridge, where it is associated with gneiss and granite. At this place it is a dark gray porphyritic rock, and contains epidote. In North Carolina it forms a wide belt, running northeast and southwest in Randolph county, and on towards the narrows of the Yadkin. It is a black tough rock, in which hornblende predominates, and in which quartz is only sparsely scattered through it.

The rock quarried at Quincy, Cape Ann, and at other places in the eastern part of Massachusetts, and which is so well adapted for columns and the walls of buildings, is more closely allied to the true granites. These are not tough and difficult to be quarried, because the hornblende is never in excess.

To the foregoing varieties there might be added a pyroxenic sienite—a kind in which pyroxene forms a perceptible part of the rock. Sienite is more closely related to greenstone than to the ordinary granites, and it often passes into the former rock.

HYPERSTHENE ROCK.

§ 63. This rock is regarded as a granite. In constitution it differs from the common granites in being composed of labradorite, feldspar, and hypersthene, the last of which is allied to hornblende. The feldspar contains lime and soda. Its composition has already been given. The color of this rock is usually a smoke gray. The color of the labradorite, however, determines the color of the rock. It has the usual granitoidal structure. This results from the crystallization of the feldspar, a portion of which is fine, and represents the base, in which there is imbedded individuals of a large cleavable size. These cleavable individuals present very frequently a beautiful opalescence of bronze, yellow, blue, and green colors. In the mechanical arrangement of its particles of composition it resembles a porphyry; but the rock chemically considered consists mostly of labradorite, the hypersthene being extremely rare in it. The rock is destitute of mica, and almost of quartz; and if quartz and common feldspar occur, they are subordinate to it. The characteristics of the rock are derived from the labradorite. Hornblende and pyroxene both replace the hypersthene at certain localities. This rock, when changed by the action of the weather, becomes light colored, and resembles the gray granites. The atmosphere acts upon the rock in stripes or bands, which run in the direction of the natural joints. The action of the atmosphere, rains, and frosts is as great as upon any of the known granites. Upon the tops of the Adirondacks large masses are strewed over the surface like huge boulders, but still in situ having been quarried out by atmospheric agencies alone.

There are but few varieties of hypersthene rock which are worthy of special notice, of which the following are the most important:

1. The first is composed wholly of labradorite, though to the eye it has the aspect of being made up of two distinct minerals.

2. Labradorite and hornblende. The hornblende takes the place of hypersthene, though sometimes the latter is still present.

3. Labradorite, hornblende, and epidote. The latter, however, never occurs in sufficient quantity to change the character of the rock.

4. Granular labradorite and mica. This variety is quite dark, and resembles a trap. The mica is in tufted, radiated masses, and almost black. It occurs usually at the junction of the rock with gneiss.

Magnetic iron in grains is diffused or disseminated through the rock. It is black, with a resinous luster. Its obedience to the magnet serves to distinguish it from other dark-colored minerals.

Hypersthene rock is traversed by a double system of joints, in consequence of which it often appears in the process of separating into large tabular masses. One set of joints run S. 5° W. The separation of tabular masses is frequently in the direction of the slope or side of the mountain where the rock occurs. A separation of the masses also often takes place in the veins of segregation, and the rock also cracks into wedge-form masses.

Like other granites, this rock decomposes, and forms a clay, which is quite refractory in the fire; but it is never so white as the purest porcelain clays. In this natural or spontaneous analysis, most if not all of the granites furnish, in connection with the pure alumina, oxide of iron, peroxide of manganese, and crystallized silica.

The hypersthene rock is confined mostly to northern New York. The western part of Essex county is made up entirely of this rock. It forms all of that group of mountains in this part of the state which are known as the Adirondacks. A train of bowlders, derived from this cluster of mountains, passes through Amsterdam, thirty miles west of Albany. It extends to Orange county. Another train of bowlders range along the St. Lawrence in St. Lawrence county, New York. This train came from another group of mountains far to the north or northeast, probably Labrador. It has no connection with the Adirondacks.

This rock receives a fine polish, and would form beautiful tables and other ornamental articles of furniture. The most important mineral associated with this rock is the magnetic iron ore. Prehnite, chalcedony, and albite are found in this rock, though by no means abundantly. It is poor in minerals. It contains subordinate beds of pyrocrystalline limestone, which are rich in minerals: those, for example, which are common to it when it is associated with other rocks.

The area which this rock covers is small when compared with the common varieties of granite. In the United States it is mostly confined to the region occupied by the Adirondacks: though I have observed a few small patches in other parts of New York, they are too inconsiderable to require a notice in this place. It is impossible to determine even approximately the age of this rock. It is isolated and disconnected with fossiliferous rocks. The elevation of the Adirondacks, however, was probably subsequent to the consolidation of all the lower Silurian rocks. Upon lake Champlain the evidence of movements of a much later date are fully established. The fact that these movements have taken place since the drift, was made known long ago in the reports of the New York survey. It can not be determined whether they extended to the central mass of mountains, situated between lake Champlain and the St. Lawrence. All that portion however of the hypersthene rock which extends to the lake has been raised about five hundred feet since the drift period.

PYROCRYSTALLINE LIMESTONE.

§ 64. *Primary limestone—metamorphic limestone in part.* There can be no doubt that limestone occurs among the most ancient consolidated rocks of the globe. The investigations which I made sixteen years ago satisfied me on this point. At that time no one had entertained this view in this country.* The rock is coarsely crystalline, usually white, or gray, or greenish, rarely blue. It occurs in beds beneath granite, and

Fig. 6.

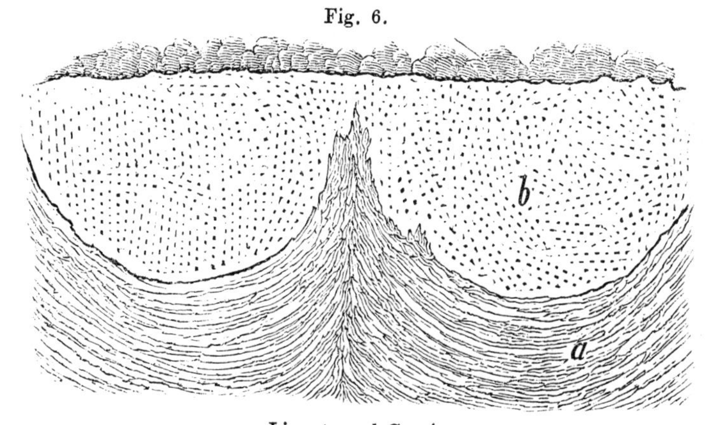

a Limestone, *b* Granite.

frequently underlies and penetrates it as in fig. 6: *a* limestone, *b* granite. The locality where it may be observed beneath granite is one and a half or two miles south of Clintonville,

* New York Geological Report for 1838, for the northern district.

New York. The line of demarkation between the two rocks is perfectly distinct. Its occurrence in veins in granite (figs. 7 and 8) is conclusive of its igneous origin; it proves that like granite it has undergone fusion, and has been injected into cracks and fissures of the superincumbent rock.

Fig. 7.

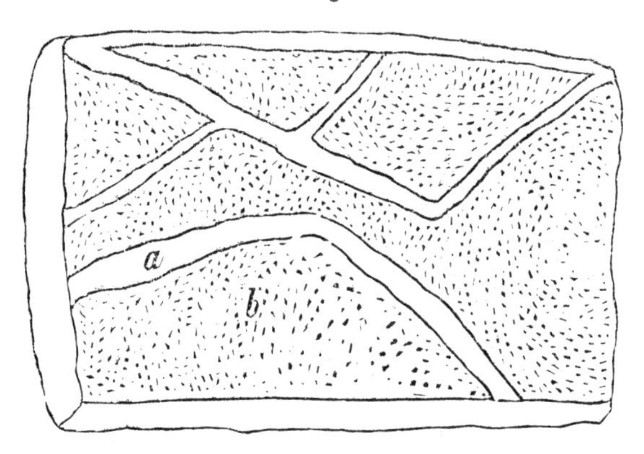

Fig. 7.—Ground plan of a system of veins in granite, as they occur at Gouverneur, St. Lawrence county, New York. These veins extend for many rods in length, and ramify in manner similar to granitic veins. Many localities occur in St. Lawrence, Jefferson, and Essex counties. They furnish an arrangement different from the foregoing, but in keeping with it.

Fig. 8.

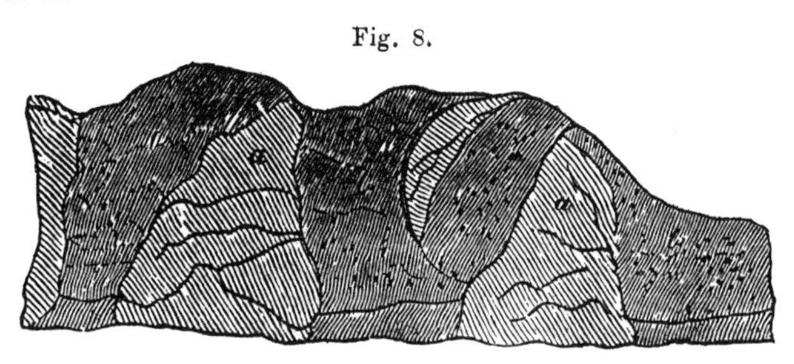

Fig. 8 shows a vertical section of portions of the same vein, where they terminate laterally in a broken ledge of granite, showing that they descend into the mass of granite, or in other words, the mass was erupted through cracks or fissures in the rock.

It contains scapolite, hornblende and pyroxene, spinelle, octahedral iron, fer-olgiste, graphite, mica, talc, phosphate of lime, brown tourmalin, serpentine, &c. The districts of this variety of limestone are limited. In St. Lawrence and Jefferson counties it occurs just within the band of the lower Silurian rocks; in Essex and Clinton counties, at Moriah and near Clintonville; also in the western parts of Essex and eastern part of Hamilton counties, in the vicinity of the Adirondack iron works; in Canada West, also, twenty miles west of Ogdensburg. This district is identical with that of St. Lawrence and Jefferson counties. The granite of the Thousand islands lies between the two districts, and is entirely destitute of limestone of this kind and the minerals it contains. In Orange county, New York, and the adjacent part of New Jersey, Sussex county, primary limestone, containing spinelles and most of those minerals already noticed, forms a band of considerable extent. The red oxide of zinc, sapphire, and chondrodite are minerals which have not as yet occurred elsewhere, excepting the latter, which is found in small quantities only in northern New York.

Fig. 9.

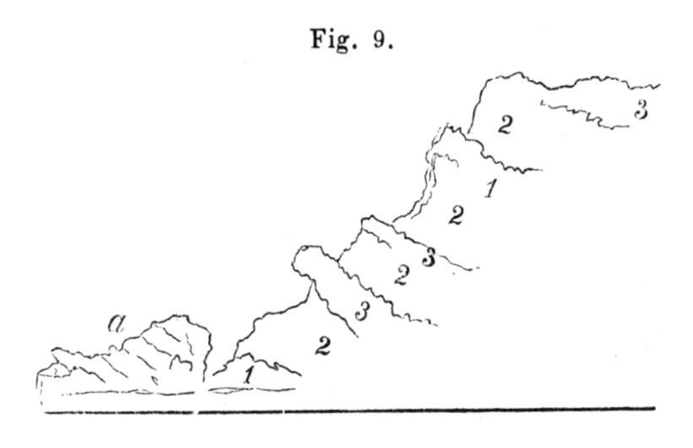

a Potsdam and Calciferous Sandrock, 1 1 Hornblende, 2 2 Limestone, 3 3 Gneiss.

In structure this rock scarcely differs from granite. It is subject to disintegration, and suffers more from the action of atmospheric agents than granite. It has been maintained that the limestones under consideration are metamorphic—the lower Silurian altered by heat. That this view is incorrect appears from the fact, that in St. Lawrence county the Potsdam sandstone actually overlies it. Fig. 9 represents a bluff of limestone and other primary rocks at Port Henry, in Essex county, New York, against which the Potsdam and calciferous sandrock reposes. The relation of these masses is far from being in accordance with the metamorphic doctrine.

Fig. 10.

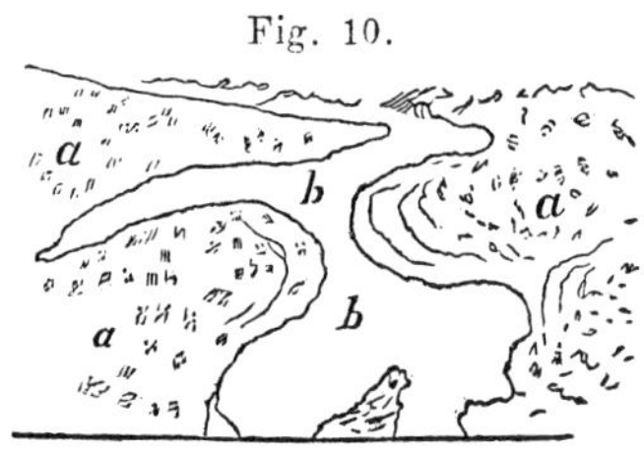

a Limestone, *b* Hornblende.

Insulated masses of hornblende often occur in primary limestone. Sometimes they appear in quadrangular shapes, and in other cases their forms are irregular, as represented by fig. 10.

The same view of the origin of limestone is supported by the occurrence of it in hypersthene rock at Long pond, in Essex county. It is an oblong mass sixty feet wide, extending nearly north and south down the face of a precipitous ledge of rock. It is rich in pyroxene and scapolite.

The metals, their oxides or sulphurets, though not very common in this limestone, have usually run out, and like the granite associated with it, it has proved an unsafe rock for mining. There is one exception, however, in the franklinite and red oxide of zinc. The specular oxide of iron, which is common and very beautiful in St. Lawrence county, is often insulated and removed entirely from its bed; and it has been as yet impossible to recover the bed or vein when once lost.

The condition of the simple minerals in limestone is worthy of special notice. The phosphate of lime, though softened with difficulty in the flame of the common blowpipe, is apparently acted upon by heat; the edges and angles are rounded or flattened, as if it had been in a pasty state since it had assumed its prismatic form. The quartz, which is often imperfectly crystallized in its usual form, has its angles and edges rounded also; but when in its most characteristic condition, it is more like a furnace slag than a crystal. Many pieces look as if while softened they had been drawn out or extended. Sometimes, again, the quartz has assumed a globular form, or the shape of a slag or of a bead. Notwithstanding, however, the similarity of the masses of quartz to furnace slag, it may not prove that those forms and conditions are due to heat; for globular quartz has been found in the calciferous sandrock, in which case it can not be referred to heat as a cause. Still, in the case of primary limestone, the circumstances are such as to favor the views I have expressed, and that the appearance of the masses referred to may be regarded as evidences of the fact.

Crystals imbedded in this rock often contain large particles of limestone in their interior. Pyroxene, quartz, phosphate of lime, and brown tourmalin are rarely free from this substance. The cavities containing the limestone are never angular, but always rounded. The imperfections of the crystals are due mainly to this cause. All the large crystals especially are subject to these faults, even the zircons, spinelles, and corundums are liable to them. I have noticed three districts in northern New York where this rock forms the most striking feature in the geology of each: 1. That of Jefferson and St. Lawrence counties, which has been already referred to; 2. Essex county, near lake Champlain, which may be traced, with very slight interruptions, into Warren county, forming a belt which runs northeast and southwest; and 3. Orange county. This last district extends into Sussex county, New Jersey, and forms a remarkable series of minor belts, which are prolonged to the southwest. In these three districts the limestone is very coarse and crystalline. Its true structure is like that of granite, and the lamination is generally obscure. Another and independent belt belongs to the Hoosick Mountain range. The beds crop out at intervals from Canada to Long Island sound. Their form is oblong, and the structure of the masses crystalline. They are in gneiss, and very frequently in close relation with hornblende. These beds are laminated, partaking of the structure of the rocks to which they are subordinate. Another belt of the same character lies west of the Highlands, extending south through Orange county into Sussex county, New Jersey. In all these belts it is accompanied with serpentine, and in all these localities the rock is massive or only obscurely laminated. At Franklin furnace, franklinite and the red oxide of zinc are largely developed in this rock.

In Chester and Lancaster counties, Pennsylvania, this rock is very common. It contains serpentine, chondrodite, pyroxene, sphene, zircon, quartz, amphibole, corundum, graphite, &c. This rock is white crystalline, and contains carbonate of magnesia in many of the localities which have been cited. It lies in wedge-form masses, and disappears after being apparent at the surface for a few miles. It occurs in belts, whose direction is southwest and northeast.

Passing into Virginia, ranges of limestone skirt the eastern base of the Blue ridge, one of which passes through Albemarle county. They may be regarded as forming several subordinate belts between Lynchburg on the west, and the region of the marls on the east. They are composed of oblong or wedge-form masses, as in Pennsylvania. They are confined to the shistose and laminated rocks, as talcose and mica slate and hornblende. These beds are exposed to a much greater extent than at many other sections of the state, in consequence of the winding of the rivers which intersect the formations. In North Carolina the limestone rocks are extremely rare; two ranges, however, traverse the state from northeast to southwest. Beginning in Stokes county, it is found crossing the Yadkin, passing onward to Lincolnton in the direction of Kings mountain into South Carolina. Another belt belongs to the Blue ridge, and has been observed in Burke and Marion counties, Buncombe and Hayward counties.

The ranges of pyrocrystalline limestone which have been very briefly, and probably imperfectly, traced through Pennsylvania, Virginia, and North Carolina, belong to the laminated and schistose rocks; and they are less coarse and crystalline than those of New Jersey and New York, which contain some of the rarer minerals, as zircon, sapphire, spinelle, and brucite. Still all these limestone beds must be regarded as belonging to the eruptive class. Those which are found in the first three states mentioned, resemble the beds in the Hoosick range, which pass through the eastern part of Berkshire and western part of Hampshire in Massachusetts. We should at any rate not confound them with the Vermont and Berkshire marbles, which belong clearly to the sedimentary series, and which are continuous and persistent through areas of great length. It is perhaps not easy to distinguish them in hand specimens, or in the cabinet, but their associations in the field attest the formation to which they belong. Geographical position often obscures the relations. For example, the beds of dolomite in Dalton and Washington are pyrocrystalline or primary limestones, while those of Pittsfield, only four miles to the west, belong to the Taconic system, or to the sedimentary class. The rocks can not be distinguished from each other by their lithological characters, and both are not unfrequently regarded as metamorphic; but the former stand in the same relation to metamorphism as granite and gneiss. The latter have no doubt undergone a change in their lithological characters, but it is not necessary to infer that the agent which induced the change was heat. Those geologists who refer all changes in structure and texture to heat, take only a partial view of the forces which act, and which have acted upon the earth's crust. A comprehensive view of the cause of structural change in rocks is of great value in geological reasoning, and we are thereby enabled to account for those changes when collateral facts forbid the agency of fire. The view which I have presented of the origin of those masses of limestone so common in St. Lawrence, Essex, and Orange counties, New York, in modified forms, two of which are prolonged into southern states, is consistent with known facts. 1. The position of many of these masses is such that they can not be referred to the lower Silurian limestones, as has been attempted by several eminent geologists. 2. From the great extent of the sedimentary limestones it will follow that these must have been formed from preexisting limestones, which once belonged to the original constitution of the earth's crust. We must go back to their primary condition and position in the earth's crust. No one at the present day pretends that limestone is an organic product in the strict meaning of the word. Is it not better, then, when we find a limestone occupying those relations which forbid the adoption of the view that it is a changed rock, to place it with those masses with which it is associated? And such is the position and relation of all those masses of limestone which I have described under this head, that they can not be referred to the Silurian system without doing unnecessary violence to the relations which they naturally sustain.

SERPENTINE.

§ 65. It is green; the variety of shades being numerous, passing into black on one side, and on the other into very pale green. It is sometimes brown. Its grain is always fine, and in this respect there is a very great uniformity in all its varieties. It is never coarse like certain varieties of limestone, and if columnar, as at Lowell and Newfane, Vt., and Middlesex and Cummington, Mass., the individuals are extremely slender, passing into asbestus. The rock is homogeneous, and is both massive like granite, and laminated like gneiss, and hence belongs to both divisions of the pyrocrystalline rocks. The massive kinds occur at Lowell and Newfane, Vt., and Middlefield, Chester and Blanford, Mass.; or it may be it is sometimes obscurely laminated. The distinctly laminated kind in Macon county, N. C., is of a dark green, where its lamination is more distinct than that of gneiss. The same variety is found at or near Port Henry, Essex co., N. Y. In Middlefield and Chester, it forms a range of hills some five or six miles in length, and less than half a mile in breadth. The serpentine of the Bare hills, near Baltimore, resembles that of Chester, and is probably more extensive. This rock is remarkably distinct from other rocks; it passes into steatite, but very rarely, if ever, into other rocks. The evidence of its igneous origin is less than that of primary limestone. I have never seen it in narrow veins and dykes like greenstone, neither does it occur resting upon other rocks. It rather appears to have been protruded between other rocks, as at Middlefield, where on one side it is bounded by hornblende, and on the other by mica slate. Chromite of iron, with many varieties of chalcedony and jasper, are among its associates. All the localities which I have named, furnish it. The serpentine of Troy, Vt., near the Provincial line, is traversed by a wedge-form vein of magnetic iron. Like primary limestone, it is an unsafe rock for mining.

Serpentine, when largely mixed with limestone, does not contain chromite of iron. It appears to be absent in the calcareo-serpentines of Canada, the St. Lawrence and Champlain districts, and I believe also in Sussex, N. J., and Orange county, N. Y. Mica and talc, however, in crystals, are common. Large plates of bronze-colored mica occur in the serpentine of Gouverneur, N. Y., and hexahedral tables of a deep green talc at Troy, Vt., associated with arragonite, and octahedral iron with brilliant faces.

The occurrence of so large a quantity of silicious minerals not unlike silicious sinter, furnishes some evidence that beds of serpentine may have been connected with ancient hot springs. Macon county, N. C., especially furnishes immense quantities of sinter-like deposit, in connection with serpentine.

Serpentine is one of the constant associates of this kind of limestone, in New York, Canada West, and New Jersey. It is frequently disseminated through it in small grains, but sometimes in large masses of an irregular form and rough surface, and again in fibrous masses. The grains and masses stand out in relief, the limestone weathering more rapidly than the serpentine. This mixed or compound rock, takes a very good polish, and might be used for a variety of purposes when the rock is sound. The serpentine is arranged in the form of coarse agatized bands, with a shape more oval than round, as in fig. 12.

Fig. 12.

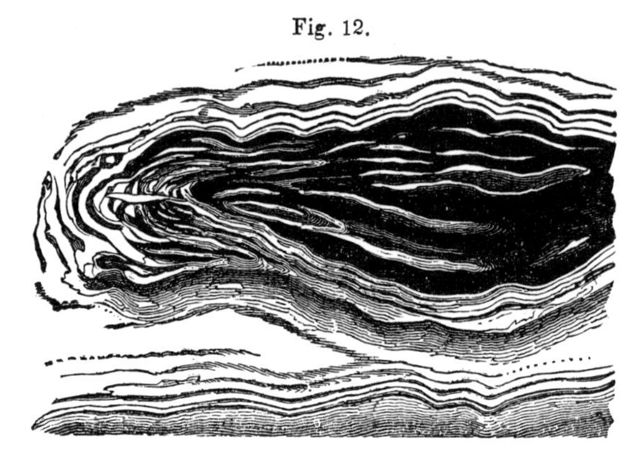

When serpentine is intermixed with limestone, it is often rich in other minerals. The primary limestone, though mixed with serpentine, is frequently pure; sometimes it is magnesian. That of Jefferson county is composed of

Carbonate of lime,	98·24
Alumina and peroxide of iron,	0·88
Silica,	0·88

The defects of this serpentine limestone, arise from an intermixture of silica, pyroxene, and hornblende, which appear in masses disseminated through it.

A serpentine of a pattern somewhat different from the foregoing, occurs in the state of Connecticut, at Milford. It has been described in the Geological Report of that state, by Mr. Percival, as a serpentine marble. It consists of two short ranges; the one includes the New Haven quarries, and the other the Milford. The rock is remarkably fine-grained, and is distinguished for the beauty of its variegated colors; blue, green and yellowish green predominating. Pyroxene, epidote, magnetic iron, picrolite, and chromic iron, are associated with it.

The serpentine marble of Milford and New Haven, is more intimately blended with the limestone than that of northern

New York and the British provinces, clouding, as it were, the mass, while in the latter places it is in distinct grains and masses, which are perfectly circumscribed, and may be detached by a blow, when aided by previous decomposition.

The serpentines of St. Lawrence county, and which are often associated with the earthy oxides of iron, contain angular pieces of quartz, from a tenth to half an inch in diameter; the quartz is not incorporated with the rock, yet it is closely invested with serpentine, and is perfectly separated from it by disintegration.

The serpentine of Cornwall, Eng., is associated with diallage rock, and is traversed by veins of this substance, as well as by granite. Our serpentines very rarely contain diallage. The dark green serpentine of Westfield, Mass., contains schiller spar. It appears then that it is traversed by other rocks, but I have not seen that the different geological writers have observed serpentine traversing in this mode, other rocks, except in Tuscany. Yet it may have been observed, at many other places, but regarded as not worthy of attention.

In the serpentine belt of Port Henry, the mixtures of serpentine and other minerals decompose and leave a scoriaceous mass, like calcareous tufa, as if there was first a deposit or formation of a very porous mass, which was afterwards filled by infiltration with carbonate of lime. I know of no true serpentine so connected with sedimentary rocks as to give a clue to its age, excepting that associated with the waterlime near Syracuse. In St. Lawrence county, iron ore and serpentine are somewhat blended with the Potsdam sandstone. As it occurs in this country, it must be regarded as one of the most ancient of our rocks. The specific gravity of this mineral is 2·55.

Serpentine is a hydrous bisilicate of magnesia. It is composed of

Silica,	41·89
Magnesia,	40·24
Oxide of iron,	3·38
Water,	15·20

Serpentine, though it can not be regarded in itself as rich in ores, yet it is often associated with, or rather in near contiguity to a great variety of minerals. Chromic iron is one of its most constant associates, and occasionally magnetic iron traverses it in veins, as at Troy, Vt., and in some parts of Europe, it is rich in copper. In St. Lawrence county, N. Y., all the beds and veins of specular iron are contiguous to serpentine, and this is the case also with the large rocks of magnetic iron in the Adirondack, in Essex county.

At the well known Parrish mine in St. Lawrence county, N. Y., the serpentine is protruded beneath the gneiss and specular

Fig. 13.

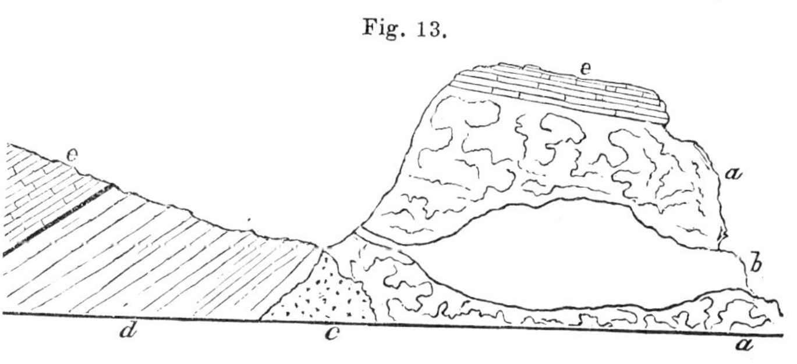

iron, as represented in fig. 13, thus: *a a* is a mass of ore, rather silicious, *b* an adit in the mass of ore, *c* protruded serpentine, *d* gneiss, and *e e* Potsdam sandstone. The serpentine in this instance, seems to have been the rock of eruption which elevated and broke up the sandstone. So also in a contiguous vein known as the Kearney ore bed, a similar dislocation is known. Near Theresa the relations of the rocks are the same, of which fig. 15 is a section: *a* serpentine, *b b* specular iron ore. Instances of the same kind and character might be multiplied.

Fig. 14.

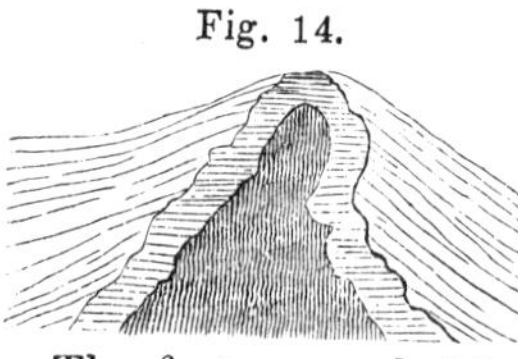

Fig. 14.—Limestone intermixed with serpentine, appears in the gneiss on the east side of the harbor at Whitehall. It has disturbed the superincumbent Potsdam sandstone.

The facts revealed by the relations of the associated rocks, support the view that surpentine is truly an eruptive rock, and belongs to the same class as granite and sienite. This view is also sustained by its distribution and mode of its occurrence, the latter of which has been spoken of. Its distribution is more or less in belts or lines, whose directions are well indicated when we compare the position and relations of the masses at distant points. This will appear on comparison of the distribution of the rock along the Green Mountain range. Beginning at the extreme northern boundary of Vermont, in the township of Troy, and following its range south, we shall find its masses distributed along a north and south line. From Troy it extends into Canada East, but southerly it is met with at Lowell, Newfane, Vt., Windsor, Middlefield, Chester, Blandford, Mass., and finally on nearly the same range as the Milford and New Haven quarries in Connecticut. The serpentine of this belt is alike. That of Milford and New Haven is more calcareous, and its colors are lighter and more blended with yellows than at the northern localities. The belt is also chromiferous, and more or less ferriferous. If the Green mountains should be regarded as a part of the Highlands, and as a prolongation of the Alleganies, we shall find the serpentine arranged in a northeast and southwest line, forming a belt of this rock along the whole eastern slope of this range of mountains, as far south as Georgia. It is coextensive with the Blue ridge, and is chromiferous through its entire extent.

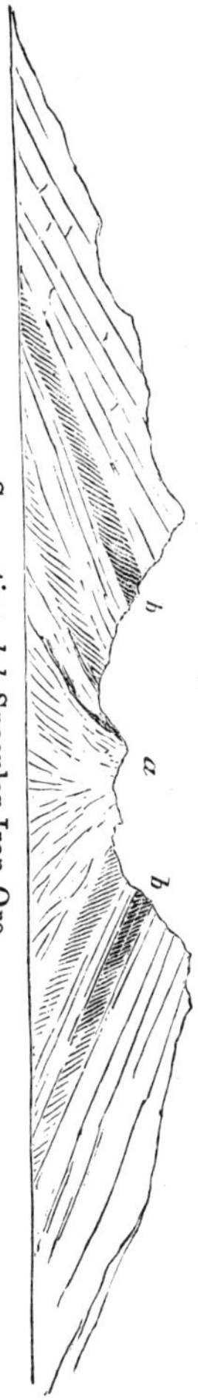

Fig. 15.

a Serpentine, *b b* Specular Iron Ore.

I have already spoken of the fields of serpentine associated with the pyrocrystalline limestone of Warren, Essex, Jefferson, St. Lawrence and Orange counties, N. Y.; to which may be added that of the district of Johnstown in Canada West. These districts or fields of serpentine differ from that which belongs to the Green or Hoosick Mountain and Allegany ranges. Pyroxenenas, amphiboles and graphite, are common to the former, but very rare in the latter; that is if they occur in proximity, they belong rather to other rocks, and not to the serpentine. The silicious minerals, as chalcedony, chrysoprase and agates, are associated with the former ranges, to which may be added from the magnesian minerals, schiller spar. In Maine, Dr. Jackson mentions only one locality of serpentine, that of Deer island, which seems to have been erupted through granite. This mass may be connected with that of the Grand Menan, on the northern coast of Maine and Nova Scotia.

Serpentine is extensively developed in Pennsylvania, Virginia, and Maryland. I may cite the serpentine rock of the Pine Barren range. It extends from East Nottingham into Maryland, after crossing the Northeast creek. It contains chrome ore. It belongs to the Chester and Lancaster county belt. It contains also silicate of magnesia, a mineral which is worth some three or four dollars per ton. This belt continues onward west by south into Harford county, Maryland, crossing the Susquehanna near Fraser's Point. The belt is prolonged to the southwest, embracing the Bare hills near Baltimore. Its continuity is interrupted in many places, still the belt extends through Pennsylvania into New Jersey in the direction of Easton, Northampton county. It is rather remarkable that serpentine, though it forms by itself hills of a moderate elevation, yet does not appear in the higher parts of the Appalachians. It is highly chromiferous through Maryland. In North Carolina, in those counties which are adjacent to Virginia, it is not so common as in the more northern states. It reappears, however, in great force in the southwest, particularly in Franklin, Macon, and Cherokee counties. The same belt extends into Georgia. Of this rock, then, it may be said to extend from Canada to Georgia in a belt which skirts the eastern base of the Hoosick and Appalachian chains. In the Appalachians its direction is nearly northeast and southwest; in the Hoosick range nearly north and south. The external characters are very uniform through the whole distance, and it is very constantly associated with certain minerals. These have been already referred to. Certain minerals, too, are as constantly absent, as galena, the sulphurets of copper, iron, molybdena, and zinc.

Notwithstanding the wide range of serpentine in this country, it occurs so rarely among the rocks of sedimentary origin that its age, even approximately, is left undetermined, in which respect it is in the same condition as the granites. In St. Lawrence and Jefferson counties, New York, the serpentine has evidently disturbed the Potsdam sandstone in numerous places, and it seems highly probable that both the serpentine and pyrocrystalline limestones were erupted subsequent to the commencement of the Silurian epoch. There is another instance of the occurrence of serpentine of a still later date: it is near the epoch of the consolidation of the waterlimes. The locality is in the vicinity of Syracuse, Onondaga county, New York. Here it is evidently a serpentine of contact, or a metamorphic serpentine. Those magnesian rocks are altered or changed into serpentine by proximity with some eruptive rock. The changes are variable. Some portions of the rock are perfect serpentines, passing into masses which are only slightly altered; and in a few cases the change has been still greater, as appears from the production of mica, forming a mass somewhat similar to granite. These altered rocks are confined to a small area. Like granite, therefore, it appears that serpentine has been the product of different periods, and to have been the product of agencies which have operated in a manner similar to those which gave origin to granitic compounds. But it appears that it is possible only in a few instances to determine the time of its eruption, in consequence of its being so rarely connected in this country with rocks of a determined epoch.

RENSSELAERITE.

§ 66. This rock has been regarded by a few of our geologists as a pyroxenic steatite. It is however perfectly homogeneous, and when the circumstances were favorable for crystalization, it is traversed by distinct joints, by which it is clearly a cleavable substance in the mass. Its hardness is 3·5–4, and hence is greater than steatite and less than pyroxene; its specific gravity is 2·874. It is composed of

Silica,	59·75
Magnesia,	32·90
Lime,	1·00
Peroxide of iron,	3·40
Water,	2·85

It resembles serpentine in the structure of the mass, but its particles of composition are coarser and more crystalline. It is white or grayish white, and tinged faintly with green. Black varieties are not uncommon. It is traversed by irregular seams of satinspar, disposed without order. It is massive, and its lamination is obscurely brought out by weathering. Exposure to the air softens, and heating it to redness slightly hardens and whitens it. It is cracked by the same exposure. The rock is tough and difficult to break. It occurs in many large beds in St. Lawrence county in the township of Russell, New York. It is also distributed in smaller fields in Jefferson and Lewis counties. It accompanies the limestone and serpentines of that district, and may be readily distinguished from the latter by its hardness. In the absence of iron, chrome, chalcedony, and the jaspery varieties of quartz, we may observe that it is unlike serpentine in its associations; still, as a mass, and in its position, it agrees with serpentine. It is not generally distributed in this country.

Rensselaerite takes a very good polish, but less so than serpentine. Its texture and grain is even, and being easily wrought, especially that part of the rock which has been exposed to the weather, it has been cut into a variety of useful articles, as candlesticks, inkstands, &c. The form of the crystal can not be distinguished from pyroxene, but the faces being too dull to admit of the use of the reflecting geniometer, it may still differ from it in its dimensions.

As a mineral, it was regarded by the late Prof. L. C. Beck as a mixed mineral, consisting of steatite and pyroxene. This view is apparently sustained by Beudant, who obtained from Sahla steatitic pyroxene swhich retained the form and cleavage of pyroxene. But contrary to this doctrine, it may be said that the mineral is perfectly homogeneous, at least as much so as serpentine, limestone, or any other mineral or rock usually regarded as simple. No foreign matter can be detected by the microscope, either in the form of plates, amorphous or crystalline grains. It is not uncommon to find hornblende and pyroxene combined in distinct crystalline particles, which together make up a crystal of the form of pyroxene. In this case the mixed nature of the mineral is evident to the senses, and it is as easy to refer each to the proper species, as it is the particles composing a mass of granite. I see no reason why we should assume a mineral to be mixed of two or more minerals in the absence of all external evidence. There is no objection, however, to receiving this mass into our list of rocks, although it may be found hereafter to be confined to the northern part of New York.

OCTAHEDRAL AND RHOMBOHEDRAL IRON ROCK.

§ 67. The largest bodies of magnetic or octahedral iron ore known in this country are subordinate to the hypersthene rock in the Adirondack group of mountains in the western part of Essex county, New York. The iron rock has a jointed structure, or it is traversed by divisional planes which bound large tabular masses. It is interlaminated with masses of this rock, and in some instances seems to disappear beneath it. At Adirondack one of the bodies of iron ore is between 700 and 800 feet thick. It disappears beneath the rock, and its actual limits are unknown. The degree of oxidation of the particles composing the masses of this ore seem to be unequal, and hence there is more difficulty than usual attending its reduction.

This species of iron in northern New York, the Highlands, and New Jersey, is associated with quartz, hornblende, and feldspar; and two very extensive bodies of ore are intermixed almost exclusively with phosphate of lime. But the magnetic oxide is usually grouped with one of the three first named. Quartz is the most favorable rock for reduction, in an economical point of view. The specific gravity of octahedral iron is 5·09; the rocks which are grouped with it scarcely exceed 3·00. If, then, these minerals were fused together, the iron, from its greater specific gravity, would sink through the molten mass, and be found at a lower level. In the majority of mines of this ore, the leanest part is at the surface. Particles of ore are scattered sparsely through the rock at the outcrop of the vein; but at the depth of twenty feet, and perhaps less than twenty, there is a perceptible increase of metal. The gravity of the iron ores may therefore explain the fact of their comparative absence as a rock at the surface; and it may be reasonably inferred from this and other facts, that the veins of ore are connected with much larger masses beneath than any which have found their way to the surface. Serpentine is very common among the beds and veins of this ore in northern New York. It is not in large masses, neither have I always found it in the beds of octahedral iron. It also accompanies the specular oxide of iron in St. Lawrence county, New York.

The great extent of iron ores of these two species, the magnetic and specular oxides of iron, seem to require that, in this country at least, they should be embraced in the rocks. They have hitherto been described as minerals only; but as they occur in mountain masses, occupying positions analogous to to serpentines, limestones, and granites, there can be no objection to ranking them with the subordinate rocks of the globe.

Magnetic iron occurs in masses and veins. In the hypersthene rock it is in masses subordinate to that rock, while in the gneiss of Essex and Orange counties, New York, it is in veins. The magnetic iron district of New York begins just north of the valley of the Mohawk, and occurs in all the counties in that part of the state, the veins themselves occurring in subordinate or smaller districts; or, in other words, veins of this ore appear to cluster together in certain areas, as it is rare that a single vein is found occupying a district: where one vein is found, it is usual to find several running in parallel lines. The specular ore occupies two or more districts in Jefferson and St. Lawrence counties. These are usually associated with serpentine or pyrocrystalline limestone. Magnetic and specular iron ores are composed of

	Magnetic oxide.	Specular oxide.
Oxygen,	28·21	30·66
Metallic iron,	71·79	69·34

Both species are variable in composition from foreign matter, particularly quartz. The special relations of the two species of iron ore will be given in the part relating to mining.

I have given the direction of the ranges or belts of the pyrocrystalline limestone and serpentine; and it is discovered that these two rocks skirt the eastern base of the Appalachians through their entire length from northeast to southwest. The same fact may be stated respecting the range of the octahedral iron in Orange county, adjacent to the Highlands. Among this cluster of mines I may enumerate some seven or eight extensive formations, among which are the Long, Rich, Forchee, and Wilkes mines. The formation which carries the magnetic iron passes into and through the state of New Jersey from the Sterling mines to those of Pompton, of which there are some three or four veins which are included in a gneiss abounding in hornblende—a rock which is quite massive, and is sometimes called, from its resemblance to granite, a gneissoid granite. A cluster of many veins are known, and have been worked, in the neighborhood of Dover. This seems to be a distinct belt, and pursues a southwesterly course towards the Delaware. A parallel belt more westerly is also known, and which embraces the zinc-iron ores which lie in a line, passing very nearly through the range of Scott's mountain towards Phillipsburg, opposite to Easton, on the southwest side of the Delaware. The belts of iron ore which pass from the southern Highlands of New York through New Jersey into Pennsylvania, might be described more particularly, and as running in at least three parallel belts. But it is my object at this time to speak generally of the relation of this ore or rock to the low ranges of mountain chains, which may be called outliers of the Appalachians, and which, taken as a whole, constitute one great belt of iron rock coextensive with the great mountain ranges of the Atlantic slope.

The magnetic ores of Pennsylvania are confined to the southeastern part of the state. They are quite limited in extent when compared with the great development in New York and New Jersey. The few veins which do occur, however, lie in the same ranges as those of the states just named. Thus in Burke county there are several parallel veins at mount Pleasant. At Durham, on the Delaware, is another district of this ore.

In Virginia, the ores of iron are abundant in the eastern section of the state, extending from a little south of Fredericksburg to Carolina; but it is an interesting fact, that most of the beds and veins have been changed into the hydrous brown oxides. They occupy the same relative position as the magnetic ores. A southwest range, however, of magnetic ores passes through Buckingham, Patrick, and Franklin counties, following closely the Blue ridge in its southwestern prolongation.

In North Carolina the iron ores, skirting the subordinate ranges of the Blue ridge, are equally abundant; but they follow two or three lines, one of which passes through Granville, Orange, Guilford, and Chatham counties; another west of the Blue ridge, through Ash, Yancey, and Buncombe and Cherokee into Georgia. The condition of the ores, however, is much the same as in Virginia, the magnetic having been changed to the hydrous brown oxide. From the foregoing it appears that all that part of the United States which lies east of and upon the Appalachians, is supplied most abundantly with this valuable ore of iron. It skirts this great range for more than a thousand miles; and though not by any means continuous, still it occurs at convenient intervals, and at such points as can not fail to supply the wants of over six millions of inhabitants.

OF THE LAMINATED PYROCRYSTALLINE ROCKS.

§ 68. *Lamination and cleavage planes.* Much has been said with a view to elucidate the efficient cause which has operated in the production of planes of lamination, or planes of cleavage. A phenomenon which is universal, is not to be attributed to local influences. Local influences are adjurants, but not the efficients of change. The wide-spread derivative matter, on the ocean's bottom, consisting of fine sand, clays and lime, mingled together, are a mere mechanical mixture, mingled together without order. But it is found that slates which are the results of such mixtures, have undergone, in process of time, very great changes. But the rocks referred to slates, differ much in the amount of change which they have suffered. Some are hard and ringing, others soft and fragile. Those which belong to the first, do not usually occupy their original position, but they are inclined, and appear to have been acted upon by mechanical forces, to a much greater extent than the latter. Pressure, therefore, must be recognized as a force which has had something to do in converting them into hard and firm slates, and in developing the peculiar structures of slaty masses. But pressure is an adjurant to an efficient cause, and this efficient cause must be referred to some of the essential properties of matter or to original endowments. This original endowment is probably crystallization. I have had occasion to speak of this property before. I have also employed the term, molecular force, a term which I have used where the result is the formation of spheroids, or rather nodular masses, while crystallization produces parallelograms upon a large scale. In the formation of the planes of parallelograms, pressure aids the efficient force by bringing the particles near to each other. Pressure unaided by this force, is insufficient to develop planes of any kind.

The production of the planes of cleavage or lamination, has been attributed, by Prof. H. D. Rodgers, to an electro-galvanic or electro-thermal agency. This, to be sure, must be regarded as one of the universal properties of matter, or a force made sensible through the medium of matter. The theoretical explanation is founded on the supposition that the slate planes, after flexure, are alternately hot and cold, and hence are the generaters of an electrial power analogous to that of the thermo-electric pile. It is difficult, however, to understand how these arrangements operate in the production of planes; moreover, it requires the preexistence of planes of some kind, in order that the analogies may be made out. It seems, that in this explanation the effect preceded the cause. If each molecule of matter has polarity, all we have to do to secure a symmetrical arrangement is, to bring the particles composing a rock within the sphere of each other's attraction. Adjustments will then take place. Compression or pressure operates in this way.

The existence of planes of lamination or cleavage in gneiss, mica slate, etc., receives an explanation at least partly from the foregoing principles. The efficient cause may be stated in different words. We may recognize the polarity of the molecules of matter, or we may use the word crystallization, and in some cases molecular forces; for in the use of the latter term, it seems we recognize a greater change of the particles in space, even entire strata are formed by this force. Concretions, too, are gathered or formed from similar particles, and from comparatively wide spaces; where the matter is insufficient to form a stratum, for in the mass of mud which ultimately forms slate, the lime and silex were intermingled without order; but now we find the lime in nodular bands, or distinct nodules, as in septaria, which could not have been in that state in the original deposition. We call it then a concretionary or molecular force from the effects produced; it is a modified crystalline force.

In sandstones and limestones molecular movements often obliterate the planes of stratification or deposition. These movements result in the formation of spheroids, or the forms represented in fig. 17; a general illustration of the kind of molecular movement, which may be observed in many sedimentary rocks, and also in rocks of igneous origin, as serpentine.

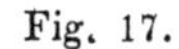

Fig. 17.

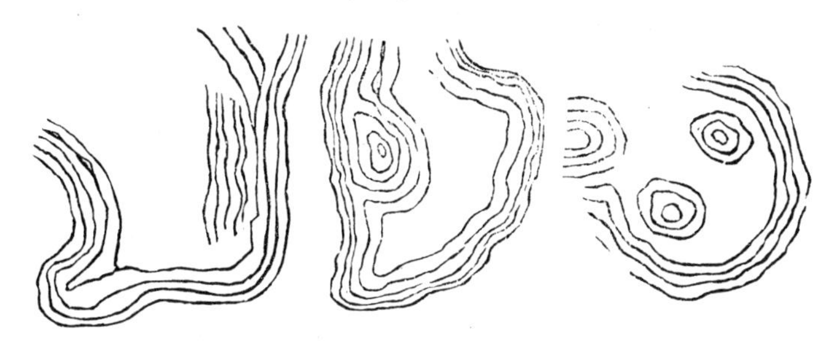

We may recognize, too, in this phenomenon, one of the efficient causes of metamorphism, a cause which whenever the spaces between the particles of a mass are charged with water, or possess from any other cause a partial fluidity, is free to operate.

The importance of recognizing the essential properties of matter as efficient causes of change in rocks, it seems to me has not been perceived, and hence has not been investigated so fully as it should be. Having stated the foregoing views respecting cleavage planes or lamination, I proceed to speak of the characters of the laminated rocks, gneiss, mica slate, &c.

§ 69. *Gneiss.* The rocks of this class, in whatever part of the globe they occur, are all alike and undistinguishable. A mass of gneiss from the Alps or Pyrenees, can not be distinguished from a mass from the Alleghanies. The mica slate of the Alleghanies differs, in no respect, from the mica slate of the Rocky mountains. The same remarks may be extended to granite, sienite, and indeed to all the eruptive rocks. Fire has left the same impress upon all of them, in all quarters of the globe where they occur. In this class of rocks there is no order of superposition. In some districts gneiss may occupy a position contiguous to granite; in other districts mica slate may occupy the same position relative to granite, and gneiss may appear in many instances overlying hornblende rock, talcose or mica slate. The Hoosick Mountain and the Appallachian ranges, furnish many examples of the varied and variable collocations of these rocks. They may, therefore, be said to belong to one epoch.

It has been stated, already, that gneiss differs from granite in its structure. Its particles are structurally parallel. Fig. 18 represents this structure. This parallelism has been regarded, by distinguished geologists, as due to water. Of this statement more evidence is required: indeed when we consider the extent of the change required to convert any sediment into gneiss or mica slate by heat, and over such wide areas, we can scarcely fail to feel that the origin assigned is at least extremely doubtful, especially when it is considered that the crust had already cooled so much as to admit of the condensation of water upon it, and the formation of oceans in all quarters of the globe. While the metamorphic view is regarded as inapplicable to this class of rocks over wide-spread fields the world over, I do not call in question those instances of local metamorphisms which occur in the Alps and other districts, where it is evident the disturbances and changes by heat are very remarkable. But who has observed in this country, sandstones, conglomorates, slates, and limestones, which have been converted into gneiss and mica slate, or hornblende rock. It is true rocks are altered. Chalk has been changed into a hard crystalline marble, but the area over which this change can be traced, is quite limited. Clay, too, has been baked, and under that process has become hard, and firm enough

Fig. 18.

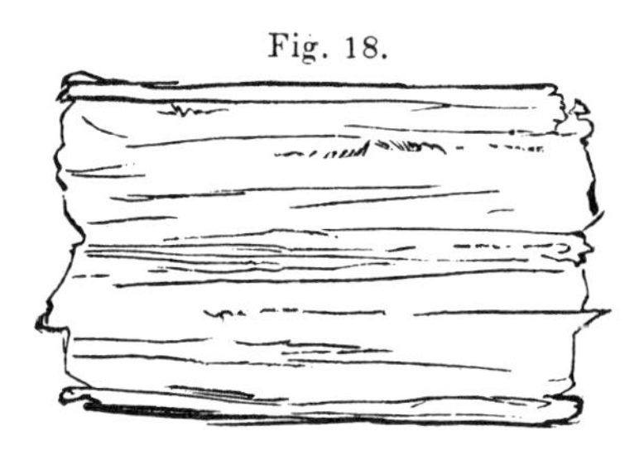

to ring like metal when struck with a hard body; but it has not become gneiss or hornblende. The extent of all the individual rocks under consideration, together with their identity of structure the world over, is to say the least, indicative of their early consolidation, and that it took place at a period when water existed only in a vaporous state.

These remarks, it will be perceived, apply to this country. The metamorphic gneiss of the Alps is admitted by eminent geologists. It is, of coarse, local, and we can not set bounds to extent of local changes. Still the metamorphic gneiss of the Alps originating in oolite and eocine rocks, must furnish by analysis, a difference in the proportion of their elements from that which exists in our normal gneiss or mica slate. It is now admitted that a parallel structure of itself is no evidence that the rock was a sediment. I pointed out, in my New York Geological Reports, that the porphyry of lake Champlain was laminated, and described it as stratified rock, notwithstanding the indubitable evidence it furnishes at the locality, that it was erupted from fissures in the shales of the Hudson River group. Darwin describes an eruptive red granite of Chili, which exhibits a decided parallel structure in many of its parts. The gneiss of Bahia, according to the same author, contains regular fragments of hornblende; hence gneiss may be regarded as a pyrocrystalline or eruptive rock, at many localities. We should subject the question of metamorphism to two tests: 1. Proximity to agencies competent to effect the change observed. 2. The continuity of the changed to the unchanged mass.

The only proof, therefore, which we can obtain, of metamorphism, is in the local change which may have been produced in a part of the rock. There will then be gradations, which may be traceable from zones of the greatest to zones of the least change. We may trace the harder, ringing, reddened, or whitened mass, to those parts which retain their original properties. Beyond this we can not go. To this extent it is useful to admit the metamorphic theory.

Gneiss is somewhat variable in structure and composition.

In structure it approaches granite in many districts in the United States. In composition the mica may be wholly or in part replaced by hornblende, and a part of the feldspar of the rock crystallized in large particles. We have then the granitoid gneiss, hornblendic and porphyritic gneiss. If mica abounds in it, it is either said to be schistose or micaceous gneiss; or if quartz, quartzose gneiss. These terms are employed to express its structure or composition in any observed locality. Gneiss passes into mica slate by the loss of its feldspar.

The foregoing considerations seem to establish the doctrine that gneiss, mica slate, hornblende and talcose slate are really contemporaneous formations. They are so blended in all the districts of our country, that it is extremely difficult, if not impossible, to define their boundaries. Even in the talcose slate of the gold region, gneiss frequently appears in wide areas. I shall not attempt, therefore, to give the boundaries or extent of either of the foregoing rocks, being content with pointing out the regions they occupy in common.

§ 70. *Mica slate.* This rock consists of mica and quartz. The laminæ are thinner than those of gneiss, and it has more mica in its composition. The feldspar, when it occurs in this rock, is in the form of seams or segregation, intermixed with a coarser mica than that which composes the body of the rock. The quartz is gray, the mica usually black, and the particles of both are fine.

Numerous varieties might be pointed out, which are due to variable proportions of quartz and mica, or to the thickness of the laminæ, or to the intermixture of other minerals, as hornblende or talc. These varieties are not so important as to require detailed descriptions. It is sufficient that the student should know that mica slate passes lithologically into other rocks without changing materially its structure.

§ 71. *Talcose slate.* It is composed of talc and quartz. Its laminæ are thinner than those of mica slate. Its color is gray, and its luster is more silken than mica slate. Its laminæ are undulating, curled, or crisped. Its texture is fine, and its feel soft. Varieties occur in which the quartz is the principal mineral, and its structure then resembles a fine-grained sandstone. On the other hand, when talc predominates it becomes steatitic or a perfect steatite. Mixed largely with scales of mica, it becomes a talco-micaceous slate. Like other rocks of this epoch, it passes into one or the other mineralogical mass, with the necessary exchange or substitution of the mineral which characterizes them. It is associated with hornblende rather than mica slate.

§ 72. *Hornblende rock.* Its color is green, light green, or blackish green, and its composition is either an unmixed hornblende, or else it is mixed with feldspar and quartz, the particles of each being arranged in parallel bands. In this last particular it differs from sienite. Hornblende is exceedingly tough, and the mineral is always crystalline. The crystals are interlaced with each other. Its composition is variable. It preserves, however, a great uniformity of character when associated with other laminated pyrocrystalline rocks; it is more variable in a trapean region.

The laminated pyrocrystalline rocks lie in proximity to each other. It is rare for a mountain to be composed exclusively of one of these rocks, and it frequently happens that gneiss, mica slate, and hornblende form an alternating series, in which they are separated by short distances only. On the sides of mountains, and in valleys, their planes of lamination incline steeply to the horizon, while perhaps upon the crests of high ridges the laminæ are nearly horizontal. This seems due to an upward thrust, by which the upper parts of the rock being unsupported, fall into an horizontal position. At the point of flexure the mass is frequently broken, when the lower portion of the rock is left highly inclined to the horizon, and the broken part is nearly prostrate. Such is the position of the talcose slate of Table mountain in Burke county, North Carolina (fig. 19). The body of the mountain is composed of strata highly inclined to the west, but the summit is quite flat. The porphyritic gneiss of the Swannanoegap of the Blue ridge, in North Carolina, is quite flat upon the crest, but quite steep at the base. But in the mountains of New England the laminæ are steep to the summits. Probably diluvial action has swept off those flattened crests which exist in the southern states.

Fig. 19.

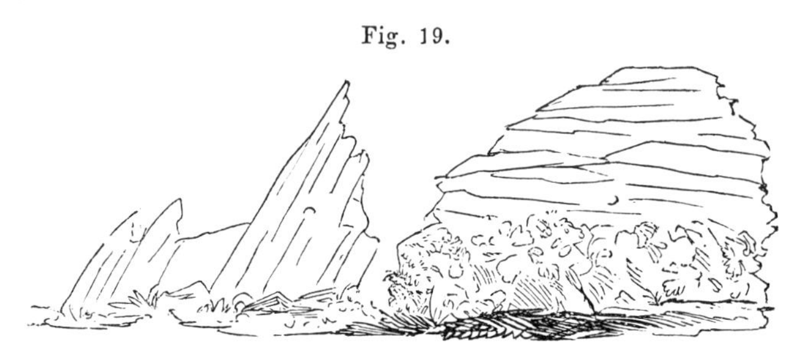

§ 73. *Chlorite slate* is a green fissile, or slaty rock, with a soft feel. In order to make out this rock, it is necessary to recognize chlorite intermixed with quartz, and a structure similar to talcose slate. It may contain feldspar and mica. It is often associated with gneiss and other schistose rocks. The specific gravity of chlorite is 2·72. Its composition is subjoined:

Silica,	26·0
Magnesia,	8·0
Oxide of iron,	43·0
Alumina,	18·5
Water,	2·0
Potash,	2·0

§ 74. *Clay slate.* This rock is a hardened clay or shale, and is for the most part exceeding fissile. Its colors are green, bluish green, and reddish or brown and purple. The red colors are variable, from a pink red to a deep brick red. The laminæ are distinct.

It is doubtful whether clay slate should be admitted as a member of the pyrocrystalline rocks. I should not regard it as an eruptive rock, and place it in this connection, were it not generally placed among the primary rocks, and were it not also quite common in proximity with veins in granite in North Carolina and other places. It is, however, possible that this variety of slate rock, which passes for clay slate, may be a variety of chlorite slate. Neither the mineralogist, geologist, or miner, has regarded the inquiries relative to the composition of rocks, veinstones, &c., of much importance, and hence they have been neglected.

§ 75. *Laminated limestone and serpentine.* Where limestone and serpentine are associated with gneiss and mica slate, they exhibit a parallel structure. They have been acted upon by forces of the same kind and nature. We are not obliged to adopt the opinion that limestone and serpentine are metamorphic when we can detect a parallel structure. The facts in each particular locality must speak for us. For a full notice of this rock I refer the reader to § 65, the serpentine marbles.

§ 76. *Quartzite.* I apply this term to a massive rock associated with the auriferous slates. It is smoke gray, light or dark, breaking with a conchoidal fracture, and hence resembles flint or hornstone. It appears to be a simple substance; but it becomes white by weathering, and hence it is probably compound. The quartzite contains sulphuret of iron, which may be auriferous. Some varieties are agatized coarsely. There is considerable evidence that the rocks associated with it are stratified, and hence it may be ranked among the metamorphic products. This question must remain open for further observation.

DISTRIBUTION OF THE PYROCRYSTALLINE ROCKS.

§ 77. The pyrocrystalline rocks are blended and interlaminated so frequently in every district of the United States, that it is impracticable at present to trace either of them separately through the range of country which they occupy. The White mountains in New Hampshire, mount Ktaddin, Maine; Monadnock, New Hampshire; Hoosick mountain and Black mountain, and the culminating point of the Appalachians, are gneiss and mica slate. Many of the passes over the Blue ridge in North Carolina and Virginia, are talcose slate; Pilot mountain in Stokes, and Table rock in Burke counties, are talcose slates;

but still they are associated with gneiss and hornblende in the vicinity. A long list of localities might be made out of the kinds of rock prevailing at different points in the Union; but it is impossible to show, at the present time, that there exist important relations between any two distant points. At the most distant points of our country these rocks preserve a great similarity of structure and of character, which renders it impossible to recognize by specimen the part of the country they represent. They are traversed also by granitic veins, the composition of which exhibits everywhere the same variations. There are districts, however, in which trap is much more abundant than in others; and these districts furnish us with a greater remove from the common character of the country at large. I shall speak of the peculiar rocks of certain districts when I have occasion to take up the subject of mines and mining.

PYROPLASTIC ROCKS.

§ 78. The first section embraces those rocks which are supposed to have been erupted through fissures in a molten state, and to have cooled beneath the ocean, and hence I have denominated them *submarine*. The greenstones amygdaloids, basalts, traps and porphyries, are comprehended in the section. The second section embraces those which have been erupted from craters, and have cooled in the open air, and hence I have called them *subærial*. They embrace the modern lavas, of all kinds: the vesicular lava, obsidian, volcanic arks, &c.

1. *Submarine.*—The rocks of this section are lithologically the same as in all parts of the globe. A traveler who speaks of trap in the greenstone or basalts of Australia, is understood by us in America. The greenstones of the Hudson river scarcely differ from those of Connecticut or Nova Scotia. Greenstone is massive, vesicular, columnar, and porphyritic. The first is a heavy black or grayish black rock, either occupying fissures in other rocks, or lying upon them, having been forced out from beneath in a molten state, and in a condition to overflow the region adjacent to the fissures. The term trap seems to be re-

stricted to that form of greenstone which is inclosed within the veins of a fissure. It is commonly called a trap dyke. It is a stony and not a metalliferous vein. Fig. 20 represents a dyke intersecting two parallel veins of granite. The forms of the pyroplastic rocks were determined by the condition under which they cooled. We can not always determine now what that condition was. It has been demonstrated that slow cooling restores to the mass the common characters of a rock, free from vitrification. So under certain circumstances the mass separates into columns of five or six sides. The tendency to the columnar condition is distinct, while the columns are often imperfect. Their terminal outline is visible, but the adhesion of their sides still remains. The columns are vertical, as at the Palisades of the Hudson, or they may be horizontal, as when enclosed between the walls of a fissure.

Fig. 20.

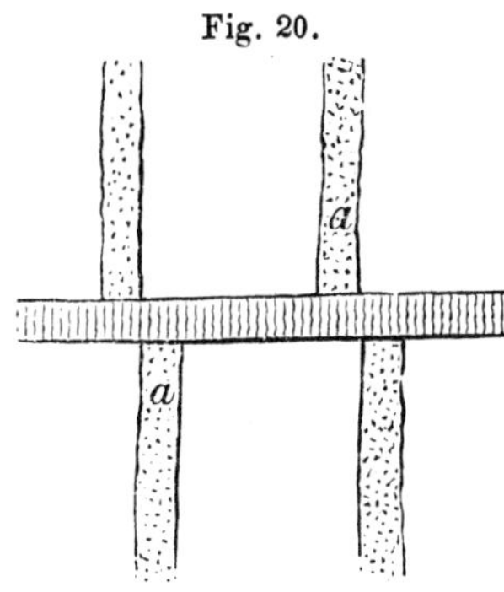

Fig. 21.

a Sandstone, *b* Columnar Trap, *c* Trap injected between the strata of sandstone.

Fig. 21 represents the rude columnar structure of the Palisades, which repose upon the lower members of the trias. The molten mass was also injected between the strata of sandstone. Greenstone is sometimes vesicular, but the vesicles having been filled with foreign substances, and remaining in relief after the rock is partially disintegrated, this variety is often called amygdaloid, those foreign matters appearing like almonds, in the rock. Both greenstone and amygdaloid have a granular structure, and feldspar may often be distinguished in small particles, disseminated through the mass. Greenstone is also porphyritic, the crystals of feldspar are distinctly formed during the process of cooling.

§ 79. *Basalt* is a black compact rock, occupying the same relations to other rocks as greenstone, just described. It is both massive and columnar. Structurally it differs from greenstone, in its perfect homogenity. Frequently our trap dykes are perfectly homogeneous and compact, and hence are basalts. Basalt is not a common rock upon the Atlantic slope. The rocks which have frequently been called basalt, are really greenstone, as either by the unassisted eye or by means of a single lens, particles of feldspar are visible.

§ 80. *Porphyry* is a rock in which crystals of feldspar are tolerably well defined. They are embedded in a compact paste. This paste is often reddish or greenish, but the color is variable, or it may be any color.

Fig. 22.

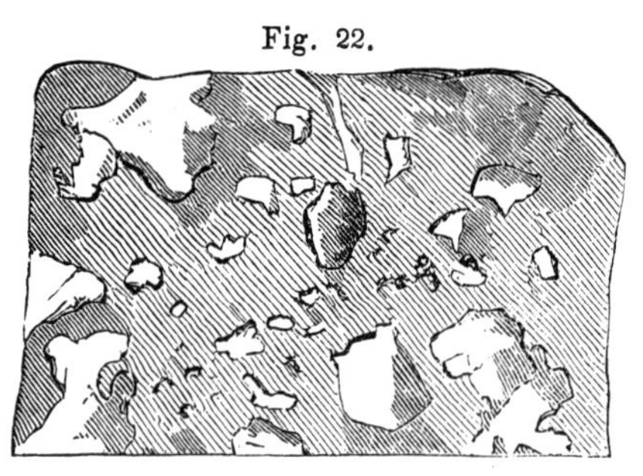

Fig. 22 represents a porphyritic mass, which is common over a large area, in North Carolina particularly, in a belt passing southwesterly through Granville and Chatham counties. The porphyritic structure is rarely perfect. The masses of feldspar in the paste, though tolerably well defined, have rarely straight edges or even planes.

The feldspar is rather concretionary than crystallized. The white spots are frequently quartz.

§ 81. *Trap dykes* are those black stony masses which are interposed between the walls of a fissure. For a limited distance they pursue a straight course. The fissure is perfectly defined, and the filling always perfect. Where more than one dyke intersects a rock, they may be parallel, as represented in fig. 23, or they may intersect each other. In the first case the dykes are probably of the same age, but of a later date than the masses which they intersect. The rock traversed by these dykes is hypersthene, and all the subordinate masses are igneous products.

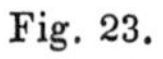

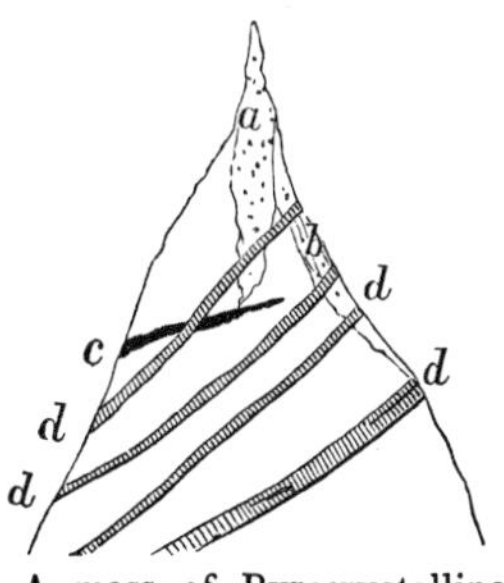

a A mass of Pyrocrystalline Limestone, *b* of Granite, *c* a vein of Magnetic Iron Ore, *d d* four parallel Trap Dykes, which may be traced a hundred yards.

DISTRIBUTION OF THE SUBMARINE PYROPLASTIC ROCKS IN THE UNITED STATES.

§ 82. In the eastern section of this country three belts of greenstone are well known. They belong to the eastern or Atlantic slope of the Appalachians and the Green or Hoosick mountains, and are coextensive with them. Two of these belts are parallel, and were synchronously erupted. The eastern belt begins in Rhode Island, and extends entirely across the eastern part of Massachusetts into New Hampshire. The belt is prominently exhibited in Weston, Watham, Lexington, Woburn, Wrentham, and onwards to Ipswich in New Hampshire. The direction of this belt, upon the whole, bears to the eastward. It is not, however, in a distinct belt or ledge, but a broad area in which these eruptive rocks are common. It is intimately associated with sienite, another eruptive rock which accompanies it through its whole route. The two masses form, as it were, a large patch of rocks, encircling in part Massachusetts bay.* This mass has disturbed the coal measures of the eastern section of this state, and may perhaps have been erupted at that epoch, as it emerges from beneath them. It may be called the trap of Massachusetts bay. The second belt belongs to the valley of Connecticut river, and passes entirely across the states of Connecticut and Massachusetts. Its direction is north and south, and it has disturbed the trias beds along its line of bearing. It is both massive and columnar. Mount Holyoke and mount Tom are conspicuous eminences in this belt. It may be called the trap belt of Connecticut river. The third belt of trap or greenstone occupies a part of the valley of the Hudson river. It is well known in the southern part of the valley, where it is called the *palisades*. This locality may be regarded as a typical representation of trap. It is both columnar and massive: the former constitutes the prominent feature of the rock. This belt may be called the Hudson river belt. It appears to terminate in a point south of the Highlands, but it is prolonged through New Jersey, and may be traced, with a few interruptions, through Pennsylvania, Virginia, North and South Carolina, and into Georgia. It is associated with the trias of those states, or it may be the permian, inasmuch as there is evidence that the coal of North Carolina is of that age, instead of the age of the oolite or trias, as has been maintained. This range of trap is not continuous the whole distance in the states I have named. It extends through New Jersey in nearly continuous ridges, the eastern parts of which lie between New York city and Newark. It passes into Pennsylvania, and forms ridges in the permian or triasic sandstones, but is more conspicuous in the northeastern counties, arranged on the line of Burke, Montgomery, and Chester counties. The Coneaeaga hills are trap.

In Virginia, upon the same belt of sandstone, the trap ranges between Fredericksburg and Buckingham county, pursuing a southwest direction into Rockingham county, North Carolina.

* See President Hitchcock's Massachusetts Geological Reports.

It continues in the direction to the Yadkin. This branch diverges to the southwest. Another begins near Oxford, Granville county, and accompanies the trias and permian into South Carolina, where it apparently terminates again in Chesterfield district. It has been asserted that the Permian system of sandstones appear in Georgia, about halfway between Savannah and Macon. If so, there can be no doubt of their being accompanied by this belt of trap.

The interesting feature of this formation is the great extent of country it traverses. It is comparatively a narrow belt of rock, and hence it seems to have been ejected through a very long or continuous fissure; and it is not improbable that the fissure extended far beyond the visible belt of trap. The palisades are, for instance, upon this north and south line of fracture, which extends northerly through the valleys of the Hudson, Champlain, and St. Lawrence, in the range of Montreal and Quebec. The trap rarely appears on this line between the Highlands and head of the Champlain valley. At this point trap begins to appear again, and with frequent repetitions down to port Kent. From this place onward to Montreal the disturbance of the rocks is much less, but at the latter place the phenomena justify us in regarding it as the center of a highly disturbed district. It may be traced onward to Quebec.

It does not necessarily follow that this belt was fractured for 400 miles north of the Highlands, in New York, at the time the eruption of trap forming the palisades took place, yet it probably was. This erupted mass ranges along this fractured belt; and if this belt extends to South Carolina, it is one of the longest lines of eruption east of the Rocky mountains. Admitting the fact of the continuity of this long line of fracture, we are led to look for some cause which determined its extent and direction. We have found a part of this belt to be occupied by trap and greenstone, and to form a very striking feature in its geology; but upon other parts of the belt, though the rocks are fractured, and very much disturbed, yet the eruptive rocks do not appear at the surface: for example, between the Highlands and the head of the valley of Champlain. This part of the belt, together with the more northerly part of it, between Montreal and Quebec, is upon a line of junction between two systems or formations, and the juncture or belt in proximity with it, is made up apparently of the thinnest masses of the systems, and hence is a line of weakness. If this position is true, then, and if it has been one of great tension, it explains the fact of the fracture and disturbance upon the line.

Again, upon lake Superior several distinct parallel belts of trap range in a northeast direction, taking, however, a curvilinear course. These traps consist of compact greenstone, amygdaloid, and basalt; they are also chloritic and ferruginous. The compact variety passes into granular semi-crystalline trap, in which feldspar is visible. The rock is regarded as a product formed by the fusion of labradorite and hornblende. The range extends from the extreme limits of Kewaunee point to Montreal river. Isle Royal is composed of materials similar to Kewaunee point. The geological investigations prove that the trap belongs to the oldest Silurian period, as it alternates with the Potsdam sandstone. Eruptions of trap took place while the sandstone was in the process of formation. This turns, it is true, on the correctness of the determinarion of the age of the magnesian limestone, which rests upon the sandstone. If this is equivalent to the calciferous and chazee limestones of New York, there can be no doubt respecting the age of the underlying sandstone.

It will be observed that the trap is not intruded between the layers of sandstone in the mode represented by fig. 21, where the trap is in wedge-form masses, and which penetrated the rock subsequent to its consolidation. In the lake Superior district, the trap overflowed the sandstone in sheets, which subsequently were covered with another bed of sandstone, and then another eruption covered the preceding with another sheet of molten trap. There were alternations therefore of melted rock and sediments; the two processes were going on at intervals, but during the same epoch.

The ranges of trap in the far west begin to make their appearance just beyond the verge of the crests which divide the waters of the Missouri from those of the Columbia and Colorado. The Rev. Mr. Parker many years since described the basalts and greenstones upon the western slope of the Rocky mountains. These eruptive rocks appear not far from the head waters of the Columbia and Colorado upon the Pacific slope. A large proportion of Oregon, indeed, is basaltic. At the head waters of the Salmon river the gneissoid granite begins to give place to the pyroplastic rocks—greenstone, porphyry, amygdaloid, and basalt—the latter of which walls up the Columbia at many points, as at the Cascades, Willamette falls, Grande Coulé, Walla Walla, the Dalles, Cape Horn, Smoke river, &c. The boundaries of this great region of trap remain, however, undetermined. It is the largest and most interesting geological field of this formation in the United States.

SUBÆRIAL PYROPLASTIC ROCKS.

§ 83. These rocks are erupted from craters, which in their perfect forms are perforated cones formed over a tubular aperture leading to the subterranean incandescent and melted matter beneath. The beginning of a cone is a fissure, simple or stellated, which in giving vent to the melted matter is modified in form and condition. The line of fissure will be filled by erupted matter, while the orifice is rounded by the exit of incandescent lava. The formation of cones has often been observed. Their shape and structure are very uniform. The material is derived from rocks which are melted and partially altered at unknown depths, and being forced upward, escape through the vent, either at the apex or at one side of the cone. A cone will consist in part of consolidated erupted matter collected around the vent. If all the substances formed beneath consisted of fused or liquid rock, flowing out at one side, the symmetrical cone would not be formed at all: but as ashes and cinders are ejected, they fall around it, and in the course of a few hours a

15

cone is built up about the orifice of the loosely coherent materials.

Volcanic action is not accompanied with the same phenomena at different times; neither do the different volcanoes eject the same kind of material at different epochs of eruption. At one time it is a thick heavy lava, which pours over one or two sides of the crater, and slowly flows down the mountain. The volcanoes of the Sandwich islands are boiling pools of melted rock—excavations rather than craters, and their activity is accompanied with moderately loud explosions, and the shrill hissing of steam issuing from a boiler. Sometimes, again, the volcanic products consist of ashes, which fall in part around the crater, while the finer particles are driven to distant countries by the winds. While the ordinary products of volcanic action consist of melted rock sufficiently liquid to flow, or of comminuted rock in the condition of an ash, there are still many other products which escape at certain times from different volcanoes. Thus gases and vapors are common. Among them are nitrogen, ammonia, carbonic acid, carbureted hydrogen, and sulphureted hydrogen; boracic acid also escapes in the steam in company with ammonia. Hot water, holding silica in solution by means of potash or an alkali, is a common product of volcanic action in Iceland. Bitumen and naptha also are found among those products, especially at Taman, at the western extremity of the Caucasus, and at Baku, a port on the Caspian sea. The latter are mud volcanoes or salses, the bitumen and naptha being derived from the superficial deposits of organic matter. The strongest indication of volcanic action in our own country was exhibited at New Madrid in 1811. The hot springs of Wachita, and those of California, witnessed by Mr. Forrest Shepard, must be regarded as due to a feeble volcanic action.

A phenomenon which stands connected directly or indirectly with volcanic forces, is the earthquake. It immediately precedes, and perhaps continues during the first outbursts of this force. The earthquake consists essentially of movements of the earth's surface in the form of waves or undulations, which travel with great rapidity in all directions from the focus of disturbance. The intensity of this movement, or the force of the shock communicated to the strata, is supposed to be in some way dependent upon the diameter of the vent through which gases and melted matter have escaped. The openings of Vesuvius, Etna, and the South American volcanoes, are narrow and constricted, and at the same time their coverings over and above the seat of activity are thick, and strengthened by repeated accessions of layers of rock from beneath, and braced by numerous intersecting dykes. Under these circumstances volcanic forces are confined by strong walls and narrow funnels; and hence, when the forces have acquired strength sufficient to rend asunder these walls, or force the safety-valve, it will be attended with tremendous earthquake shocks. On the other hand, where there is an opportunity for a free escape of gas and melted matter, as in the Sandwich islands, Mouna Loa for example, where the craters are rather deep and wide excavations, volcanic action of great intensity begins without endangering the surrounding country by earthquakes. The seat of activity seems to be more superficial, and the resistance to be overcome far less, than those of South America and Europe. An eruption of the volcanoes of Europe and South America, therefore, is attended with violent movements or undulations of the crust, which are propagated from the center of action in all directions. It is to be recollected, that as the crust is not homogeneous, and the material through which the impulse is propagated is not equally dense, the effects of that impulse must be modified in its transit. The impulse here spoken of generates a wave in the crust which usually moves onward, as has been already stated, with great velocity in all directions from the center which receives the first shock of the explosion. The surface rises and falls like waves of the sea; or, in other words, the undulations travel onward with great speed in obedience to the ordinary law of a force propagated through a resisting medium. The undulation is modified, however, by the position and condition of the resisting medium. In its progress, a direct undulation may be converted into a gyratory one by an increased resistance in its course, or into a vertical one at the place situated immediately above the point of impulse. An instance of direct earthquake undulation, extending over a large portion of Europe, a part of Africa, the Atlantic ocean, and West India islands, is well known in the remarkable visitation of Lisbon in 1755; of the gyratory, or of the vertical movement, which took place in the great Calabrian earthquake in 1783, by which two obelisks at the convent of St. Bruno were twisted on a vertical axis without falling. Of the vertical movement, a striking example occurred during the earthquake at Riobamba, 1799, where a sudden rise of the ground took place, which hurled the bodies of men to a height of several hundred feet.

Considering earthquakes as earth-waves, it is evident that when those waves are generated in the ground beneath the ocean, their impulse must be communicated to the water above, whose motion will partake of the same character. Waves will therefore be generated therein, which will travel onward in directions which the impulse communicates; but from the nature of the medium, the water-wave will travel with less speed than the earth-wave. In consequence of this fact, a person upon a shore where the two waves are tending, will experience first the earth-wave, and soon after the water-wave will follow; lastly, another wave will be recognized through the medium of the air. In each of these cases the rate of transit depends on the nature of the medium receiving the shock. Experience proves that the intensity of the shock is very variable in volcanic districts. It does not, however, seem to depend directly on the activity of volcanic action, for according to Dana, the activity of Kileaua is not exceeded by that of any other volcano; yet earthquakes are rare and feeble, even when a force immeasurable by us in its power is manifested in some of the results of this action, especially in rending the earth for twenty-five miles without having produced an earth-wave worthy of notice; while in other cases the visible activity is much less, but is accompanied with disastrous and terrific effects.

A fact should be stated in this place, which is probably the most important one which attends volcanic activity—it is the change of level which a country often suffers during its paroxismal throes. The coast of Chili, for example, in 1822, was permanently elevated for one hundred miles, in some places more than ten feet, in others less. Large areas in all countries furnish many facts in proof that they have undergone a similar change of level. It is not determinable now whether those changes occurred during a single paroxysmal effort; but where a coast has been stationary a long time, and then appears to have been stationary again at a higher level, the probability favors the paroxysmal view. But wide areas are elevated slowly, and apparently uniformly. Where the change is going on slowly, as in Scandinavia, and perhaps on our own coast, it may be due to the expansion of rocks by heat.

CAUSES OF VOLCANIC ACTION.

§ 84. Much has been said and written of the cause or causes of volcanic action, and for the solution of the question many ingenious and indeed philosophical reasons have been proposed. Among the causes assigned, chemical action, excited by electro-magnetic arrangements, has had many advocates. Known analogies are favorable to this theory. We may arrange our apparatus, or we may devise in the laboratory the needful conditions for imitating nature's processes within the earth, hence its favorable reception; and it is not strange that chemical forces have been regarded as the prime agencies of volcanic action. When we take, however, a larger view of the facts and phenomena which constitute in the aggregate the volcanic forces, we can hardly refuse to admit that the chemical actions which no doubt go on on a magnificent scale during the eruptive periods, are effects and not causes. We are therefore driven to the necessity of going still farther back in order to find the primal cause; and it is no disparagement to the intellectual power of man to acknowledge, that respecting the primal cause we see only the hand of infinity who kindled the once blazing fires of the universe of matter. How or when, are questions too deep for us to answer. We may therefore regard the primal cause as the remains of that original incandescent state, and it is the prolonged activity only of the burning mass which has but just retired to the deeper parts, above and around which the crust has become scaled as it were by simple cooling.

As terrestrial volcanic action is to us the most interesting of geological phenomena, this circumstance alone has led both geologists and astronomers to scrutinize with great care the only heavenly body which admits of examination, in order to ascertain if our satellite shows indications of the same agencies of which I am speaking. The result of these examinations has clearly proved that the moon has been the theater of intense volcanic action. This luminary, which shines with such silvery light, and appears so plane and even, becomes under the telescope studded with rough and rugged mountains, whose tops are crateriform, or its planes have the semblance of deep excavations, in which are standing sharp conical peaks, perforated like the cones of Vesuvius, Etna, and Cotopaxi.

Our satellite, then, is but a smaller pattern of the earth, exhibiting, an intenser volcanic action than that of the earth—a fact which is probably due to the absence of water, an agent which upon the earth has modified its surface so far as to conceal in part beneath the sedimentary rocks its original volcanic nature. The moon, however, presents its face covered with ancient eschars, which time never has healed, and which are destined to remain in all their original roughness and rigidity.

In illustration of lunar volcanic phenomena, I have presented the student with a crystallotype of her surface, which was taken by Mr. Whipple with the Harvard telescope. It will be observed that her surface is studded with prominences which represent faithfully its structure. The strong shadows, which are visible under the telescope, are but faintly perceptible in this likeness. But what is particularly worthy of notice, are the lunar craters, which are represented by the circular projections upon different points of her face. The stellated fractures or fissures produced by soulevements of her crust, are among the remarkable features of the portrait.

In order that the student may locate the lunar volcanic peaks, I have copied a diagram from the Penny Cyclopedia, fig. 24, upon which the relations of many of their points may be observed at leisure. But I would particularly recommend the study of the moon's surface by means of a good telescope.

The arrangement of lunar volcanoes, like those of the earth, is very nearly in lines, which indicates their connection by means of passages beneath. Considering the mass of the moon, and the rapidity with which heat escapes into space, it is probable volcanic action has long since ceased, and that its primal fires were long ago extinguished.

The moon has neither an atmosphere or water upon her surface. The temperature of the moon is different from that of the earth. She has fourteen and three-fourths days of sunlight, and of course the same number of days (terrestrial) of darkness. Intense heat and cold succeed each other. The moon's mass is $\frac{1}{80}$ of that of the earth, and the average density of her material 0·615 or $\frac{6}{10}$ of the earth. Hence a body weighing six pounds at the earth, would weigh one pound at the moon, if each weight retained its terrestrial gravity. Some of her circular or cusp-shaped mountains exceed one and a half miles in height. I have already spoken of their volcanic character.

Fig. 24.

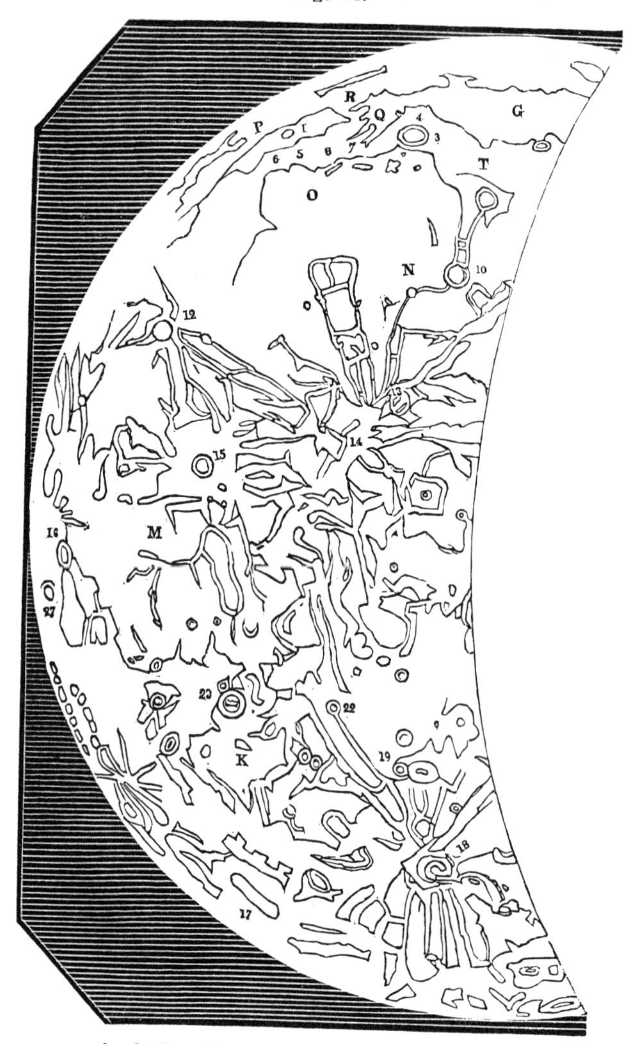

An Outline Plan of a part of the Moon's Surface.

The following are references to the most important points upon the moon's surface: 1 Pythagoras, 3 Plato, 4 Aristotle, 5 Hercules, 6 Atlas, 10 Archimedes, 14 Copernicus, 15 Kepler, 18 Tycho, 23 Gassendus, 19 Pitatus, 16 Helvetius; G Mare frigoris, M Oceanus procellarum, K Mare humorum, T Terra grandinus.

SECTION OF THE EARTH'S CRUST.

§ 85. This section is designed to illustrate the structure of the earth's crust at the parallel of 40° or 45° N. If a segment of the earth were cut off at this parallel, we may suppose the lower part of the consolidated face to consist of pyrocrystalline rocks intersected by veins of the same material, but of a date posterior to the original consolidated matter. These veins consist of the newer granites, dykes of trap, serpentine, veins of iron, copper, lead, &c.; and as there is no country which has been explored, which does not furnish clusters of veins and dykes ot some kind, we may regard the crust as constituted of pyrocrystalline rocks, penetrated everywhere, and traversed by a network of veins. As Vesuvius, Etna, and South American volcanoes are but ejected matter in beds, which are fissured and dyked in every direction, so the general crust may be regarded as equally fissured and traversed by the more recent of the erupted rocks in the form of veins and dykes. It has been stated in the preliminary remarks, that the bottom of the ocean is not a plane, but an irregular surface of the same character as the dry land—sinking in places to the most profound depths, and rising again in mountain peaks, some of which reach the ocean's level, while others peer just above it. All those parts of the earth's crust which lie beneath the area, represented as America, Europe, and Asia, were once beneath the oceans. The vast amount of sediments which are accumulated upon these continents show the vastness of the time during which they were covered with water; and as the extreme height of the mountains of these continents nearly equal the profoundest depths of the oceans, the vertical exposure of the crust, or the depths to which our observations may extend, rather exceed 60,000 feet.

The only points to which I wish to call the attention of the student, on the physical map of the world, are the general directions of the old and new continents, both of which were determined by the direction of the upheaval of their principal mountain chains. In the old world the direction is nearly east and west, and in the new, north and south; both continents are therefore prolonged in the direction of these mountain chains. The description of the subordinate mountain chains will be furnished in a subsequent part of this work. The geological column is designed to show the relation of the systems of rocks to each other. The sediments or hydroplastic rocks are represented as reposing upon each, according to their age; and though the pyrocrystalline are represented as lying also one above the other, this is by no means the constant order of arrangement. Fig. 1 represents granites of different ages, penetrated by a volcanic funnel or tube, terminating in a crater. Figs. 3 and 4 a trap dyke and vein intersecting the laminated pyrocrystalline rocks, which consist of masses of gneiss, mica slate, hornblende, and talcose slate, upon which the oldest of the sediments, the Taconic system, reposes. At different places, however, the several systems of sediments are found reposing directly upon the primary and crystalline rocks. Thus the permian system in North Carolina; the new red of the Connecticut valley rest upon the pyrocrystalline rocks, proving that the places where those rocks are spread out and deposited, were subærial until these geological periods. The relative lengths of the geological periods are approximately represented in the column. It shows to the eye that the paleozoic periods were immensely longer than the messozoic and kainozoic. These periods are the great triads of geologic history. The paleozoic, is the period remarkable for its duration; the other, for the rank of the organisms entombed in their strata.

APPLICATION OF GEOLOGICAL FACTS AND PRINCIPLES TO THE BUSINESS OF MINING.

§ 86. Mining can be conducted safely only by employing the principles of geology as guides in conducting its labor. It is, therefore, a practical application of the observations of geologists which constitutes scientific mining. Mining, as commonly understood, is the extraction of ores from the beds and veins which they occupy in the earth. Its signification, however, may be extended so as to embrace the removal of rock as well as ores from their beds; embracing, also, what is usually known as quarrying.

Though mining is strictly the application of labor, as I have just defined, yet it has its theoretical part which is really of great importance, especially when facts form the foundation of those views. These constitute the philosophical part of the business. The skillful laborer will not disregard this part of the subject; it will even aid his mechanical labors in detaching the useful parts of the mine or of a quarry from their beds, and assist him in bringing to light the riches hid in the earth's bosom.

Mining, in the comprehensive sense in which I design to use the term, may be treated of under the following heads:

1. The theory of the formation of those depositories which contain the metals and ores.

2. The structure of those depositories.

3. The changes which the mineral undergoes in depth.

4. The best modes for extracting their metals and ores, including those which have been devised for raising blocks of rock from their beds.

5. The expense attendant upon different kinds of work in mining and quarrying.

6. The value of the products of mining.

1. In the foregoing part of this treatise I have already had occasion to speak of points which belong to, and which also

illustrate, the subject under consideration. Notwithstanding this, it will be necessary to recapitulate certain facts and principles which have an intimate bearing upon the subject.

In the first place, we must look upon all the repositories of the ores and metals as ancient arrangements, by which they are made accessible to us; and that those arrangements are the necessary result of the constitution of the globe. They are by no means to be regarded as accidents, arising from conditions which might have been otherwise. They, too, are general results, confined to no limited scale; and when the forces and plan were determined, upon which to form and fashion the earth, the results of which I have spoken became an essential part of those causes, and it would have required special instrumentalities to have prevented their operation just as we now see them to have operated. It is for this reason that the formation of repositories for the metals has been controlled by law, by which certain constants may always be looked for. This being the case, the miner has not overlooked the plainest of these results, but is constantly referring to them in his operations with confidence.

It is not necessary that we should connect these laws with the early conditions of the globe, in order to understand them; but as facts it is proper that they should be borne in mind. What was that original condition, then, which gave birth to the repositories of the metals? It was that incandescent state of the crust of the globe, of which I have already had occasion to speak. We have no occasion now to inquire what gave birth to that incandescent state; the fact is attested in the phenomena everywhere visible in those portions of the earth's crust which belong to its earliest epoch. The most important effect of this state is the expansion of the crust, or the occupation of a larger space for the time being. But the earth, situated in space, and in a colder medium than itself, has necessarily lost that primitive heat which belonged to its earliest stage of existence. It has cooled, and the most important result which interests mining, is the consequent contraction of the cooled part. Contraction has severed the bonds of the continuity of the strata; and those fractures which are the result of the cooling process have been made in comparatively straight lines, or in given directions; or we may regard the causes of fractures simply as subterranean, but due to general conditions, and which must necessarily affect the whole of the cooling envelope.

The fact that fissures may be, and probably are thus formed, is agreeable to all that is known of cooling bodies; and observation which has been directed to those fissures proves, in the general at least, that the fissures are made in lines of bearing quite constant. We may not infer that a mechanical force is applied beneath a stratum, and has erupted those strata in the lines I have spoken, as a previous step in the formation of a vein. A cooling state has given rise to a state of tension, which increases in the direct ratio of the diminution of temperature, which is finally too great to be borne, when the continuity is broken. It is true that a subterranean force is often operative in the mode represented, and by which strata are uplifted and fractured; but it is more consonant to facts to suppose that vein fissures are the result of cooling and the great tension which arises therefrom. As a general rule the direction which a fissure has taken was in a line of the weakest part of the stratum; but it is easy to conceive that a greater strain may be made upon a stronger part, so as to form a fracture in a line which is apparently along the strongest part of the stratum. In crystalline rocks the planes of lamination must be regarded as weaker planes, and hence it is that a very large proportion of our veins of magnetic iron lie along those planes.

Having alluded very briefly to the force which has been operative in the formation of vein fissures, it is proper that I should speak of the manner and force by which they are filled. It is in the first place admitted that the matter which fills those fissures was liquid or semi-fluid at the time it passed into them; and furthermore, that the vent produced by over tension extended to the liquid or semi-fluid mass below. It is inconsistent with known facts to suppose the fissures to have been filled with matter in a solid state. The filling of fissures, then, is supposed by many to have been effected by subterranean forces analogous to the forcing of fluids upwards, or in any direction, by the elastic force of vapors acting upon a molten mass; or a new way being opened, it is forced by the elasticity of vapor into that way. It would be difficult to disprove such a view of the manner, but under certain conditions it is unnecessary to bring to our aid the elastic force of any fluid; for a fissure being opened so as to communicate with a semi-fluid matter below, would necessarily fill instantly, in consequence of the vacuity of the fissure itself. It would take place in the same way that water rises in an exhausted receiver of an air pump; or it would rise up and fill the vacuity by what is sometimes called suction. This view comports with the remark, that means would have to be devised to prevent the filling of fissures under the present arrangement. Again, it is evident that veins are not always opened to the width we find them by one single operation of the force of tension. In the regular cooling of the crust, which goes on subsequent to the formation and filling of a fissure, it is evident that as the cooling may still go on, the tension or strain will begin anew. Now, under these circumstances less force will be required to widen anew the former fissure; for we can scarcely suppose that the filled fissure will unite the torn edges of rock so as to equal at all its former strength of attachment: the fissure will therefore run along the line of the old one with ease and certainty, because that has now become the weak part. This will result in the addition of new matter to the vein; and it is not at all improbable that in the extent upheaved, as well as in the line of bearing, it may be increased.

I am disposed to adopt the foregoing view of the manner, as well as that which relates to the force by which fissures are filled. I do not reject, however, the view which brings in the instrumentality of an elastic force of a gaseons fluid, by which the semi-fluid matter is forced upward into a fissure. Both modes have undoubtedly been operative under different circumstances.

But, again, there are other kinds of veins—those which have no communication above or below with a fluid or liquid matter. They are fissures which begin, and which end in the rock, and yet these fissures are filled. In furnishing an explanation of cases of this kind, we must recognize the existence of the same forces as those which I have alluded to. The fissures are produced by tension in a cooling or drying mass, and when they terminate in the rock the fissures are absolute vacuities—each fissure is a vacuum. The filling of such fissures is effected in a mode similar to that already intimated—by fluids charged with lime, or any matters contained in the rock which are soluble under the circumstances; for towards the fissure soluble matter will tend, and crystallization will take place, and the fissure will ultimately be filled with it. It may be inquired, how it happens that veins possess such a uniformity of width. It may be answered satisfactorily in this way: a given rock, mica slate or gneiss, possesses a great degree of uniformity in texture, and hence the different parts of the mass expand or contract alike by equal increments of heat. A fissure may proceed from above downwards, as the outer surface will cool faster than the inner; but its subsequent extension through this uniform mass of matter will form a fissure of nearly the same width throughout, though it may be successively formed. We may justly suppose, however, that in case a fissure proceeds from the outer surface inwards, the resistance of tension will be less in the interior and lower parts of the rock; hence a fissure may rend the lower stratum, as it were prematurely to a great depth, the tenacity being proportioned to the state of consolidation.

We can with difficulty resist the conclusion, that as fissures are formed by the cooling of the surface, the width of a fissure must necessarily be wider at the surface than at considerable depths in the interior. The true mode of representing fissures or veins in diagrams should be in accordance with this view.

The depth, however, to which mines penetrate the earth's crust is exceedingly small compared with the earth's semi-diameter, or even with that of fifty or one hundred miles. We have no occasion, therefore, to attempt to illustrate this principle by a different mode than the one which is usually employed.

But another fact may require a word of explanation. A fissure or vein is shifted, or is jogged out of its line of bearing. We may suppose in such a case, that while cooling, the stratum is subjected to an unequal tension of its parts, or to a tension in two directions. Hence, we may infer that a shift in the position of a part on one side of the fissure may take place at the moment the tenacity of the rock yields to this force in another direction. Or the shift may take place at a period long subsequent to the first fissuration, by the tension in an opposite direction to the first; the shift taking place by an unequal support of the mass. I can not conceive that the force of the entrance of the matter of the vein, tends to the displacement of the stratum. Its entrance acts equally upon the sides of the fissures; and though it is evident that there is considerable friction upon the sides by the striation of the walls and the vein-stone, still, it may be due to the weight of the mass resting upon an unstable foundation.

If a fissure opens to the surface from a great depth, but does not extend to the molten mass beneath, it becomes a water course, a *drainage* fissure, upon the sides of which incrusting matter will be deposited. This is called veinstone, or the gangue, and with its metallic associates forms the vein. The upper part of a vein fills, or is filled, in part, by veinstone intermixed with metal in specks or small lumps, which are diffused through it very sparingly at or near the top, but with an increase of depth increases in quantity.

In addition to the function of drainage, fissures may become *galleries of sublimation,* in which the sulphurets, chlorides, &c., will be deposited. Metallic zones and stripes of metal will traverse the gangues wherever they are penetrable by subterranean exhalation. The exhalations passing upwards through the crater of a volcano, carry up with them ferruginous combinations, which condense on spicula of lava, or colder pieces of rock, which project into the passage. The sulphuret of lead melts in the furnace, and in cooling returns again to its original sulphureted state. Both lead and copper volatilize by furnace heat, and may be condensed again; so subterranean exhalation will condense upon the cold surfaces of fissures. These facts leave, however, other facts observed in mines unexplained. The Rossie mines furnish large, fine crystals of sulphurets as well as crystalline masses enclosed in crystals of calc spar; a fact which seems to sustain the view that the materials were mingled together in the great furnace beneath, and were ejected bodily into the vein. In all those cases where lime and the sulphurets are intermingled in this way, it is evident that their fusion took place under great pressure, otherwise they would both have been decomposed, and it is highly probable sulphates would have been formed; as it is, we have reason to conclude that the minerals fused without parting with their sulphur or carbonic acid.

From the foregoing views, we are justified in the belief that vein fissures are not filled by one mode only, but that they may have been filled by two or more modes conjointly; the upper portion by endosmosis in part as a drainage fissure, and the lower by injection, or by pressure, or sublimation. The copper veins of Cornwall rarely contain copper at less than a hundred feet from the surface; yet there is a fissure with its veinstone. I have said nothing of the electro-magnetic force as an agent, for I conceive that the detection of this agent is not proof that it has been operative in the modes assigned to it.

I have probably presented the simplest view of the formation of metallic veins; and if no other agencies were operative than that of the cooling of the earth's crust, the business of mining would be less complicated and more certain than it is. It is to be recollected, however, that consolidated rocks have been subjected to many disturbing agencies at different times, and it is well known that a mining district is always one in which those agents have been particularly active. An undisturbed district is never a mining one, though it does not follow that all disturbed districts are rich in mines. But without doubt all mining districts have been the seat of great disturbances of the strata. They have been subjected to chemical as well as mechanical forces; strata have been crushed and displaced by faults, and along with those displaced strata their mineral veins have suffered displacement also.

CONTINUITY AND PERSISTENCE OF VEINS IN DEPTH.

§ 87. The expenditure of capital in mining, is warranted only where there is a good degree of assurance of the persistence of the vein in a downward direction. While it must be admitted that each particular case should have its own evidence of its continuity, still that evidence is based on general facts and principles. We refer in the first place to the theoretical views we have offered relative to the origin and formation of veins. If that theory is sustained by observation, we are carried a great way towards a satisfactory establishment of a persistence of veins in a downward direction; subject, however, to an interruption, it may be by the operation of local causes, which may have deranged or interrupted the formation of the vein fissure. We may not, however, neglect the teaching of facts as they are being furnished by the workings of the oldest mines. To the mining records of other countries where it has been the business for centuries, we may refer with great satisfaction. The veins in the mining district of Cornwall, England, have been pursued successfully to the depth of 1800 feet. It was thought by one class of geologists that they had become less rich at the depth of 600 to 1200 feet; still they appear to retain their richness to day with very little or no abatement, to the depth of nearly 2000 feet. There are still deeper mines in the districts of the Hartz mountains. Those of Clausthal and Zellerthal have been pursued to the depth of 1920 feet. Those in the district of Andreasberg, 2400 feet. In the neighborhood of Freyberg, Saxony, the mines have been worked to the depth of 1800 feet. The Mexican silver mines, have been explored to the depth of 1500 or 1600 feet; and the quicksilver mines of Almaden, in Spain, have sustained their richness beyond the depth of 900 feet.

The continuity of veins in the direction of their strike, has not been determined with exactitude. Some of the veins of Freyberg are known to extend, in length, from 12,000 to 24,000 feet. Another is known to extend 36,000. But in the direction of their strike or bearing, there is a great variation in their condition. An auriferous vein of quartz in Maryland, extended only 20 or 24 feet in its line of bearing, plunging down at each extremity on a rapid slope; and yet there is little doubt that its depth continuously is as great as any of the auriferous veins of our country, whose line of bearing may be traced for one or two miles. The records of mining furnish many curious, as well as interesting facts. Veins are often explored for several years without returning a compensation for the labor and expense incurred, when they are abandoned. After many years of rest, an enterprising miner or capitalist, who is acquainted with the history of its former working, having faith in its value, reopens the mine and pursues it with great success. Occurrences of this kind are common in all mining countries. They show in the first place what superior skill may accomplish; or, indeed, the subsequent success may have arisen not so much from superior skill in working, as from having opened a richer part of the mine. All experience in mining proves that the wealth of a mine is variable; some zones are rich, while others are comparatively poor; and this fact is one which should be universally known; it is one which the capitalist should be prepared to meet in any mine, however rich it may be at certain points of exploration. This remark is applicable to veins of lead, copper, silver and gold, rather than of iron.

Veins have been observed to terminate in the direction of their strike, in many thin branches, which appear to be lost in the rock. It is not, however, determined that those smaller branches are always the extreme ends of those veins, for the fissure may have opened wider beyond their apparent terminations, and hence be prolonged still farther. Vein fissures extend beneath the soil farther than the indications upon the surface warrant us to expect. In the direction of the strike they may plunge down beneath the surface on a gentle or steep slope. The outcrop is lost—the fissure does not extend to the surface—the lamina of rock are in contact—and yet the vein is prolonged, and may reappear in an outcrop, several miles from the points where it is known. The argentiferous galena vein of Davidson county was struck in an excavation six miles northeast of its principal workings, and yet the surface gave no signs of its presence beneath. In instances of this kind, the vein plunges downward to an unknown depth, when it takes an upward movement and ascends to the surface. The fissure may be filled, however, with veinstone only, and hence excites no attention. The foregoing statement proves the existence of vein fissures which do not reach the surface. The same fact has been observed in dykes; they penetrate the inferior layers of a rock, but the fissure not extending to the surface, and the stony matter having no power of itself to form a passage, stops at the upper limit of the rent. Branches of veins, too, are often cut in sinking shafts at many levels, which have no connection with the surface. These facts illustrate the direction and mode in which those vein fissures have been filled, and clearly point to their igneous origin.

As the quantity of metal which a vein carries is variable at different points on its line of bearing, so it is also variable at different depths. It may be variable in consequence of the diminished amount in the vein, which at the bottom preserves its ordinary width, or it may be diminished by a contraction of the whole vein in width. It may be pinched out, and the veinstone, together with the metal, becomes a mere trace or string, retaining its position between the walls. The variation of the latter kind is well illustrated by the terminal outcrop of the Rossie lead mine, which is exposed in an uplift of the mass of gneiss in which the vien is enclosed, as is shown in the annexed cut, Fig. 25. It will be observed that while the vein is exposed for about forty feet, it may have an average width of eighteen inches, it bulges out and contracts at many points upon the line of exposure, and is four feet thick at one of the places of dip. It will be observed, also, that the vein is not perpendicular, the rent is not even or vertical, but one which is inclined; or in other words, it has a dip slightly variable at different points, but which, when followed down for one hundred or two hundred feet, or more, is found to be constant, so as to conform to those of the district. The direction of dip may be depended upon, and shafts are often sunk two or three or even four hundred feet, with the expectation of cutting it at one of those depths. Sometimes a vein which dips eastward at the surface, is found to dip westward at 80 or 100 feet. The indication that such a change may be expected consists in the character of the disturbances at the surface. The dip at the surface in those cases is entirely at variance with the common dips of the veins of the district, and the existence of dykes and faults, show that something unusual has taken place. The veins of a district usually dip in the same direction; and when a vein dips in a contrary one, under the circumstances I have stated, there is ground for expecting that at a certain depth this vein will change its direction.

Fig. 25.

J. T. PEASE del. R. H. PEASE Sc.

STRUCTURE OF MINERAL VEINS, TOGETHER WITH THE KINDS OF VEINSTONE WHICH CONTAIN THE ORES OR METALS.

§ 88. A metallic vein is originally a fissure of an unknown depth and length, bounded by two walls of rocks, whose composition differs from the contents of the vein fissures, and from which the latter easily cleaves. When the fissure or vein is not vertical, the wall against which the vein rests is called the foot wall, and the other the hanging wall. These walls are sometimes called the floor and roof of the vein.

The structure of a vein is simple, and we have no occasion to multiply varieties. A dyke is the simplest and least complicated of that class of veins. It is a fissure filled with stony matter, with specks of sulphuret of iron disseminated in small quantities through its substance. It seems to have been formed and filled at one time. A mineral vein consists of stony matter and the oxide and sulphurets of the metals, and sometimes with pure metals; as the native copper and silver of lake Superior, and the native gold and copper of North Carolina. It is necessary to learn the character of the veinstone, before an opinion can be formed of the probable value of the vein. If the veinstone is solid and compact, the judgment will be unfavorable to its productiveness. If it breaks easily, or if it is porous and seamy, the judgment will be favorable for its productiveness. In a solid veinstone or gangue containing disseminated metal in isolated particles, and which do not run together as it were, the encouragement for a valuable vein is small, as long as it continues in this condition; and a vein of this description rarely assumes a favorable character. If, however, the veinstone is striped vertically with metal, or if the masses are elongated and run together, the prospect is favorable; the metal is assuming a vein-like character. The metal may possess only the disseminated character at the top of the vein, and may pass into the veiny character below. The beginning of this change is always regarded as a favorable indication, and an encouragement at least to pursue it to a greater depth. A vein is often laminated. The metal and rock being arranged in parallel stripes, it possesses the parallel structure of gneiss. This structure is regarded as one of the most favorable changes,—one which will ultimately lead to the formation of a productive vein. The lamination of a vein is unlike the lamination of gneiss and mica slate; it is always vertical. The structure is sometimes developed so prominently that a question arises respecting the periods of its formation. The indications suggest the probability that the vein was filled by successive openings. At each opening it received an accession of new matter from beneath. I have already stated that the first fissuration may not have communicated with the metalliferous matter, and hence is merely a *drainage fissure*, a receptacle of stony matter. For example, we often find that an auriferous vein is composed of two parts; one of copper, which lies next to the hanging wall, and which is most stony, and a lower mass, which lies against the foot wall, and which is metalliferous. In an example of this kind the fissure may have been widened by the same force which produced it in the first instance. It is upon the foot wall that we look for the greatest quantity of metal. But the metal may pass from the foot to near the hanging wall; and it is frequently central, as in the Rossie lead mine. From the foregoing remarks it will not escape the notice of the reader, that a vein consists mostly of stony matter or gangue, the metal being distributed through it in elongated branches, which run together; and there may be two or more such confluent ranges of metal arranged in parallel stripes, and which are separated from each other by an excess of stony matter.

The foregoing statements may be taken as a general description of the structure of all productive veins, without regard to the kind of mineral which forms the gangue.

INDICATIONS OF A GOOD OR BAD VEIN, JUDGED OF BY THE CHARACTER OF ITS WALLS.

§ 89. The structure of a vein furnishes important information respecting its value, as I have already intimated. The walls, however, are not to be passed by unheeded. In the first place, we look for clear and well defined boundaries between the vein and wall; and if the walls are marked with what the miners term slickensides, which are polished striations more or less vertical in direction, the indications of a well-formed vein are sufficiently satisfactory. Sometimes, however, the walls are not equally well defined. We look in this case to the foot wall, and miners are generally satisfied if they find this hard or firm, and distinct from the veinstone, even if it is not marked with slickensides.

As regards the indications of a well-formed vein in its structure and character of its walls, we find that a porous veinstone furnishes the best indications, especially if the gangue, with the metal, is arranged in parallel stripes, the bunches elongated and confluent, or become more so as the shaft increases in depth. On the contrary, where the ore or metal is sparsely disseminated in a hard, compact, tough gangue, with scarcely any tendency to become confluent, in a veinstone of this description, the indications of a well-formed and productive vein are unfavorable. So also where the vein, though tolerably distinct in parts, is mostly incorporated with the walls, or what are taken for walls, the indications are unfavorable. I have said nothing of the judgment we should form of the increase of metal as the shaft increases in depth, for it is sufficiently plain, that in this country, there should be an increase of metal within forty feet of the surface. The kinds of veinstone are numerous; and how much the value of a mine is dependent upon them, is not well determined, excepting the general fact that some are much softer and more workable than others, and require less expense in working them.

The veinstone in which gold, silver, lead, copper and zinc are found, in this country, is generally quartz, or in the language of miners, flint; and we have numerous examples of rich and poor veins in this kind of gangue. I know of but one lead vein in North Carolina which has a calcareous gangue. The following kinds of mineral form the veinstone of sundry mines in this country: Quartz, calcspar, pyroxene, hornblende, feldspar, phosphate of lime intermixed with a small quantity of hornblende, prhenite, and magnesian carbonate of lime. Either of these minerals may form good stoping ground, or either of them excepting the calcareous may be bad or hard. Much depends upon the connection of the gangue with the wall; if this is such that a gad or pick can be employed in taking down the vein, it belongs to the kind called easy or good ground. Quartz sometimes partakes of the mineral character of a hornstone; it is then an exceeding tough rock; it is an expensive vein to take down, and though it may be rich, yet the expense attending its working is so great as to consume the expected profits.

CHANGES WHICH A VEIN UNDERGOES NEAR THE SURFACE.

§ 90. Every miner has observed that the part of a vein near the surface, differs from that below. The change takes place at that depth where water always remains. The difference between the part of a vein near the surface and that below water, consists in two particulars. There is first a mechanical difference in these two parts of the vein. The veinstone is porous, and the metals are oxides, in loose, slightly coherent masses. I have in mind the gold veins of the south. In the second place, below water, the veinstone is more solid, and the metals are sulphurets, disseminated in the gangue. The upper part is brown, or reddish brown, and the quartz is thus stained; while below, the gangue is blended with specs of metal, or perhaps with the elongated masses. This change in the character of a vein is due to the action of the atmosphere, aided by the alternations of dry and wet states to which the vein is exposed. In consequence of the changes of which I am speaking, many intelligent miners believe the vein has become poor; and in this opinion he is sustained by the fact that as usually worked, the profits have very materially diminished. But this result is due to what nature has done for the upper part of the vein, by detaching the gold from its most intimate combination with the sulphurets, and no process has hitherto been invented by which all the gold has been separated from the mineral. Experience proves, however, that by repeated pulverizations, by means of stamps and chilian mills, the aggregate amount of metal which can be obtained, is about as great below as above water.

The changes referred to in the foregoing paragraphs are true of the auriferous veins, in connection with the sulphurets of iron and copper of this country; while those containing galena are far less chemically changeable. When to the sulphurets of iron and copper arsenic is added, as in the Ducktown mines in Tennessee, the changes are still more striking.

The question whether mines are richer or poorer above than below water is not perhaps fully settled. For myself, I believe that the facts, when well determined, go to prove the undiminished value of the vein below water. But it is a question whether means sufficiently simple and cheap, can be devised by which mines can be made to pay a profit. Some will pay for forty, fifty or sixty feet, but ore may not be really rich enough to pay a profit below those points. The depth at which a vein is changed by atmospheric influence is variable. It depends undoubtedly upon the depth of drainage. It is not a point which can be determined beforehand; we may reach water in twenty-five or fifty feet, or it may not be reached in sixty; but the mean is about forty-five or fifty feet.

NOTE.—It is not designed to intimate that water has any influence in increasing or diminishing the amount of what metal it carries. The reason why the expression above or below is used is sufficiently obvious from the explanations of the text.

CHARACTER OF SOME OF THE METALLIFEROUS VEINS OF THIS COUNTRY.

§ 91. *Veins of magnetic and specular oxide of iron.* We have not as yet placed a due estimate upon the value of the iron ores of this country. The increased and increasing use of iron itself, together with the demand which grows out of our increasing population, are facts sufficiently positive and absolute to prove that the wants for this metal will soon more than double its present consumption. We have seen that our ores are abundant and favorably situated to supply the wants of the country. I need not dwell upon this topic. My object is to illustrate the formation and structure of the repositories of these ores, and in doing this I shall compare them with some of the oldest iron mines of the old world.

The magnetic iron veins are upon a magnificent scale, especially those in northern New York, to which I shall direct the attention of the reader.

The repositories of this ore seem to be of two kinds. Of one kind I have no hesitation in saying that they are veins according to the definition usually given. Of another kind there may be doubts whether they are veins or beds, but I have regarded them as rocks or masses of magnetic iron, inasmuch as their boundaries with the rock are indefinite, and they are upon so large a scale that they are worked like quarries of marble or granite.

The structure of the veins of iron scarcely differ from those of other metals. The ore itself is crystalline, and there is no doubt but that it is pyrocrystalline. It is not subject to great changes in its composition. The veinstone, however, is often stained red or brown, by a change of the protoxide into the peroxide; such a change is regarded as a favorable indication of the quality of the ore.

In veins of magnetic iron the rich part of the ore forms a belt parallel to the walls, and it occupies very frequently the centre of the vein. The width is variable, a mass frequently extending beyond the usual limits of the vein. The Pendfield mine in Essex county, N. Y., (fig. 26), swells out 160 feet.

Fig. 26.

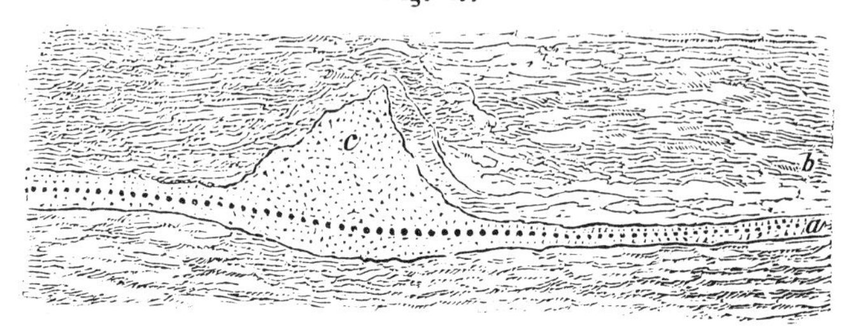

The ground plan takes in this immense expansion of the vein, showing in addition the rich belt of ore by a dotted line which extends along the middle of the vein. The average width of the vein is 40 feet. Its line of bearing is northeast and southwest, and has been exposed by the removal of its covering about 20 rods, but it has been traced over half a mile.

The ore adjacent to either wall is largely intermixed with quartz, while the middle is a solid mass of unmixed ore. The rock of this section is gneiss dipping S. E.; this vein of ore coinciding with the lamination of the rock.

The iron manufactured from this ore has a high reputation. It was tested at the Navy Yard at Washington, and was found more suitable for cables and chains than any iron of the country, which had been submitted at that time to the necessary tests. The superiority of this ore depends much upon the veinstone. Quartz or flint forms an admirable flux for iron, when the quantity is too great it is removed by water. There is no pyrites or phosphate of lime intermixed with it,. and the forgeman has to deal with a pure oxide of iron. This ore is uniform in quality, is of dull black color, rarely bright, but crystalline and strongly magnetic. This vein has furnished ore for thirty years, but has been worked only in the forge and into blooms. The supply will be equal to any demands which are likely to be made for ore, inasmuch as it is prolonged beyond the limits I have stated, on its line of bearing, more than half a mile from the original opening upon the tract, and another mine has been worked for twenty years, and the prospect for the permanence of ore is greater than could have been anticipated. The width of vein increases with the depth, and no fact is yet brought to light which indicates its discontinuance.

Veins of magnetic iron are distributed over limited districts. Several veins traverse this district in parallel ridges and they may be known to belong to separate and distinct veins by the dissimilarity of their ores, or of their gangues. Those which are only a few feet asunder possess unlike qualities. In deciding upon the number of veins it is necessary to guard against deception, as a fold in the strata or an anticlinal axis may place the parts of veins in such relations that they may appear to be two veins when there is only one. Fig. 27 illustrates the im-

Fig. 27.

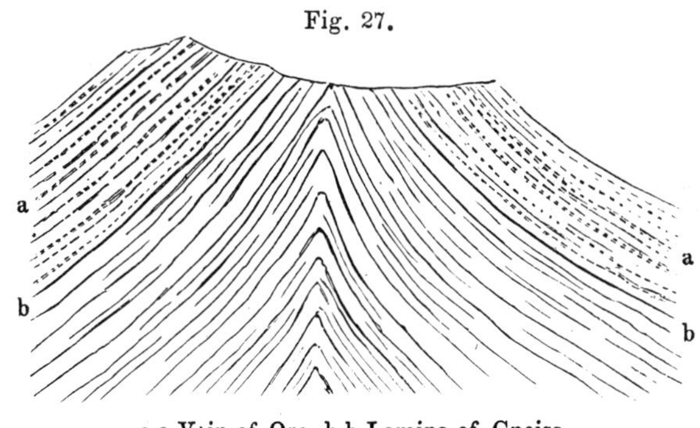

a a Vein of Ore, b b Lamina of Gneiss.

portance of being on the guard. b, b, The folded lamina of gneiss supporting a lean vein of ore, a, a, and indicated by the dotted lines. This instance is a plain one, but others of a more obscure character will not mislead the observer, provided due caution is exercised in his examination. An arrangement of veins approximating to the former occurs at the Cook mine, where an anticlinal axis seems to occur, and which might at first be regarded as a repetition of two veins, admitting that the narrower vein is divided; but the quality of the ore and the thickness is quite different. Fig. 28. The wider vien is fourteen feet thick, and the others six, three, and two feet. These are known as the Cook veins. They traverse a north and south range. The rock is gneiss and the lamina of the planes nearly vertical.

Fig. 28.

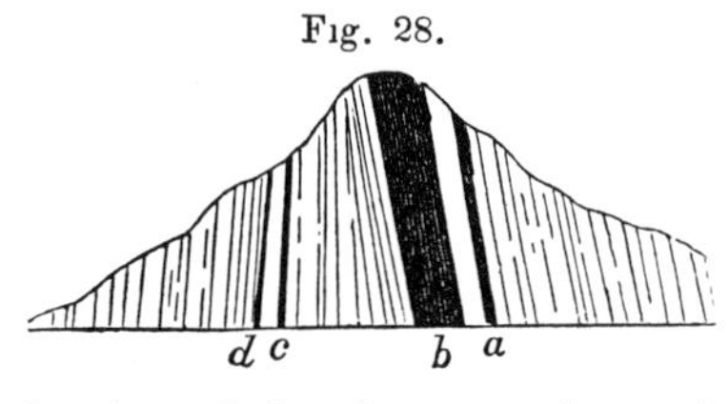

The experience which has been acquired in mining in northern New York, has now become valuable. The exploration of the magnetic ores was prosecuted at an early day, and in consequence of the adventurous spirit of the owners of mineral lands in this district, trial shafts were sunk at points which were not promising at the surface. The rock was observed to be charged with particles of ore which were found to be persistent and to have a direction correspondent with the veins which had been proved. On sinking however a shaft upon those stripes of rock and lean ore, it was found that the iron increased, and at a moderate depth, a productive vein of metal was established.

A mine widely known as the Palmer vein, is a good example of the change from a very lean ore to a rich one. A vertical section of the mine, fig. 29, illustrates the fact under consider-

Fig. 29.

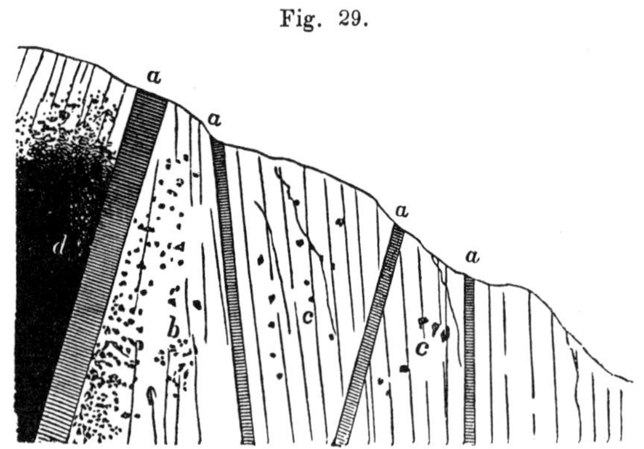

a Dykes, *b* Increased Ore, *c* Disseminated Ore, *d* Mass of Ore.

ation, and what is quite as interesting at this locality in Clinton county, is the intersection of the rock by several trap dykes. The surface of the rock is denuded and shows particles of ore disseminated through it, particularly between the dykes; a, a, dykes, c, c, the spaces in which ore is sparsely disseminated at the surface, at b, the ore increases, and at d, on the other side of the 14 foot dyke, it is a solid mass of ore, with scattering grains of gray quartz. This wide dyke cuts the veins obliquely. The ore was lean on the east side, but much better upon the west, as an adit on being cut through it, disclosed a mass of ore on the opposite side, seventy feet thick.

Fig. 30.

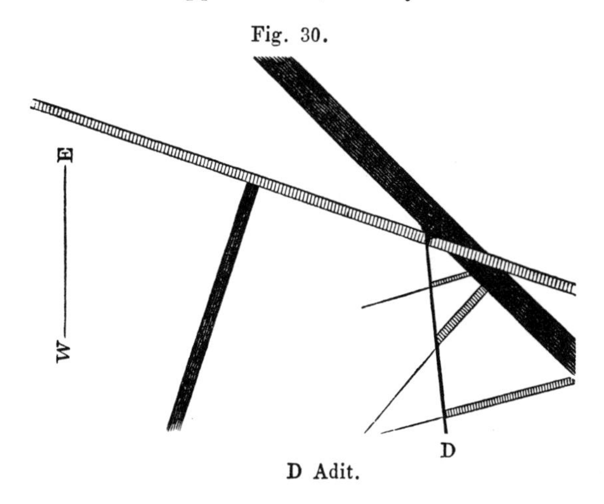

D Adit.

Fig. 30, is a ground plan of the dykes, showing their relations to the ore. These are not parallel, and as they intersect each other, they are clearly of different ages. The widest intersects the vein of ore, but the narrower ones are intersected by the latter. A, vein of ore running N. E. and S. W. B, vein of ore which runs nearly east and west. The latter is a rare example of a vein pursuing this direction. D, adit. Veins of magnetic iron often contain masses of rock in the midst of the ore which under certain circumstances have the semblance of walls. They are generally pure rock free from the ore, and as they are arranged in a direction parallel with the true walls, may, when not examined with care be mistaken for the walls. An examination will lead to the detection of their true form which is that of a wedge. It is scarcely necessary to add that they will disappear in the progress of mining.

Fig. 31.

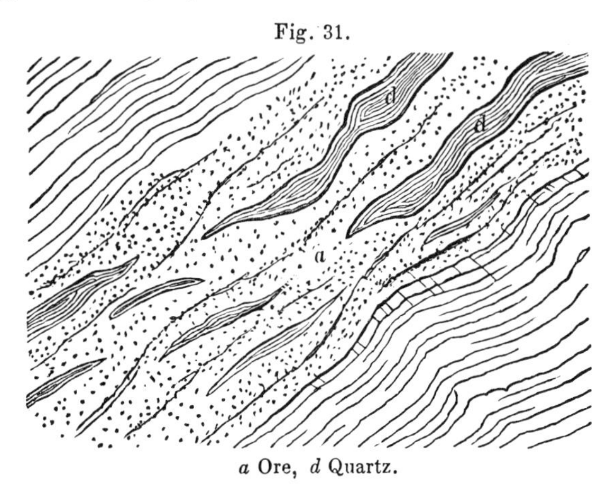

a Ore, *d* Quartz.

Fig. 31, illustrates the fact I have just stated. It is a section of a part of the Hall vein, in Moriah, Essex county, the parallel masses of rock, are, d, quartz, a, ore; but they may consist of hornblende or feldspar. An interesting instance of the same kind, is now being exposed at the Old Sanford ore bed, where masses of trappean rock and feldspar, are intruded into the midst of the ore. Fig. 32. In this instance the strike of the vein is not determined, and hence it is impossible to say whether these masses are parallel with the walls or not. Each of those masses might be regarded as limiting the ore, but on cutting through them, equally good ore is found in the spaces between the dykes, as at any other part of this remarkable vein.

Since the foregoing was penned, I have ascertained, by an examination of the wall, that three of the dykes have disappeared, and the three obliquely placed masses of rock are entirely removed, and there now appears a breast of solid ore 146 feet long and twenty-five feet high, traversed by a single dyke

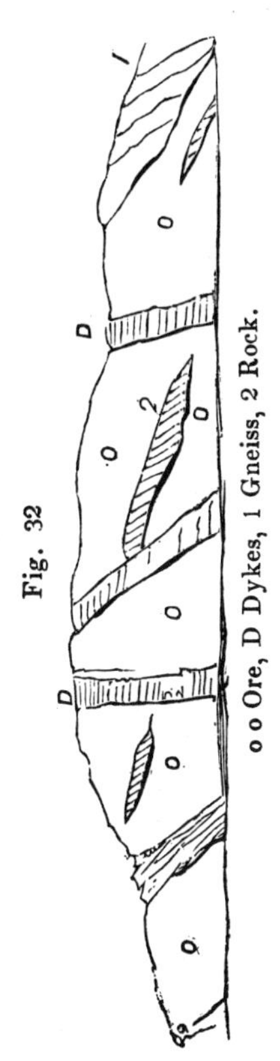

Fig. 32

o o Ore, D Dykes, 1 Gneiss, 2 Rock.

twelve inches thick. The dykes are composed of a greenish, foliated crystalline feldspar. The disappearance of a net work of nonmetallic veins, is by no means an uncommon circumstance in mining in Northern New York, and hence there is encouragement in mining of this description, that the dead work will diminish as it progresses. The gangue of this ore is phosphate of lime, in small reddish particles, imbedded in the grains of the oxide. They are usually of the size of a kernel of wheat. When the phosphate is separated, the iron made from this ore is good, but if manufactured with the phosphate it is brittle. This mineral constitutes about one-fifth of the mass—but in parts of the vein it is equal to one-half of the mass of ore. The phosphate is now separated by magnetic separators, is ground and prepared for use in agriculture.

Adventurers in mining are often startled on finding the vein diminishing in width. In iron mining, however, those fears are

Fig. 33.

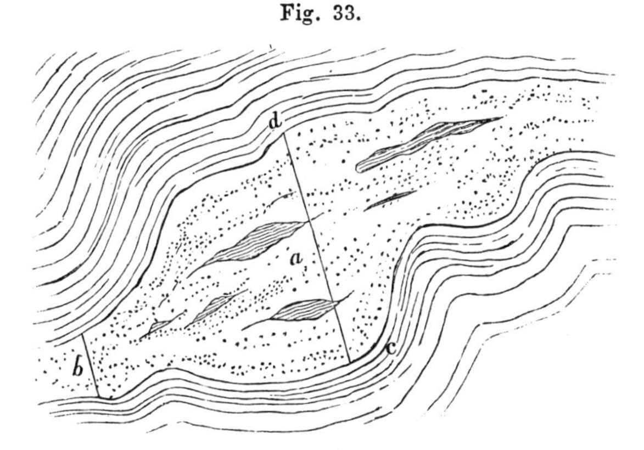

b **Constricted Part of the Vein,** *c* **Gneiss.**

groundless. When the Hall mine was first opened, it enlarged to eight feet. Afterwards it diminished rapidly to less than four feet. Fig. 33, is a section of this part of the vein. It was eight feet wide at a, a, and when I examined it in 1837, it was pinched out to four feet. At this stage of the working it was a question what would become of the vein. On pursuing it farther and into the constricted part, it began to widen again, and has proved to be one of the most valuable mines in the Moriah mineral district. It has been worked to the depth of a hundred feet. The vein appears inexhaustible. A transverse section of this district, about one hundred yards, furnished at least three parallel veins. The annexed section was made in 1837, fig. 34, a, a, veins, b, b, gneiss. All of these have been proved

Fig. 34.

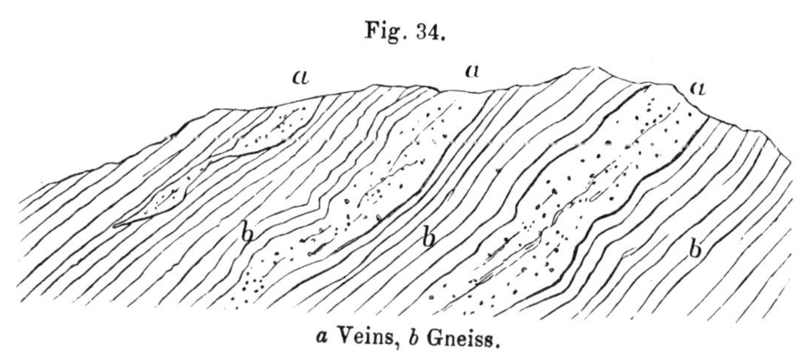

a Veins, *b* Gneiss.

to the depth of seventy or eighty feet, the ore increasing in richness with the depth. They have been traced about one and a half miles. Their line of bearing is northeast and southwest, and they dip with the plane of lamination of the gneiss in which they are inclosed, and half a mile west of the Sandford mine, they have been extensively worked, and yield a pure black granular oxide. Phosphate of lime is not present in the three parallel veins, although in the immediate neighborhood of the Sandford mine which is so rich in it.

The veins of magnetic iron contain ore whose quantities differ materially from each other. No two veins, however near, supply the same kinds of ore. A good example and one in point, are the well known veins in the Clintonville mining district. Four veins have been worked to a depth from one hundred to two hundred and sixty feet. These occupy a high hill four miles west of Clintonville. The first vein which was opened contained a beautiful blue and irridescent ore, both soft and granular. The next vein which is parallel with the first, furnishes a black ore, and the others a gray ore. The blue ore probably makes the softest iron of any ore in this country. The others a harder iron. The first is from four to eight feet wide, and the direction and dip of the four correspond, being north-northeast, and dip west-northwest, at an angle of 70°. The ore of all of these veins has been changed from that of a protoxide to a peroxide, as they all give a red powder, but the change is more descisive in the blue vein. The gangue is a blue gray quartz in the gray veins.

These veins have been shifted simultaneously by dykes in a

Fig. 35.

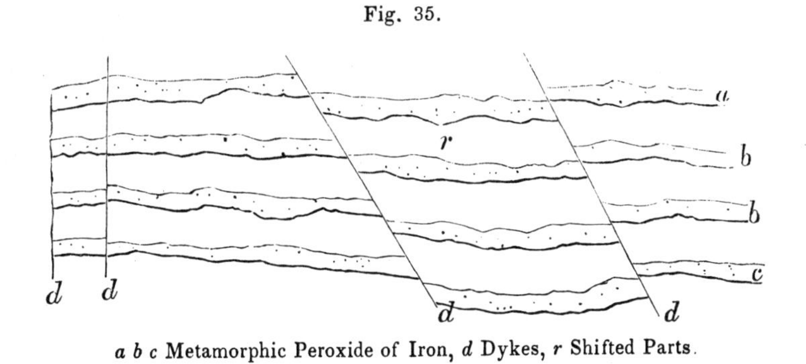

a b c Metamorphic Peroxide of Iron, *d* Dykes, *r* Shifted Parts.

mode represented in diagram No. 35, d, dykes, a, b, b, c, four parallel veins of metamorphic peroxide of iron.

About half a mile north of Clintonville, a mine worthy of a passing notice was opened at an early day. It is known as the Winter ore bed. The ore is hard, but being pure, it made a good iron, and as it was situated conveniently, it was desirable to make the most of it possible. The vein reposed upon the top of the rock, and it excited attention from the novelty of its position. It was in the form of a flat superficial mass gently inclining to the north; this mass was about a hundred feet long and forty wide. It was all quarried out, and when this was done, the miners lost all trace of its direction. Several shafts were sunk in hopes of striking some parts of the mass or veins which it was supposed might be prolonged beneath and between the lamina of the rock. These were all unsuccessful. The following diagram illustrates the relation of the vein to the rock, fig. 36, a, a, a, a, masses of ore, a is the

Fig. 36.

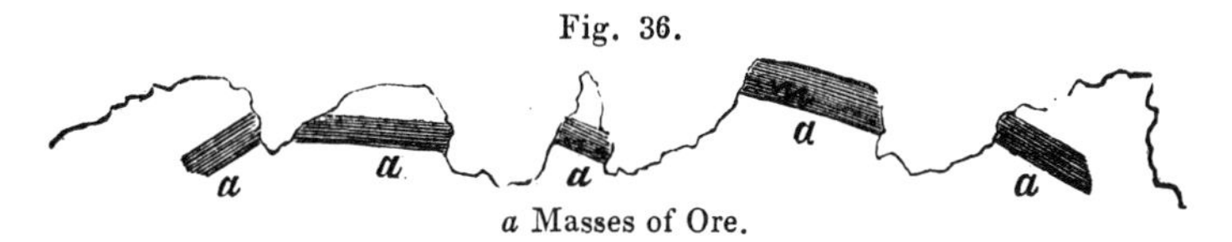

a Masses of Ore.

large mass already referred to. These five masses are regarded as parts of one vein, interrupted and broken at the time of upheaval, presenting a very imperfect anticlinal axis. The mass a forms the crown, but being unsymmetrical and the others being lean their true relations had been overlooked. The dip on one side is to the east, and on the other to the west. The plan proposed for recovering the mine or vein was to tunnel from the eastern slope with a view of intersecting it, two hundred feet below the surface. The plan was adopted and the vein recovered. The complication was increased by numerous trap dykes. No less than seven crossed the mining tract in about two hundred feet. In diagram 37 the dykes and

Fig. 37.

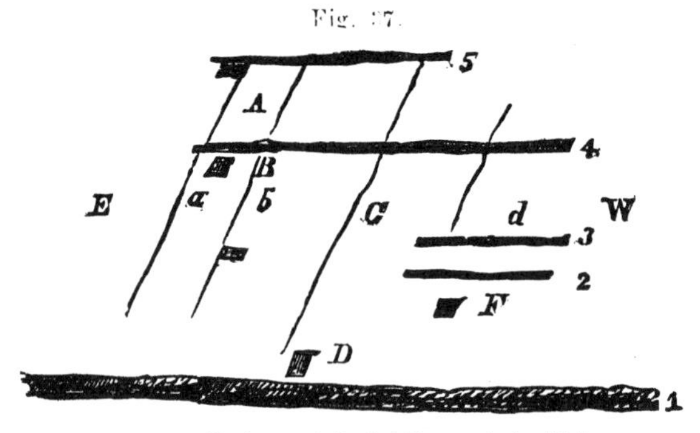

1 2 3 4 5 Dykes, *A B C d* Parts of the Vein.

parts of the vein are represented. 1, 2, 3, 4, 5, are the parts of the dykes which appear to have produced shifts in the vein, or rather the irregularities, we may suppose, in the anticlinal axis are due to their presence. The dyke No. 1, is twenty-seven feet thick, and in tunneling it was proposed to cut it along this dyke. The cost of tunneling was five dollars per linear foot, with an adit six feet high and five feet wide.

A vein at the outcrop sometimes has a less dip than in the earth; and when the vein is flat, as it may be, the inexperienced are very likely to overestimate the width of the vein. The Averil vein at Clinton prison, was nearly flat for forty feet. It had been worked from its easterly outcrop to the west, and along its strike about one or two hundred feet in length. At a point about forty feet to the west, it showed at first, an inclination to the west, and then began to dip more rapidly—when uncovering it still farther west, the hanging wall was discovered. For more than forty feet this wall had been removed, having a flat mass of ore from fifteen to twenty feet thick. The position of the vein was calculated to deceive those who were or might be interested in the property respecting the quantity of ore at this mine, or near the surface.

The illustrations and remarks which I have made respecting the iron ores of Northern New York, are applicable to all those which lie upon the base of the Appalachian range, from north to south. I have already referred to the numerous veins of this ore, and have spoken in general of its value and extent. The resources for the supply of iron in this country have never been properly estimated. We may, however, expect that for the future the attention to the manufacture of iron will be turned more exclusively to our means and resources, and that necessity, as well as interest, will soon establish for the iron manufacture what has already been established for cotton.

The iron ore of the western part of Essex county, N. Y., lies in immense beds, or rather constitutes rocks of no mean extent. We are unable to discover that it is enclosed in regular walls, though rock bearing a resemblance to walls is often encountered in the midst of the ore. The ore of Sandford lake, at or near Adirondack, may be cited as an example of a huge mass of magnetic ore in the midst of hyperthene rock, whose boundaries have not been determined. It is between 700 and 800 feet over, measuring from east to west. It extends still farther, but the debris and soil is too deep to admit of removal for the single purpose of disclosing the extent of the quarry of iron rock.

The masses of ore at Adirondack frequently extend into the adjoining mass in branching or ramifying veins, which usually terminate in threads or strings, and are therefore lost in the rock. At certain points, as I have already remarked, the ore rests against the rock, which appears like a wall, but it has been observed in many instances, that on penetrating beyond this apparent barrier, the ore is found extending beyond it. The extension of veins into the hyperthene rock is illustrated in fig. 38. Garnets border or fringe the veins lying between

Fig. 38.

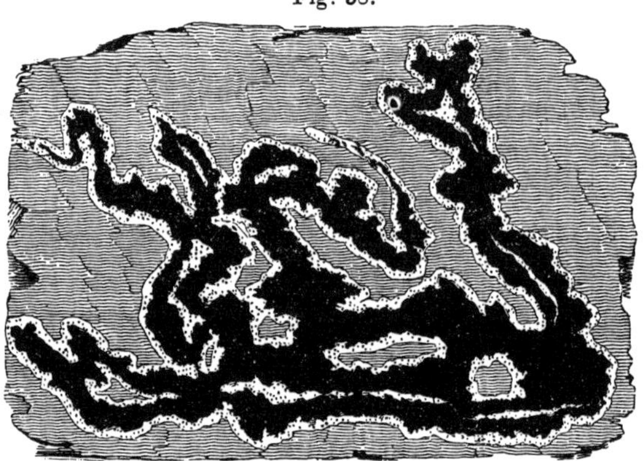

the rock and ore. These garnet borders seem to have been produced by the influence of the ore upon the rock.

The ore of Adirondack produces a remarkably tough iron. In the Catalan forge the result is too uncertain, and it does not furnish when reduced in this simple way, an iron the qualities of which are constant; yet bloomers of intelligence have succeeded in making from it iron whose qualities were equal to the best Swedes iron. Many samples of this iron have been tested at different times and by different modes, all of which tended to confirm the favorable opinions entertained respecting the value of this ore. The late Prof. Johnson's tests were the most satisfactory which were made. The iron could be drawn to an half inch bar under the common trip hammer, and bent when cold without breaking the fibres, or producing cracks. The bent bar represented in fig. 39 illustrates in the most satisfactory manner, the quality of the iron which may be made from this ore, both as regards toughness, and durability. The outer curve of the bar is 3·8 inches in length, the inner only 2·15 inches. The length of this bar before it was bent, was 2¼ inches; the width 1·29, and the thickness 0·59.

Fig. 39.

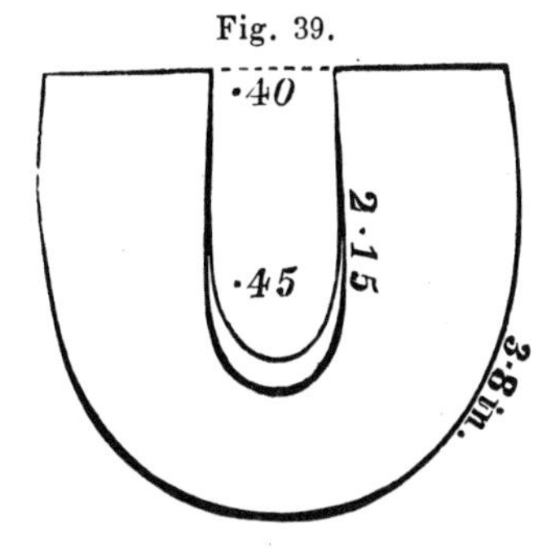

The specular or peroxide of iron takes the place of the magnetic, in Jefferson and St. Lawrence counties, N. Y. The associated rocks of the St. Lawrence side of the mountains differ somewhat from those of the Champlain side. Gneiss, granite and pyrocrystalline limestone belong to both sides, but on the Champlain side the former rocks contain the magnetic ores, while the latter on the St. Lawrence side contain the specular. Serpentine is also a constant associate of the latter. There is really no difference in the mode of occurrence of the two oxides, both are primarily in veins. The specular oxide, however, may appear to be an exception, as a very large proportion of the ore of Jefferson and St. Lawrence counties has been obtained from beds in the soil. The ore, however, is derived from veins. Its constitution favors its disintegration. The associates of the specular ore are crystallized quartz, mostly in dodecahedrons, carbonate of lime and magnesia, carbonate of iron, sulphate of barytes, sulphuret of iron, cacoxenite, sulphuret of nickel. The last two are rare. The ore, except when broken down or disintegrated, is in crystals and crystalline masses, with brilliant surfaces—and in this condition is very pure and free from sulphuret of iron. The veins lie in parallel position with the lamina of rock, where they can be traced to rocks of that class; when in serpentine or limestone the ore is apparently in masses, and may be removed entirely from the rock. The specular ore is not confined to pyrocrystalline limestone; several veins, as the Polly and Tate ores are subordinate to gneiss, but maintain a connection with serpentine. The Kearney and Parish veins are important ores, and extend for two or three miles in a northerly direction.

Magnetic ore occurs at one or more localities in St. Lawrence county, in the township of Chaumont. The ore is rich, and being situated upon the Oswegatchie, and in a well wooded forest, will in time become a valuable location for the manufacture of iron.

From the foregoing remarks and illustrations, it is apparent that iron ores do not preserve those geological relations which are absolutely similar in all parts of our country. While we may observe, however, considerable diversity in their occurrence, still there is such a general similarity in those respects, that we may avail ourselves of the use of general principles in conducting the necessary mining operations. This general similarity extends also to the conditions of the great masses and veins of ore which are best known, and which have been the most extensively worked in foreign countries, particularly those of Norway and Sweden.

The annexed diagram, fig. 40, copied from a valuable article on the mines of Arendal,* would illustrate our own mines of magnetic iron. It is a ground plan; M M, masses of ore prolonged in the direction of the vein, but variable in width and apparently interrupted; *gn* gneiss, *g* limestone. This vein branches out into the walls of the gneiss. It occupies a middle part of

*Annales Des Mines, Tome IV, 1843.

the fissure, a fact which is quite constant in our own mines. A vertical section of the mines of Arendal exhibits similar arrangements to the mines of the Appalachian range. Indeed, the phenomena are so constant that only a few illustrations are required to exhibit all the most important facts relating to the veins of magnetic and specular oxides of iron. The relations and circumstances under which these ores occur, seem to point to their igneous origin. In numerous instances minerals of contact are developed, such as garnet, hornblende, epidote, &c. In this respect the action of these ores resembles that of other pyrocrystalline rocks. Garnet and epidote appear to be the most common minerals which are developed under circumstances similar to those under consideration. Of the two, epidote is the most common: it accompanies the numerous varieties of porphyries, appearing in different stages of development, from a mere yellowish coloring of the rock in patches, to perfect nests of crystals. The slaty rocks of sedimentary origin, frequently furnish this mineral in all the different stages referred to. In the case of garnet, its development usually takes place contiguous to the mass of igneous matter, while epidote is found more extensively and widely diffused in the rock.

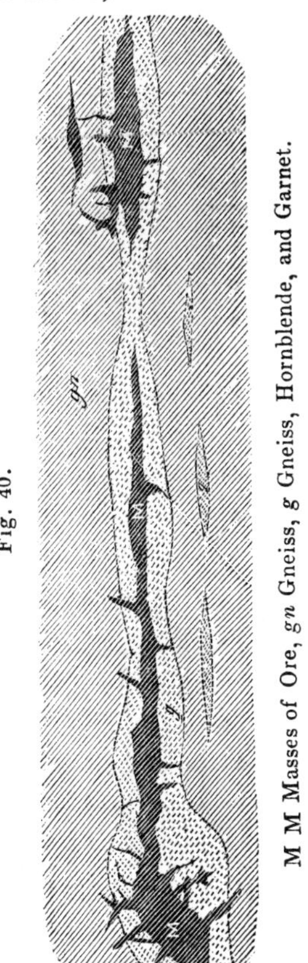

Fig. 40.

M M Masses of Ore, *gn* Gneiss, *g* Gneiss, Hornblende, and Garnet.

LIMONITE OR HEMATITIC ORES OF IRON.

§ 92. The geological position of this ore in New England and New York is not well determined; but when in beds it seems to be enclosed in materials very similar to drift. But the drift beds have been derived from distant places; and as the rocks are rarely uncovered, it has been up to this time an unanswered question, where this ore belonged. At Adams, Berkshire, Massachusetts, the ore has been traced apparently to a vein in the quartz rock, near its junction with mica slate of the Hoosick mountain. Farther south, in Maryland, Virginia, and North Carolina, limonite occurs in veins. The most conclusive facts bearing upon the geological position of limonite have been disclosed in Cherokee county in North Carolina, in the vicinity of Murphy. It occurs upon the surface, rising up like mounds which cover several acres. When it had been observed at the copper mines in Ducktown that this kind of ore covered the black oxide of copper, those mounds were explored by cuts and shafts in expectation of finding copper. Though these explorations terminated in disappointment, yet it was discovered that those mounds of iron ore surmounted a vein which, near its surface, was composed of gossan mixed with sulphuret of iron, but no copper. In all those cases where gossan was cut entirely through, the back of the vein was found intermixed with sulphuret of iron only. I was quite satisfied that the prospect of finding copper was too uncertain to warrant the prosecution of the object when the expenses were likely to become considerable.

The accumulation of mounds or masses of the hydrous peroxide of iron or gossan many feet beneath the surface, is owing, first, to the ordinary decomposition of veins of sulphuret of iron, and the breaking down of the rock as the process proceeds; and the debris remains, as the country in this latitude has never been subjected to causes which disperse and scatter it abroad after the decomposition. At the north, however, embracing the Canadas, New England, New York, and Northwestern States, as far south as Cincinnati, diluvial action has removed the loose materials south, southeast and southwestward. Hence the hematites are buried deeply in drift, often having been lodged in protected positions. The bog ores are accumulations of the hydrous peroxide of iron in swampy depressions. Instances of bog ore upon hillsides and in dry places, are not exceptions to this statement: they are beds which have been elevated by subterranean forces. Examples of this kind occur in Jefferson county, New York. They contain the leaves, trunks, and roots of trees now living. In the neighborhood of the elevated beds of bog ore the rocks have immense pot holes; and connecting these facts, the conclusion is forced upon us, that the country has been elevated during the modern period. Bog ore is also formed from the decomposition of the sulphuret of iron, which is disseminated in rocks which undergo disintegration and decomposition. Water bears the materials into the depressions, where, in time, thick beds accumulate.

THE GOLD-BEARING ROCKS OF THE UNITED STATES.

§ 93. In this country, both in its mining districts and in its general geology, we have sought to discover and make out similitudes or analogies with Europe, rather than to discover the grouping and relations of our rocks, minerals, and fossils. Our attempts have been made rather to compel facts to bow and obey what has been represented to exist abroad, rather than to give them the independence they deserved, or to honor them by the deductions they sustain. Our gold, copper, and lead are associated with minerals differing from those of Cornwall or the Hartz. Our copper of lake Superior has no parallel in European mines. Every day's experience proves that the metals have associations more general than has been represented; that they are by no means confined to one rock and one geological position. It has been supposed that the gold of this country was confined to talcose slate. This is now known to be untrue; for, in addition to talcose slate, it is found in mica slate, hornblende, granite, and limestone. It has a wide range of relations, and it would seem that the metals have geographical rather than geological boundaries; that is, in a given district all the rocks may be expected to contain them; or, in other words, that there are certain types of association which belong to certain geographical districts, and that in order to pursue successfully mining interests, we must study that type. The doctrine, however, holds good, that a certain rock is commonly the bearer of a given metal rather than any other.

In this country quartz is the common associate of gold. But other associates may exist: the massive talcose slate, hornblendic gneiss, hornblende, granite, mica slate, limestone, and even serpentine. In talcose slate, gold may be disseminated in the rock for from forty to sixty feet in width. It is a mass of rock, or a mass mine, which bears no semblance to a vein, except that belts or strips of it are richer than others; but it is impossible to detect walls or defined boundaries to the auriferous parts. The gold of the mass is mechanically mixed with particles of a brown substance, disseminated in the rock sufficiently abundant to impart red and brown tints to the mass. These are sometimes of a rosy or peach-blossom color, indicating the presence both of iron and manganese. There is no doubt that these brown and reddish particles were sulphuret of iron which held the gold in mechanical combination. Where the auriferous mass is inspected, it is found to consist of fine grains of quartz, which are only slightly coherent, intermixed with silvery talc, which at the first inspection seems to predominate, but on closer observation proves the predominance of quartz in its composition. This kind of rock might with propriety be called *talcite.*

Gold is also disseminated through a pure saccharoidal quartz, which contains neither sulphuret of iron or talc; but this white and pure quartz may be contained in a vein or mass subordinate to that just described. It is found again in milky and granular quartz in veins, carrying the sulphurets of copper and iron and sulphuret of iron by itself, together with galena, and sometimes sulphuret of zinc. These veins may traverse argillaceous chlorite and talcose slates, or all the other rocks which belong to the mineral district, as gneiss, hornblende, or granite. In the last place it may be contained in a false vein, or rather beds, as the depositions are parallel with the strata or lamina. They are always thin, rarely exceeding twelve inches in width. The metal in this case is confined to a well defined strip of rock, but not limited by distinct walls. The eye can not detect a difference between that portion of the rock which is auriferous and that which is not. These false veins or beds being narrow, may be lost, unless the miner resorts constantly to the use of the pan for testing the presence of gold. These strips are usually very rich, and often yield from twenty-five to one hundred dollars per bushel of ore.

The gold of the soil is derived from rocks broken down by the ordinary atmospheric agents. In its dissemination in the soil, it has obeyed the same law of distribution as all the bodies which have great weight. As its gravity exceeds that of all other substances which may be expected in the soil, it will be sought for at the lowest planes of subsidence.

Leaving out of the account the gold deposits of the soil, it is most frequently found in the regular veins, associated with the sulphurets of copper and iron. It is probable that the sulphurets should be regarded as the true matrix of gold, though not the only matrix. The veinstone is quartz, though in a hundred auriferous veins of quartz containing sulphurets, one or two are known in limestone, and occasionally also it has been found in serpentine.

The veins of auriferous quartz do not differ in structure from other veins; and they are subject to the same irregularities in width, and the same varieties as to richness in parts of the vein, both when examined in the line of bearing and in depth.

OF THE CHANGES WHICH TAKE PLACE IN THE AURIFEROUS VEINS OF SULPHURETS OF COPPER AND IRON IN DEPTH.

§ 94. The condition of the materials at and near the surface differs materially from that below. Those sulphurets which are gold-bearing seem to be more subject to decomposition than those which are not. At a variable depth, then, the matrix of gold consists of a porous quartz, colored brown upon the surface by a superficial deposit of oxide of iron. This partially fills the cavities. It is a friable dry powder, or it may have become consolidated, in which case it resembles the oxide of iron with a resinous luster, or the common brown hydrous oxide of iron, or brown hematite. This condition of the iron is due to the decomposition of the mixed sulphurets of iron and copper. The sulphur is discharged, and usually disappears entirely, and leaves the rock; but not always, for in a few instances it is retained in the cavities, and has crystallized in octahedrons with rhombic bases. The decomposition proceeds from the outside to the inside, as the inside of a mass of oxide often contains an undecomposed one of sulphuret of iron or copper. Where the decomposition is complete, the gold is attached to the quartz in irregular plates, or is also retained in the midst of the oxide in grains or thin scales. But it is a still more interesting fact, that the gold occasionally crystallizes, and appears under the form of octahedrons and dodecahedrons, or in skeleton crystals, the general form of which is developed, but the faces are deeply striated. Beautiful productions of this kind were quite common for a time in the Ward mine in Davidson county, North Carolina. The figure of one of these striated or skeleton crystals appears on the title page of this work. Regular octahedrons have been obtained in the mines of Burke county in the same state. But crystals of gold are extremely rare in North Carolina.

ORIGIN OF GOLD—THE GEOLOGICAL POSITION AND RELATIONS OF GOLD.

§ 95. If we attempt to account for the origin of gold on facts and principles which are inapplicable to the origin of other metals, we entirely lose ourselves in conjecture. Indeed, the phenomena which accompany the auriferous quartz veins are by no means unlike those which accompany lead, copper, and iron. Its subterranean origin should therefore be admitted.

The gold of North Carolina is connected with three divisions or sections of rock: 1. With granite and associated rocks; 2. Gneiss and its laminated associates; and 3. With a series of slates which I am disposed to regard as sediments. The immediate repository of this metal is the ordinary constituted vein, differing in no respect, in its structural relations and origin, from those of the other metals and ores. It will be observed that I have passed unnoticed the deposits of gold in the soil and grits of decomposed rock for obvious reasons.

The slates are soft, greenish or reddish rocks, intersected by quartz veins and trap, and appear to repose upon granite so as to admit the outcropping of low and long ridges of this rock where the slates have thinned out. The rock, however, which I have noticed under the general name of slate, is really made of a series of rocks, which furnish a series of subordinate beds which have a wide range of lithological character. The following is a list of the most important kinds: 1. Soft green chlorite slate; 2. Soft red, reddish, and purple and purplish slates; 3. Soft talcose slates, which contain, however, quartz in fine grains, and which are also reddish and purplish. This variety might receive with propriety the name of *talcite* as already proposed. Alternating with the foregoing soft slates are the harder masses, which consist of

1. Quartzite, a mineral which in all respects resembles chert or hornstone. It is whitish, green, bluish, passing into black, and often coarsely agatized. It breaks with a flat conchoidal

fracture: it has sharp and translucent edges, and is usually very tough. When mixed with a small proportion of argillaceous matter, it forms also a hard tough rock, which appears above the surface of the ground in hatchet-shaped projections. Sometimes these rise, with rounded, sharp edges, seven or eight feet high, and extending as many feet in the direction of their strike, and then disappear in the soil. A succession of such outcrops continue for miles, forming a peculiar feature in the geology of the district.

2. A quartzite porphyry. This rock is quite common. Its porphyritic character is obscure; and it happens not unfrequently that it takes on the character of a breccia. Fig. 22 is designed to illustrate its porphyritic character; but it is often more obscure, and I am well convinced that I have found a few rolled quartz pebbles imbedded in the mass.

The quartzite or chert is in beds, and in one or two, and probably more districts, composes the largest part of the rock in a belt half a mile wide. It appears homogeneous in the protected part of the mass, while the weathered surface consists of a white or gray fine granular substance, similar in its condition to tripoli. This change extends several inches into the rock, and forms a well defined border around the unchanged parts within the mass. Although the slates bear strongly the indications of sediments, no fossils have as yet been found in them. In their general features they bear a strong resemblance to the slates of the lower half of the Taconic system. I do not propose to attempt to give a full and minute description of this formation of slates of North Carolina, and of the adjacent states, at this time. I shall proceed to the consideration of certain questions relating to the gold as it occurs in veins, and other relations to the rocks which contain it.

COMPARATIVE RICHNESS OF THE UPPER AND DEEPER PARTS OF THE VEIN.

§ 96. The opinions of miners, as well as geologists, do not agree on this question. The facts sustain apparently the doctrine that the yield of gold diminishes where the vein is beneath water. When the facts, however, are carefully examined, they are not so decisive of the assumed position. The mine which is of all others most reliable in North Carolina, is the Gold Hill mine. Its depth is about five hundred feet; the vein quartz, bearing the sulphurets of copper and iron, and the rock, a greenish slate. The upper forty or fifty feet of the vein was decomposed, and of course the gold easily obtained. Different companies had leased parts of the vein. One company worked upon a level between one and two hundred feet from the surface; another company worked a three hundred feet level down to four and five hundred feet. Now, while the vein did not yield profits at all large at the upper level, the lower was very profitable; even twice the amount of gold was obtained in the deepest part of the mine.

The opinion that a gold mine is less rich in the lowest levels or below, arises from the state of the metal, or from its combination with the sulphurets; and experience has proved in thousands of instances that it is extremely difficult to separate the gold by the mechanical processes usually resorted to. After the ore has been subjected to one operation, and all the gold apparently obtained from it, and this fine material is allowed to remain a year or two exposed to the air, and then worked again, it will usually yield by the second process nearly as much gold as was obtained by the first; and so the same ore may be worked five or six times, and made to yield profitable returns of gold. These facts are especially applicable to the auriferous sulphurets. It appears then from experience in North Carolina, that the auriferous veins are equally rich below and above water, to use the common comparison, and that there is no diminution of gold in the deeper parts of the vein. Like other veins of metal, it is variable in quantity—rich and lean places are met with in the same mine; but the Gold Hill mine, the McCulloch mine, Conrad Hill, and all those which have been skillfully worked, yield as much gold now in the aggregate as when they were first opened. It must, however, be acknowledged that though the quantity of gold obtained below water may equal that above, yet the increased expense of mining may consume all the profits of the business, and, in many cases, prudence would suggest the abandonment of deep mines.

DIRECTION OF THE AURIFEROUS VEINS.

§ 97. The extremes in the direction of veins lie between N. 10° W., and N. 70° E. The ordinary limits lie between N. 20° E. and 70° E. Their direction of dip is northwest in North Carolina, and the angle of dip varies from vertical to 20° at the surface, while where it is as small as 20°, the dip becomes greater in its progress downwards. The ordinary dip is such that a vein will make to the westward seven or eight feet in seventeen or eighteen feet perpendicular descent.

OF THE AURIFEROUS VEINS WHICH PASS FROM ONE ROCK TO ANOTHER, AND WHOSE COMPOSITION AND AGE ARE DIFFERENT.

§ 98. In North Carolina a belt of granite underlies parts of Guilford, Davidson, Cabarras, and Mecklenburg counties. The east side of this granitic belt is bordered by a fine greenish slate. It is the opinion of well informed miners and geologists that the auriferous lodes pass from the granite into the slate. The observations have been made where the slate is thin, and overlaps the granite. Fears were entertained that certain lodes would become valueless for gold when they entered the granite, agreeably to a law which is well established in other mining districts. My opportunities for observation are too limited to warrant the expression of an opinion; but I have been informed that in two or three instances the lode has not deteriorated after it has left the slate and entered the granite. The question, however, it appears to me, is not yet satisfactorily settled.

Of the age of this slate I have as yet been unable to form an opinion. In certain localities its character favors the view that it is a sedimentary rock. But so far as I can speak from observation, it is not fossiliferous, and it is doubtful whether it contains rounded pebbles. The only slate rocks which resemble it are situated above the quartz rock of the Taconic system.

AGE OF THE AURIFEROUS VEINS OF NORTH CAROLINA.

§ 99. The idea which has gained a few supporters, that gold is of recent origin, does not seem to be sustained by facts. President Hitchcock,* for example, quotes approvingly the opinion of Sir R. A. Murchison, that gold is of a recent origin; as late, for example, as the tertiaries. Opinions of geologists whose reputation is so widely spread should not be set aside for slight reasons. We find, notwithstanding the high authority to the contrary, that the permean rocks of North Carolina contain the debris of the auriferous quartz veins: the gold itself may be obtained from the quartz in the usual way. Whether the auriferous rocks of North Carolina and Virginia belong to the same period as those of Australia is not determined. But that the Carolina gold rocks and those of California belong to the same period, there is scarcely a doubt. An interesting fact connected with this subject should be stated in this place. In Burke county, North Carolina, E. Emmons, jr., discovered pottery and implements supposed to be of Indian manufacture, such as arrow heads, in the auriferous quartz grit seven feet below the surface. The grit is overlaid with a stiff clay. So the mammoth remains are found in Siberia in the same alluvia that contain gold.† These facts do not prove the recent origin

* Geology of the Globe, p. 31.

† Idem, p. 31.

of gold, neither can they be employed to prove that a change of climate took place about the time the Mammoths of this country became extinct.

I have already had occasion to state the facts relative to comparative richness of the gold veins at considerable depths. This question is economically one of great importance. Capitalists in this country are now pursuing their schemes and plans for working gold mines at all accessible depths. They make no distinction between auriferous and cupreous veins, neither do the auriferous veins appear at all analogous to the staniferous veins of Cornwall, which were said to be staniferous upon their backs, but cupreous in the main parts of the lode. Here the gold accompanies the yellows, as the sulphuret of copper was called in Cornwall, without giving place to it as the tin lodes referred to.

AGE OF THE GOLD BEARING ROCKS OF THIS COUNTRY.

§ 100. The talcose slates, mica slate, hornblendic gneiss, etc., which are traversed by auriferous veins, belong to a period which preceded the deposition of all the hydroplastic rocks of this country. The evidence which sustains this position is of two kinds. 1. Evidence of composition and derivation. 2. Of superposition and succession.

1. The oldest sediments belong to the Taconic system. The talc and mica of the gneissoid granite enters into the composition of the limestone. A peculiar bluish tinted quartz of the same rock enters into the composition of the quartz rock at the base of the Taconic system. These minerals are distinguishable from other varieties of the same species. So it is not less certain that the materials of the Taconic system are carried up into the Silurian. Thus a stratified limestone is found in the Potsdam sandstone at Chazy. The Taconic slate enters into the composition of the lower part of the calciferous sandstone.

2. The quartz rock of the Taconic system rests unconformably upon the gneiss of the Green Mountain range, or its gneiss-

oid granite, at many places; and the Potsdam sandstone, or in its absence the calciferous sandrock, rests unconformably upon the Taconic slates and other members of the same system. The auriferous rocks, therefore, are inferior to two systems, and the lines of demarkation between the systems are so distinctly drawn, that they should not be overlooked. There is no evidence that the lower Silurian are metamorphic rocks, which contain the gold of this country, though this veiw is taken by a distinguished geologist in the governmental survey of Canada.

Several rich auriferous veins traverse a hornblendic gneiss in Rutherford, N. C., at the eastern base of the Blue ridge. On the west side of the same ridge gold is derived from mica slate, four miles west of Sowannanoe gap. I entertain the opinion that we have no facts which sustain the doctrine that the rocks of the Blue ridge are altered Hudson river sandstones and shales, and yet the Blue ridge, where it is auriferous, is identical with the Green Mountain range.

DISTRIBUTION AND EXTENT OF THE AURIFEROUS ROCKS EAST OF THE MISSISSIPPI VALLEY.

§ 101. We can not with certainty determine the extent of the auriferous belt of the Atlantic slope. Its width in North Carolina is at least one hundred miles on a line, passing through the state from southeast to northwest; but the area over which gold has been found, is equal to 10,000 square miles. It is found in this state, in all the counties lying west of Wake, and is from fifteen to thirty miles wide in Virginia. One gold vein only is known in Maryland; but the slates which usually contain gold are about ten miles wide. The gold vein in Maryland is at Brookville, in Montgomery county. Gold is not found in this direction until it appears in Somerset, Vt., on the east side of the Green mountains. All that is interesting at Somerset is the fact that gold occurs in the same relations as at other places in the South. Still farther north, in Canada East, it reappears again, at Sherbrooke and La Chandiere river. It is associated with quartz, as usual, but the most productive locations are the deposit or branch mines. The indications bear a favorable aspect in the Canadian mines. From North Carolina, the gold belt extends south through South Carolina and Georgia, disappearing at last beneath the tertiary of the Atlantic slope.

OF THE PYRITES AND AURIFEROUS COPPER LODES OF THE ATLANTIC SLOPE.

§ 102. There are classes of gold mines in which copper is rarely if ever found. There is first the disseminated gold of the soft slates. When the metal is widely diffused in the rock, the produce amounts to from ten to thirty-five cents per bushel of ore. These masses of slate may yield gold through a width of one hundred feet; some places are richer than others. They may be called mass mines. In auriferous rocks of this kind, the gold is combined with very small particles of sulphuret of iron, which are decomposed for thirty or forty feet in depth, but below the water level, the sulphuret becomes visible. Copper is never found in this species of auriferous deposit. It is absent again in the thin auriferous belts, which rarely exceed six inches in width, in slates of a similar character. These are false veins, as they are destitute of walls. So copper is absent in some of the veins of auriferous iron pyrites. But a very large proportion of the veins which carry gold carry copper also; and some of them have become the best copper mines in North Carolina. The upper part, or back, as this part is sometimes called, is simply a decomposed iron pyrites; but when the lode is penetrated thirty or forty feet, the triple compound of sulphur, copper and iron, comes in, and increases in quantity with the descent, or in other words, it becomes a copper lode. A few years ago a change of this kind was dreaded, as it was invariably deemed necessary to abandon them, for no other reason than that it cost too much to separate the gold from alloy, which would be found with the copper. At this time many of these abandoned gold mines are regarded as the best and most valuable mines for copper. As lodes or veins they furnish nothing new as to structure, or add to those facts which are interesting in the eyes of a geologist. The gold is sacrificed, or at least no successful attempts have been made to save it; although it has been proposed, first to separate it by the usual mineral process, and afterwards smelt the remaining cupreous material.

While there is a general similarity in the structure of the veinstone, and arrangement of the materials constituting the lode, there are a few peculiarities belonging to individual veins, which are worthy of a passing notice. The McCulloch vein, in Guilford county, N. C., for example, is remarkable for its width, and the extent or depth to which the decomposition of the ore extends, which can not be less than one hundred feet. The vein expands to seven and eight feet in width, and contains, notwithstanding this great expanse, very rich ore; and the proprietors did not work ore which yielded less than one dollar per bushel. It is a magnificent vein.

The Gold Hill veins have also been very successfully worked. Their greatest width is seven and eight feet, and the principal vein has never been less than two feet. This vein has been worked to the depth of five hundred feet. It is an auriferous quartz and copper pyrites, in a green fissile slate. It often yields five and six dollars per bushel of ore. The Reed mine, Cabarras county, contains but little pyrites, but it is remarkable for the large masses of gold it has furnished; one of which weighed twenty-eight pounds, another sixteen, and another nine, and so on in respectable lumps, which have been very numerous; and another of sixty pounds: though it requires to be more fully authenticated, still the evidence of the fact is by no means slender.

The Fentriss copper mine in Guilford county, North Carolina, has furnished a large amount of ore. This vein is forked, or consists of a flat vein, rising at an angle of ten degrees towards the surface. Its veinstone is a coarse quartz, carrying large lumps of sulphuret of copper and iron on its inferior side.

When it unites with the vertical vein it swells out to a width of seven or eight feet. It carries also a quantity of iron pyrites which is silver white. It has not been analyzed, but probably contains arsenic and cobalt. Among the remarkable veins of the south, those of Ducktown, Polk county, in Tennessee, have probably excited the most attention. The rock is a talcose slate, interlaminated with hard layers of gneiss and hornblende, and highly inclined. There is nothing in the general appearance of the country which indicates a mining region, and accident alone brought to light the remarkable repositories of copper ore. The ores of this district are in some respects unlike those of North Carolina: they are arsenical, and probably to the presence of a third metal the peculiar condition of the copper ore is due.

A section of the Congdon vein at this place, gives all the information we desire respecting this structure.

Fig. 41.

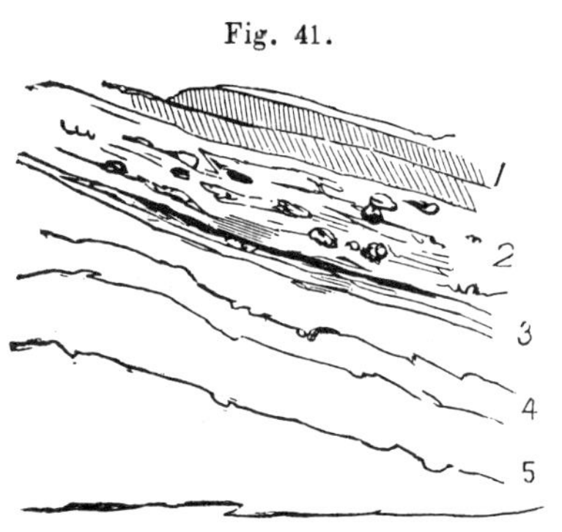

1 Talc Slate, 2 Gossan, 3 Bell metal, 4 Black oxide of Copper, 5 Mundic.

The section is vertical and longitudinal for the purpose of showing the slope or pitch of the materials of the lode, for it has a slope in addition to its dip. The following exhibits the order of arrangement: 1, talcose slate; 2, Gossan, or hydrous peroxide of iron; 3, bell metal ore; 4, black oxide of copper; 5, masses indicative of the commencement of the yellow sulphuret or mundic. The Gossan was seventy feet thick in the direction of the shaft sunk upon a hill. It is destitute of copper in the upper part of the mass, but it is present in the lower. The bell-metal ore is only from twelve to sixteen inches thick, and the thickness of the black oxide is variable in consequence of resting upon an uneven floor, but exceeded three feet. The width of this vein at the surface is five feet, with its walls very well defined. At the depth of thirty-five or forty feet, its extreme width is forty feet. No one can inspect this lode and be uninterested in the chemical changes that have taken place since the fissure was filled with metal. The first change which perhaps would be noticed, is the perfect destruction of the original lode. Its gangue, which is quartz, is dispersed through the mass of black oxide and gossan in part. The true, original walls are broken down, and the copper is extended laterally into oven-shaped cavities, in the soft adjacent slate. The smaller oven-shaped expansions extend into the rock six or seven feet, the larger still farther. The smaller are three or four feet high, and as many wide, and as they are occupied entirely with the black oxide of copper and its gangue, the contents only require to be shoveled and screened, in order to be prepared for market. Three hundred tons of this ore were taken out from this vein in one month. The chamber, after the removal of the ore, presents an irregular shape. Its roof is more uniform from the existence of the layer of bell-metal ore.

This vein presents its most remarkable feature only when we contemplate it in its original condition; when the iron, copper, and arsenic were in combination with each other, and arranged in the usual order. Now we find the copper in the form of a black oxide, occupying the lowest place in the lode. The iron forms a mass seventy feet thick overlying all the rest. The more volatile elements, sulphur and arsenic, have disappeared. The change is undoubtedly one which should be referred to molecular forces, representing in its effects the formation of nodules, septaria, and sometimes entire strata.

Ducktown can boast of five veins, rich as the Congdon, and similar in structure. Here is, therefore, a peculiar mineral district, rich beyond any which had been explored, and yet there is nothing upon the surface which would lead a geologist or miner to suspect the value and magnitude of the mines beneath it. It is true that on the surface there are mounds of gossan, or the hydrous peroxide of iron; yet copper is never found upon the surface. The slates of Tennessee, Virginia, and North Carolina abound in this species of iron ore; but so far as discoveries have yet been made, the mounds of oxide of iron do not overlay copper, except at Ducktown.

A vein three miles south from the Congdon mine, and just within the limits of Georgia, exhibits the general original form of the lodes in the vicinity of Ducktown (fig. 42). This vein was discovered by following the indications furnished by the gossan. Upon the surface this substance was observed to be rather common at the locality referred to, and selecting a place which represented the center of dispersion, a shaft was sunk almost at random. When the earth was removed from the rock a narrow crevice was observed, which contained the gossan; and on following it down twenty-five feet, the crevice expanded into a large pipe vein of the form presented in the figure. This peculiar lode was struck five feet higher on the north side of the shaft than upon the south. This pipe vein penetrates the rock obliquely. The black oxide of copper and gossan occupy the same relative position as at the Congdon vein. The Congdon mine was originally a larger pattern of the same kind of vein as the Georgia mine. At the deepest part of the shaft this is five feet wide, and a ton of black oxide was taken out of it.

Fig. 42.

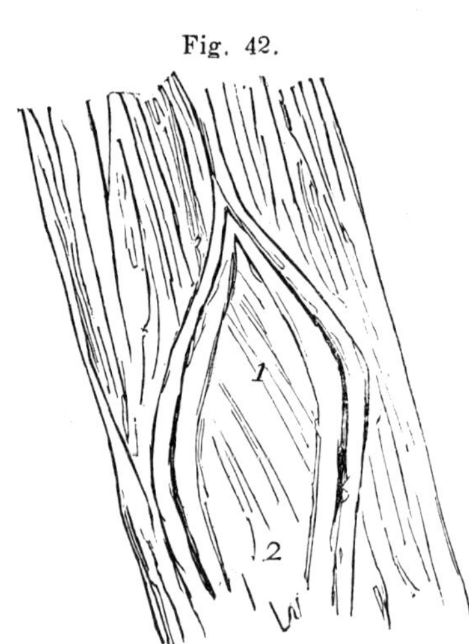

The Georgia mine is noticed for the purpose of illustrating the peculiar form of the pipe vein, which seems to constitute an interesting feature in the Ducktown mining district.

COPPER MINES OF LAKE SUPERIOR.

§ 103. The occurrence of native copper, in sufficient abundance to make it an object of commercial value, is one of the most recent of geological discoveries. The fact is another illustration of a remark, that each mineral district contains a certain type of mineral character peculiar to itself. That type does not seem to be entirely dependent upon the surface geology; for that of Polk county, Tennessee, is not ostensibly different from North Carolina. Many other illustrations of the fact might be given.

The geological position of the copper has been determined by the late Dr. Houghton and Dr. Jackson, and confirmed by other American geologists. They have shown that the native copper lies in veins in the trap rock, and that this rock forms alternations with the potsdam sandstone; and save the exception, the exclusive metallic condition of the copper, its repositories do not differ from those where it is found in other states, as that of the sulphuret or the gray copper ore.

As a general illustration of the copper lodes of lake Superior, I have copied the annexed diagram from Foster and Whitney's Governmental Report (fig. 43). In this diagram the shaded

Fig. 43.

A A layers of sandstone, D mass of trap containing the copper.

part shows where the copper has been removed by means of the vertical shafts and horizontal galleries. It will be observed that two of the horizontal galleries penetrate the borders of the sandstone, but are soon discontinued. It is proved by many observations, that an ore or metal in passing from one rock to another is greatly diminished in quantity, or the contrary. In the case before us, the copper, when it passes into the sandstone, becomes a mere thread or string; and though the fissure may exist, or may have been formed. the metal it contains bears no comparison in quantity to that in the trap rock. The fissure of the sandstone, if it equals in width that of the trap, will be filled mostly with veinstone. It can not escape the reader's notice, that this fact—the change which a vein undergoes in passing from one rock to another—is one of the most interesting, as well as important, in all mining operations where two rocks bear the same relations as those of the trap and sandstone of lake Superior.

The traps, however, are not equally productive in metal; and there appears to be as much difference in the three kinds of trap, the soft amygdaloids, hard granular trap, and the greenstone, as between the most productive trap and sandstone. The soft amygdaloids contain copper lodes, but they are thin, branching and scattered; while in the hard greenstone the veins are contracted and pinched out. In the fine granular subcrystalline trap the veins reach their maximum of excellence.

The gangue of the copper lodes is in keeping with the rock which contains them. The zeolites, prehnite, laumonite, &c., are minerals of trap rock; so in their geological position they become the veinstone of the copper, and other metals of the rock. Calcspar is also a veinstone, but it is also an associate of zeolite It appears from this fact that the veinstone is derived from the rock containing the vein.

The veins of native copper vary in width. I may cite, as an illustration of the fact, the well known Cliff mine, situated on Keewaunee point, three miles from the lake shore. The outcrop of the vein is in a cliff of greenstone. It is only two inches wide in prehnite as its gangue, but an exposure on the

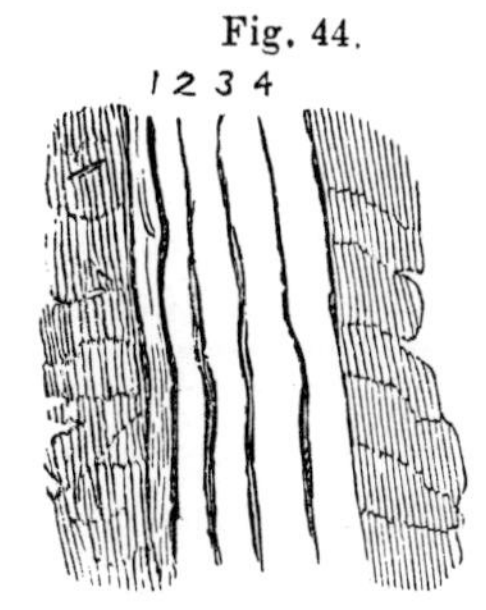

Fig. 44.

Parallel masses of copper and rock, forming the vein.

line of dip disclosed the fact that its width increased. With this encouragement, the vein at the base of the cliff was exposed. At this point the vein had increased to two feet in width. This encouragement led to the further exploration of the mine, which soon afterwards led to the discovery of immense masses of native copper, some of which weighed fifty or sixty tons. From the first discovery of these masses, the mining enterprise assumed a better aspect. The progress of discovery has kept pace with the labor and expenditure of capital; and now the mines of lake Superior take rank with the most productive mines of the world. The shipments of copper from Cliff mine have amounted to 1800 tons per annum, containing from sixty-five to seventy per cent of pure copper.

DISSEMINATED NATIVE COPPER IN HORNBLENDE.

§ 104. In the southern counties in Virginia, native copper has been discovered which occurs in small pieces in hornblende, with epidote in combination. It shows no tendency to arrange itself in veins, but is distributed irregularly in masses from the size of a pigeon shot to an almond. This discovery of copper was made in Carrol county, Virginia, and it has been found over a region thirty miles long, the breadth of which is undetermined, but not over, it is supposed, ten miles.

It appears that the discovery, up to this time; is interesting to the geologist, rather than useful to the miner or capitalist. Its origin is no doubt coeval with that of the rock. The cabinet specimens which I have seen, bear a trappean aspect.

VITREOUS OR GRAY COPPER ORE.

§ 105. In North Carolina, Chatham county, several veins of copper have been discovered, having at the outcrop of the vein and for fifteen or twenty feet more, gray copper; but it gives place to the yellow sulphuret sooner or later. There is nothing peculiar in the structure of the veins which carry gray copper. It is worthy of remark that it is massive, and that it is probably an altered yellow sulphuret.

The gray copper of Bristol, Conn., is associated with the yellow sulphuret and sulphuret of iron. It is remarkable for its fine crystals of gray copper.

The red sandstone of Connecticut and New Jersey contains the carbonates, sulphurets, red oxide and blue sulphuret of copper. The locality at Simsbury, Conn., was known in the time of the revolution. Those of New Jersey appear, many of them, to be exhausted. Mining operations have been prosecuted at Somerville, Woodbridge and Farmington. As the efforts to obtain good veins of good mines of copper have signally failed, it is a matter of interest to determine the causes of these failures. Prof. Rodgers, who has investigated the question very carefully, came to the conclusion that veins do not exist—that the copper occurs in proximity to the trap, and without veinstone or mixture of other metalliferous matter, but in ramifying strings or bunches, which are more or less blended with the adjacent rock. At the Schuyler mine, the sulphuret and carbonate occur in a sandstone twenty or thirty feet thick, in which it is imbedded and forms a band which traverses the layer in a series of offsets or steps, and which has been pursued to the depth of 212 feet below the surface. The phenomena which bear upon the origin of the copper ore of the sandstone, seem to favor the view that it was sublimed through shrinkage cracks, or imperfectly formed fissures, and which also penetrated more or less between the layers and into the porous substance of the sandstone.

In the north-western part of New York, the sulphuret of copper is found connected with the pyrocrystalline limestone, in strings and bunches. Its origin in this instance may also be attributed to a like cause.

THE LEAD BEARING ROCKS OF THE UNITED STATES.

§ 106. The only ore of lead which is found in sufficient quantities to pay a profit to the miner, is the sulphuret of lead or galena. It is found in rocks of several epochs. The Pyrocrystalline, the Taconic, Silurian and carboniferous, are each of them lead bearing. In the Alleghanies and other primary ranges, it is found in veins. The gneiss of this great range is generally the repository of it, as at Rossie, St. Lawrence county. It occurs in veins also in all the systems I have named. At Ancram in Dutchess county, it is in the Taconic system; near Spraker's, on the Mohawk, and at Martinsburgh, it is in the Lower Silurian. At Wisconsin and Iowa, Upper Silurian. In Derbyshire, England, it is in the Carboniferous system.

The Rossie lead mine has been worked to a greater depth than any other lead mine in the country, and hence may be referred to for the purpose of illustrating the general facts which pertain to the repositories of lead in gneiss, as well as the other rocks which belong to the same class. I have already had occasion to refer to the lead vein of Rossie, the structure of which is exhibited in the figure. The vein is exposed at this place in consequence of a shift in the gneiss by which it has been elevated thirty-five or forty feet above a low swampy country, which bounds the outcrop of the rock on its eastern side. The peculiarities and characteristics of this vein as well as other veins, are exhibited in the cut, Figure 25.

The middle dark broad line represents the position of the galena, imbedded in a calc spar on each side, and which fills the fissure. When the outcrop of the vein was exposed, the white gangue of spar, and the brilliant seam of lead in the middle, formed a fine and beautiful contrast with the darker and sharply defined wall of gneiss which inclosed it.

The wall of rock which rises out of the low, flat and swampy ground, seems to terminate here, and to limit the extent of the vein in this direction, as no rock appears for one-fourth of a mile. This wide interval and absence of rock is due to diluvial action, at least in part, its upper part having been removed; but the vein and rock continues in the direction indicated by the strike of the vein, beneath the low grounds; and the workings of the vein may be carried on beneath these grounds, which are frequently covered with water, which sets back from Indian river, with which they communicate.

I have had occasion to observe more than once, that it is a rare circumstance that a vein fissure is produced without being accompanied with others also. This fact I propose to illustrate again by the accompanying ground plan of veins which have been discovered at Rossie.

Fig. 45.

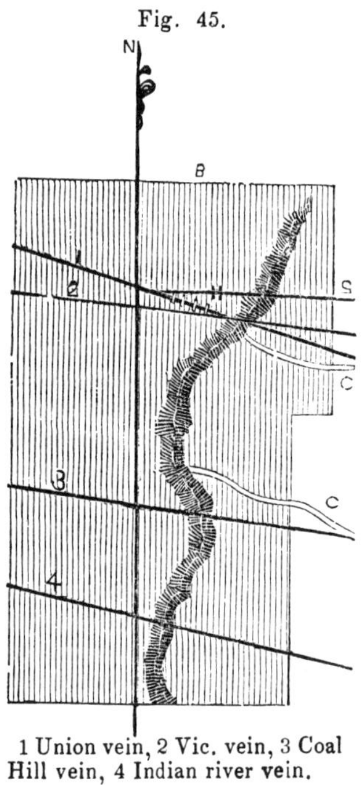

1 Union vein, 2 Vic. vein, 3 Coal Hill vein, 4 Indian river vein.

Fig. 45 exhibits four lead veins which constitute this mineral district. Thus, 1 shows the bearing of the Union vein, which is S. 73° E. It sends off a branch S. 88° E. 2, the Victoria vein, whose bearing is S. 84° E., with its diverging branch; 3, the Coal Hill vein, bearing S. 82° E. This is the vein already referred to in Fig. 25. 4, Indian river vein, bearing S. 75° 45′ E. The heavy transverse shaded band shows the boundary between the high and low grounds already spoken of. Two of the veins have been extensively worked. The great Coal Hill vein was first discovered, and has been explored to the depth of 200 feet, and has furnished $241·000 worth of metal. This mine, which was opened in 1836, although it presented a fair prospect of yielding large dividends to the owners, yet was abandoned after three or four years' trial. This arose from great extravagance and unskillfulness in mining. The work was prosecuted by opening the vein for a great distance along its outcrop, which exposed it to inundation from surface water.

To exhibit the plan pursued in opening and working this mine, I have copied the following diagram, Fig. 46, by which it will be seen by the darkly shaded parts of the figure how much galena has been taken out of the mine.

Fig. 46.

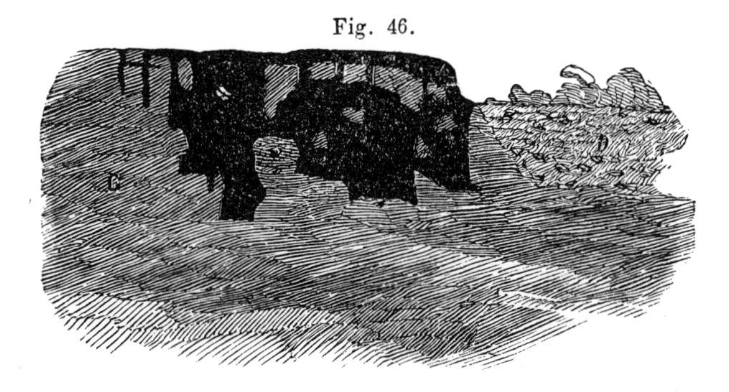

It will be observed that this shaded part extends the whole distance along the outcrop. Where there should have been at most two shafts, there is a trench extending the length of the vein as far as worked, some twenty or thirty feet deep, the effect of which is to convert what should have been closed into an open way for the ingress of water, which, in some cases, would totally ruin a mine.

The most important lead-bearing rocks of this country belong to the western states, Illinois, Missouri, Iowa and Wisconsin. The lead of these states belongs mostly to one rock, the Cliff limestone of the western geologists, the lower part of which is equivalent to the Niagara limestone of New York. Lead is also found in the calciferous sandstone, the lowest limestone of the Silurian system. In New York, this lower

23

limestone is not destitute of lead and zinc; it is in a sufficient quantity to keep up the analogy between the beds of rock of the same age at distant points. These limestones are both magnesian, though the magnesia is variable in quantity in the different layers.

The composition of the lead-bearing rock of Wisconsin was found to be as follows:

Carbonate of Lime,	47·40
" Magnesia,	40·70
Oxide of Iron,	2·40
Silex,	7·10
Water,	2·00
Loss,	40
	100·00*

The texture of the rock differs in different places; in some it is friable like a sandstone; in others, hard and durable. It is also cherty or traversed by bands of compact quartz resembling flint.

This rock is generally favorable both for the reception and retention of the metals. It has been fissured extensively, and these fissures had remained open until they received the lead or copper or other metals which we find in them.

A fresh fracture of the rock exhibits a sub-crystalline aspect, a flat conchoidal surface, a granular texture, and a light yellowish or drab color. When weathered it is brownish or reddish yellow. Specific gravity 2·65 to 2·70.

Owen divides the Cliff limestone into, 1, upper or shell beds, consisting of a white or light colored limestone, destitute of magnesia, fossils calcareous; 2, middle or coralline beds, which are cherty and magnesian, and of a yellowish color, fossils silicious; 3, lower or lead-bearing beds, color yellowish and composition magnesian. The rock presents cliffs when it appears at the outcrop, which are more or less fissured and separated

* D. D. Owen's Rep. of Geol. Expl of Iowa, Wisconsin and Illinois, 1836, p. 24.

into columns. Its stratification is obscure and its vertical fissures numerous, and which are sometimes prolonged or extended horizontally.

Wisconsin, and the states which are lead-bearing, have been subjected both to the denuding action of a rush of waters, and to a slower disintegration of rocks by the common atmospheric agencies These external influences have wrought many alterations in the rock, and have also changed the position and association of the metal which the rock contains. Originally, the galena was injected into its fissures, which it undoubtedly filled, and which it continues to occupy in part; but in consequence of the disintegrations which the rock has suffered, all the upper parts of the veins seem to have been broken down, and the lead has become commingled with a stiff, reddish clay. The galena is therefore found in veins and in beds.

LIMITS OF THE VEINS OF THE LEAD-BEARING STRATA.

§ 107. The lead is not confined wholly to the lower cliff limestone. To a limited extent, it reaches the blue limestone beneath, though this rock is not by any means so rich and productive. The lead, therefore, for all useful purposes, so far as is known, is confined to the Niagara limestone, and to the overlying debris which contains the products of the broken down veins.

The fissures vary in thickness from a few lines to fifty feet, but Mr. Owen remarks that the common width of the fissures filled with solid metal is about four inches, and rarely exceeds twelve inches. One instance is given in which the galena of a fissure was six or eight feet thick in the center, and extended thirty-five feet; it tapered to a point at each extremity. The foregoing statement leads to another still more remarkable in the annals of lead mining. Thus, fissures sometimes expand into large caverns, the walls of which are lined with galena, while this is overlaid or incrusted with spar, stalactites, etc. The galena which lines these large chambers is sometimes twelve inches thick. The enlargement of fissures subsequent to the beginning of the metallic accumulation has been mentioned already, but the peculiarity in the cases referred to, is no doubt due to the nature and composition of the rock.

The direction of the lead veins is east and west, or rather south of east. When north and south they are less productive, though exceptions to this statement do occur. The peculiar fissured condition of the limestone has produced a great irregularity in the dip of the veins. Instead of dipping regularly, and at a constant angle, the dip takes the form of several successive offsets. Thus, for a given stratum, the angle of dip may be 45°, but in passing to the south it may be more oblique or become horizontal, by following the bedding plane between two strata, pursuing a route where the difficulties and distractions are readily overcome. When, however, it has taken a horizontal direction, it may pass to the vertical. There is therefore great irregularity in the dip or lead of the vein. The fissures of the Cliff limestone usually stop short of the superincumbent limestone, and hence are capped over by the upper mass.

POSITION OF THE ORE, AND THE LEAST THICKNESS OF VEIN WHICH CAN BE WORKED WITH PROFIT.

§ 108. The broken down veins are concealed in reddish clay, ferruginous sand, masses of rock, etc. The lead is then obtained from these deposits by digging.

The position of the ore in the vein is often changed, and it has been in process of time detached from its walls, and is obtained in pieces varying in size from a pea to those of a thousand pounds weight.

The width of vein required for working may be profitable if it does not exceed half an inch, provided the ground is not very hard. It is evident the profits of working a vein depend on the character of the walls, gangue, etc.

INDICATIONS OF THE PRESENCE OF A VEIN BENEATH.

§ 109. It has been observed that where a *sink hole* occurs there is a probability of finding lead. So, when a series of them can be traced upon the surface in an east and west direction, there is probably a fissure in connection with them, which will prove to be lead-bearing. Depressions, miniature ravines, the presence of barytes in the debris, or calc spar, become indications of the existence of a vein fissure in the neighborhood. In *prospecting* for lead, the miner looks for the above indications. Mineral gravel, the existence of the Coscinopora Sulcata, Goldf., is regarded also by Mr. Owen as indicative of the presence of a vein in the vicinity.

The galena is associated with blende, calamine and sulphuret of copper. Instances occur when calamine displaces the galena and becomes the principal mineral of the lode.

EXTENT OF THE LEAD-BEARING ROCKS OF THE WEST.

§ 110. The lead region comprehends eighty townships, or two thousand eight hundred and eighty square miles, and is about one-third larger than the state of Delaware. The length ot the lead region, from east to west, is eighty-seven miles, and its greatest width, from north to south, fifty-four miles. The produce of the mines of the upper Mississippi, exceeds 50,000,-000 of tons annually, which is probably less than one-half the quantity this region is capable of producing under increased facilities and a better system of mining.

LEAD AND ZINC OF NEW YORK, MASSACHUSETTS AND CONNECTICUT.

§ 111. The lead mine of South Hampton, Massachusetts, was opened as long ago as 1810. Its value has never been determined. The vein together with the gangue is seven or eight feet wide at the surface, is vertical, and is included in granite.

This mine is now being opened again, after having been abandoned for thirty years, with some prospect of success.

Galena occurs in more than twenty places in the old county of Hampshire, and in veins which seem to be well defined, but the low price of lead has operated to prevent their exploration. They all belong to the pyrocrystalline rocks, and the ore is associated with pyritous copper, sulph. of iron and blende, together with the rarer minerals, sulphate, carbonate and molybdate of lead. The gangue is barytes and crystallized quartz. The probability is, this region is an extensive lead district which will one day become important.

Another lead region has been explored rather extensively in Ulster, Sullivan county, New York. These mines are known as the Ellenville, Ulster and Shawangunk mines. The latter is near Wurtzborough, in Sullivan county. The first occupies a transverse break in the Shawangunk grit; its direction is S. 60° E. The mineral matter filling the fissure consists of galena, iron pyrites, blende and copper pyrites. Galena and blende predominate. The gangue is quartz.

The Ulster mine is near Redbridge, upon a line of fault, according to Prof. Mather, lying between the grit and slate, the Hudson river slates having been upheaved.

The Shawangunk mine occupies a fissure in the grit of that name; a rock which is equivalent to the Oneida conglomerate. The fissure is from two to five feet wide, and was opened between the strata, but it shifts its dip in a manner resembling somewhat the mines of Wisconsin. It carries galena, blende, iron and copper pyrites, intermingled with crystalline quartz which forms the veinstone. Very large masses of galena have been raised from this mine, one of which weighed 1400 lbs. Neither of these mines have proved profitable, though they are not to be regarded by any means as valueless. The galena of the Shawangunk mine is argentiferous.

It will be observed on reviewing the principal points respecting the geological position of galena, that we find it associated first with the pyrocrystalline rocks, and second with the paleozoic rocks. In the latter it ranges from the calciferous sandstone to the Niagara; these extremes together with the included rocks contain veins of galena. In Europe, especially in England, it extends upwards into the carboniferous limestone. So far as observations have been made, this ore does not occur in veins in the Messozoic series.

The foregoing facts sustain the doctrine which many distinguished geologists have inculcated, that metalliferous veins are productive and important the nearer they are to the igneous rocks, and those districts which are rich in metals, are traversed and disturbed by them. The Wisconsin lead district, however, is much less dyked and disturbed than any other one in this country which is equally rich and productive in metals.

ON THE EXPENSES ATTENDING MINING OPERATIONS.

§ 112. It is scarcely necessary to observe that the expense of working a mine must depend upon a variety of circumstances, and that the estimate of those expenses which will probably accrue in the working of any given mine, will not be applicable to any other though the actual amount of metal may be the same in each.

In some cases a work may be prosecuted successfully without the aid of gunpowder, in others the hardness of rock may become excessive at certain depths, and trap dykes may cut off a lode, and hence heavy expenses will be incurred, in recovering it. Variability in the hardness of rocks is one of the principal causes of the difference of expense in working mines which belong to the same rock or formation.

Mining operations embrace three kinds of work—shafting, tunneling and stoping. Shafts are right-angular cuts into the soil and rock, in which respect they differ from wells. They may be perpendicular or inclined. They are necessarily perpendicular if the lode is vertical; they may be inclined and sink upon an inclined lode to great advantage by employing a proper apparatus for elevating the mineral. But they are more frequently vertical, and sunk at a distance from the outcrop of the vein so as to intersect it at any given depth which may be desirable. To do this it is necessary to ascertain in what direction and how much the vein dips, or in other words, how many feet the vein makes on the dip side in any given number of vertical feet. Thus a vein may make to the west seven feet horizontally in seventeen vertical feet. It is easy from these facts to calculate how far west from the outcrop the shaft must be sunk to intersect the vein at one hundred or two hundred feet; and so of any other ratio which the vertical plane may bear to the inclined one. It is to be borne in mind that a very flat vein, as has been already observed, will become steeper as it descends. In very flat veins, therefore, calculations of this kind can not be relied upon until the vein has reached its maximum dip.

Tunnels or drifts are horizontal cuts into the earth or rocks. They may be designed both for conveying outwards the mineral and for drainage. In that case they are made from the outside, or begin on a slope and are worked inwards, or they may proceed from a shaft to a lode which lies upon one side of it.

Stoping is a term applied to that kind of work by which the vein, or its metal, is removed from its bed. The most economical way is to stope from beneath upwards. The dead rock is readily disposed of, and placed so as to save the expense of raising it to the surface, and at the same time aid in supporting the walls. The arrangements for stoping should be such that several gangs of miners may be employed in taking down the vein at different levels at the same time. In this case they will be arranged generally in a series of steps, one above another. The cost of a shaft seven feet by five, and sixty feet deep will not be less than two dollars and a half per foot. If sunk to one hundred feet, it will not be less than four dollars per foot. A windlass will answer for the purpose of raising rock and earth for the first sixty feet, but a whimsey, or whim, is more economical below that depth. The apparatus for raising the rock and water, however, must depend very much upon circumstances. If it is a trial shaft merely, and the vein is of a doubtful character, it will not be prudent to incur the expense of a whimsey until after the vein has been proved to a certain extent.

Other expenses are incurred in mining operations before the mineral is ready for smelting, besides those which attend the shafting, tunneling and stoping. Almost all ores require roasting in order to fit them for stamping. There are, therefore, three operations, at least, to which the ore must be subjected to fit it for smelting, viz: assorting, roasting and stamping or crushing. Passing by the expense of assorting, as this must be the most variable of the three, and can not be calculated before trial, I proceed to state the expense of roasting and crushing a given amount of iron ore. The roasting is performed in kilns of a simple construction.

One kiln, containing 107 tons of ore, will cost:

1. For mining, fifty cents per ton,	$53·50
2. Five cords of wood, two dollars per cord,	10·00
3. Teaming one hundred rods,	11·50
4. Labor for filling the kiln,	16·00
	$91·00

In the case of iron, there is usually the expense of separating the rock from the ore, which is effected either by magnets or by washing. If the first method is adopted, the expense will amount to

1. 6 men, $2\frac{3}{4}$ days in crushing, one dollar per day,			16·50
2. 1 engineer, $2\frac{3}{4}$ days, at	$2·00	per day,	
3. 1 assistant, 1 "	1·50	"	
4. 1 machinist,	2·50	"	
5. 1 man, "	1·25	"	
6. 3 men, "	1·00	"	
7. 1 boy, "	50	"	
			40·42
8. 10 cords of wood for engine,			20·00
Total cost,*			$167·92

* Sandford mine, Essex county, N. Y., Lot 21.

There will also be a loss of ten tons in shrinkage, which will leave only ninety-seven tons of ore after it has passed through the several operations. As separated ore is now selling, this will be worth $253 00

Phosphate of lime saved, worth 125 00

The allowance for loss in separating by water is about twelve tons per one hundred.

VALUE OF THE MINING PROPERTY IN CERTAIN STATES OF THE UNION.

§ 113. The country is not wanting in a class of persons who decry mining enterprises and pronounce all attempts for developing its mineral wealth, schemes more fraught with expectation than with reasonable prospects of fruition. This class, though wealthy, never invest their means or their money in mines.

There is another class who may be equally unbelieving in the real value of mineral property, yet are ready to plunge into any scheme or project in which there is enough to make a bubble, something which may be inflated into consequence. This class of men have little to lose, and being reckless in representation, are ready to avail themselves of such arts as are necessary to advance their unrighteous schemes.

There is also a third class, who look at matters in a different light. They are satisfied that there are valuable mines, and moreover, they look at the world as progressing in the arts, and requiring every day for its progress, the wealth which is concealed, except to the eye of science, in the bowels of the earth. They see that for years to come, the increased wealth of the nation is to be drawn from this great storehouse. They are disposed to invest their money in this kind of property, and to be content with the fair gains of the business. They are aware that it is not without its risks, but to diminish them resort to all the means which may be necessary, in each particular case, to secure a successful result.

The course pursued by the second class has been so notorious, as to give character to mining enterprises, and to invest them with a suspicious appearance.

I have not space, however, to follow up this subject. My principal object now, is to state, very briefly, the value of some kinds of the mining property of our country, for the purpose of placing the subject in a prominent light, and thereby show that it is for its interest to foster and encourage all mineral enterprises which are undertaken in a proper spirit.

The time is not far distant when mining will form one of the great industrial pursuits of the country. Every blow of the pick, and every gunpowder blast, will add their farthings to the common wealth of the nation, for every pound of copper or iron is a real addition to its resources.

1. Northern New York. The net proceeds per annum, which may be realized from the ores of iron in northern New York, will pay the interest, at seven per cent, on $3,000,000.

The mines at Adirondack have just been sold for $500,000, a sum much below their real value. The Sandford ore bed in Essex county, can not be estimated at much less than $500,000. At this mine, from two pits alone, 21 and 23, 200 tons of ore per day have been raised at a cost not exceeding fifty cents per ton; and which, when crushed and separated, yields from five to fifteen tons of phosphate of lime per one hundred tons of ore, which is worth on the ground twenty dollars per ton, and twenty-five to thirty dollars in New York.

There remain the Clintonville and the Saranac iron districts, together with inexhaustible quantities of the specular ore in Jefferson and St. Lawrence counties, and the magnetic ores of the Highlands.

2. Pennsylvania furnishes an amount of iron which may be estimated at $5,000,000, annually.

3. Missouri, from the Pilot and Iron mountains is capable of furnishing as much iron as any part of the world. Situated in the great valley of the Mississippi, its value can scarcely be overrated.

4. The iron mountains of lake Superior are equally as rich as northern New York. There are some, perhaps, who may regard this comparison as unjust to lake Superior; but it must not be forgotten that one mine, the Sandford Lake mine, is between six and seven hundred feet thick. A square yard of ore weighs four tons.

5. Maryland, Virginia and North Carolina possess inexhaustible supplies of iron ore, which are mostly the hydrous peroxides of iron. The hematites of Vermont, eastern New York, are very extensive.

6. The brown ores of iron in the south-western counties of North Carolina, and in eastern Tennessee, are immense.

A mineral so important as iron should be widely distributed, and it appears that in the United States, every important section is supplied with it. The largest section or formations which are destitute of the ores of iron and of the metals, are the Cretaceous and Tertiary, which skirt the Atlantic coast and which form our great basins and valleys. So also, the Silurian and Devonian systems are in a great measure destitute of iron ores, with the exception of the argillaceous and oolitic ores of iron of the Clinton group.

7. I have already spoken of the value of the lead ores of Wisconsin, Missouri and Iowa. The highest estimate which I have noticed of the probable productive capacities of the lead region, is from one hundred to one hundred and fifty millions of pounds annually, having already reached that of fifty millions under unfavorable circumstances.

8. The production of copper is in its infancy. It is too early to attempt to determine the value of its mines, and yet the lake Superior copper district has already produced two thousand tons in a single year. The value of the copper which has been produced equals, at twenty-five cents per pound, $2,700,000. The copper region which ranks next in value is in North Carolina. It has been referred to. The ore is the yellow sulphuret; the country is far better adapted to mining than that of lake Superior. Indeed, it is of all others the best, whether we

consider its climate, its means of sustaining a mining population at a cheap rate, or the production of timber for shafting, tunneling, fuel etc. We do not yet know the real extent and value of its copper ores, but we have not the least doubt of the ultimate success of its copper mines.

It is not to be expected, however, that one-quarter of the veins which are now being tested will prove to be mines. Even if one in ten turn out well, North Carolina will become one of the richest mining districts in the Union.

The resources in copper in Tennessee are also remarkable, and particularly so as several mines became productive from their first trials. I allude to those of Ducktown.

9. Although gold has been obtained in considerable quantities for half a century, still the mines and deposits have not been worked in a systematic manner. Gold mining operations have been conducted in the loosest manner. Present and immediate gains have been sought for, and hence no permanent works have been erected, except in a very few instances. Within the last two years, more system and more capital have been employed, and a better and more consistent view is now taken of gold mining, and the prospect is becoming daily more favorable to the enterprise. North Carolina is the center of the gold region, and will rank in value next to California. There are no accurate returns for the amount of gold North Carolina has furnished. Of the gold of California, the estimated production is less than the actual. The Hon. T. Butler King estimated it for 1848-9, at $40,000,000.

10. Our plaster, salt, marble, granite and freestone, form other large items of mineral wealth with which the United States abound. In the list of mineral property, mineral springs should not be forgotten. They administer to the health of the people.

11. The only mines of quicksilver which are now known in the United States, are situated in Santa Clara, twelve miles from San José, in California. It is found in bunches in ferruginous clay, forming in part a hill 1360 feet above tide. It is

associated with broken down magnesian rocks. The deposit is large, but no accurate returns of the yield of quicksilver have been published. The mine is being worked in a systematic manner.

We have no mines of tin, properly speaking.

I have said nothing of coal. It is almost impossible to measure or weigh in calculation its amount. But President Hitchcock observes truly, that the whole amount in solid measure, of the coal in the United States, equals at least 3,500,000 square miles.

When it is considered that our country is destined to support its hundreds of millions of souls, that its fleets, its mines, its manufactures, its locomotives and the domestic firesides must depend upon its mineral fuel, and when we also estimate the vastness of our resources in the mineral kingdom, we can not but see that everything necessary for prosperity has been provided with the most liberal hand, and on the most gigantic scale; and has moreover been so distributed as to accommodate the many and the most distant parts of the Union. There can be no centralization of products and resources, so as to confer a preponderating influence on a few favored sections of the Union, in the south or in the north. But enterprise and industry may create any where a prosperous community by availing itself of those natural mineral and manufacturing resources which are provided throughout the land. Where these do not exist, agriculture comes in to supply the necessary elements of prosperity. Nature had no sectional favors to dispense when she grew the coal plant for nine hundred miles from north to south, and more than a thousand miles from east to west, extending over an area of a million of square miles, and at the same time distributed her iron and other metals still more widely; to say nothing of the lands and mineral productions upon the Pacific slope.

Although I have not pretended to present a statistical account of our mineral resources, and what is recorded in the foregoing paragraphs is exceedingly meagre and unsatisfactory, still enough has been said to prove that our mining interests are to become one of the great sources of wealth, and that the real additions to the wealth of the nation is hereafter to be largely derived from its mines and quarries.

STATICS AND DYNAMICS OF GEOLOGY.—CONCLUSION.

§ 114. Geologists have often employed the phrase, "*The Dynamics of Geology*," which, if it is appropriate, has its counter phrase, "*The Statics of Geology*." The first comprehends all that relates to the processes which are productive of change; the latter, all that relates to the rocks and formations as they are, without regard to the cause or causes which have been influential in the development of their present condition.

Statical geology stands first in the order of time; it is descriptive in its objects and ends, and contains a record of the phenomena which they exhibit.

When statical geology was wholly neglected, dynamical geology was ridiculous and absurd. Indeed, it is impossible to construct the dynamics of the science, without first perfecting the statics. The dynamics of our principal mountain ranges will be better understood when their statics have been more thoroughly studied. The dynamics of our great system of lakes seem to point to diluvial action as a cause, but we require more facts before that theory can be established. If our dynamics do not grow out of, or legitimately follow from our statics, we are likely to beget error rather than truth. If, on the contrary, the dynamics grow out of our statics, truth is begotten, and geology becomes an inductive science.

The dynamics of geology have too often been formed or constructed from what may be termed *the possibles*. It is evident, however, that they can not be true because they are possible; it only saves them from absurdity. The truth of our dynamics is to be tested by their conformity to the statics of a region in each particular case.

As an instance illustrative of this kind of error, I may cite the ingenious explanations which were offered to explain the easterly dipping of the quartz, limestone and slate rocks of Berkshire, Mass., beneath the gneiss of the Hoosick mountain. But this does not exist immediately adjacent to the range; there is a synclinal axis, the dip being westerly and from the range. The error in the dynamics arose from having determined the statics only in part.

Among the points in statical geology which always require considerate attention, is the derivation of the materials composing a formation or mass, when mica and talc enter into their composition. In cases when those minerals compose a rock, we might infer that it was metamorphic; especially when they exist in large proportions. But such a conclusion becomes doubtful if other facts in the statics of those rocks are wanting; for mica and talc from the form of their particles, and the nature of their constitution, reproduce the same lithological masses as those from which they were derived. Hence it is certainly safe to be cautious in pronouncing a rock metamorphic when composed of mica and talc.

There are three kinds of forces which are recognizable in the earth's crust; physical, chemical and biological. The first leaves its imprint upon its masses, and from those imprints its nature and its degree are inferential.

The second furnishes results generally complete, and activity is rather inferential, and our conlusions must be based upon our knowledge of chemical principles, and the known mutual action of bodies upon each other. Chemical forces must frequently give origin to the physical, and they no doubt become the source of some of the most energetic of this class of operations.

The third is entirely passive, leaves no mechanical impress, but they leave memorials of the highest moment. Biological forces are entirely passive when put in conflict with the dynamical and chemical, and always yield to their movements.

It is in the construction of the globe that these forces are to be studied, and it is here that the conditions are revealed under which nature has elaborated the greatest of her works, concerning which she invites our serious inquiries.

We must also consider that the conditions or the medium in which all beings have lived, whether belonging to the palæozoic or kainozoic life, for the time being must have been in harmony with their organization. In what manner or how this harmony was interrupted, so as to result in the extinction of successive faunas, remains one of the profoundest problems for geology to solve.

We have already passed that incipient period in geology when the mind follows its inherent propensity to inquire after causes instead of the laws of phenomena. Like all other sciences, its history illustrates this propensity in a remarkable manner. Its records show its attempts to obtain at a single leap the origin of the structure of the globe, without the trouble of acquiring facts. The inquiry after causes, however, need not be discouraged. Who can doubt the legitimacy and use of such inquiries, when we find the tendency of the human mind inclined so strongly in that direction? What is to be discouraged are the attempts to leap the wall at a single bound. We are to climb, and the steps are to be cut by labor. Proceeding in this way, even the essential nature of things may be opened before us. And who shall forbid inquiries into the essential nature of spirit? Step by step we climb the ascending pathway. Light gleams up in the distance. The essence of cause, the essence of God may faintly illume the horizon of our prospects. It is the goal of man's hopes and aspirations.

EXPLANATION OF THE PLATES.

Section of the earth, exhibiting an outer envelope consisting mainly of primary or pyrocrystal line rocks, penetrated by veins which may be supposed of more recent date than the consolidation of this exterior mass. As veins or fissures are supposed to arise from the cooling of the exterior mass, it would be more consonant with this view to represent the wedgeform veins, as inverted or thicker towards the exterior.

Resting upon this exterior crystalline mass, I have represented the hydroplastic rocks or sediments; their lines of stratification are sufficiently distinct to be observed. The more steeply inclined parallel lines represent the position of the laminated pyrocrystalline rocks, beneath the former.

The physical map of the world represents four different subjects: 1. The white, the land; 2. The light-shaded, the table lands of the continents; 3. The mountains, with their directions upon the continents; and 4. The direction of the ocean rivers. These may be traced from the antarctic regions, and first as flowing up towards the coast of Chili. Their several directions are indicated by arrows.

On the American continent the reader will recognize the Andean and Rocky mountain chain, the coast chains upon the Pacific, the Ozark, the Appalachian and Lawrentine chains, and the generally western and eastern direction of the principal chain of the old world.

The geologic column represents the relations of the systems of rocks, and the four geologic positions of the coal seams: 1. The great Carboniferous. 2. The Permian. 3. Triassic. 4. Oolitic. In part of the edition the Liassic or Lias was omitted by mistake.

1, 3, 4. Veins and dykes. 2. Volcanic vent.

A, B, E.—Gneiss, Mica and Talcose slate, etc.

The moon's surface presents two varieties of cones. 1. Those with flat areas within a circle. 2. A circle surrounding an illuminated apex of a cone, evidently a large crater, containing one in its central part. These cones appear as the terrestrial cones, Vesuvius, Etna and Cotopaxi, would appear from a station on the moon's surface.

The deeper shaded parts of the moon are the comparatively depressed planes, which are dark in consequence of their being less illuminated than the more prominent parts. The jagged edge of the moon will not escape attention. Here it exhibits its rough, mountainous and volcanic character. The lunules which stand out in the dark-shaded part are tops of volcanic mountains which are beginning to be illuminated by the changing of its phases.

ERRATA.

Page 55, seventh line from bottom, for "bed" read *feel*.
" 74, fifteenth line from bottom, for "are" read *is*.
" 74, fifteenth line from bottom, for "Labradorite" read *Labrador*.
" 106, end of section 78, for "arks" read *rocks*.
" 155, fifth line from top, for "often" read *after*.
" 156, eighteenth line from bottom, for "strips" read *stripes*.
" 167, sixth line from top, for "mineral" read *mechanical*.
" 174, for "Farmington" read *Flemington*.
" 180, fourteenth line from top, for "distractions" read *disturbances*.
" 192, second line from bottom, for "construction" read *crust*.

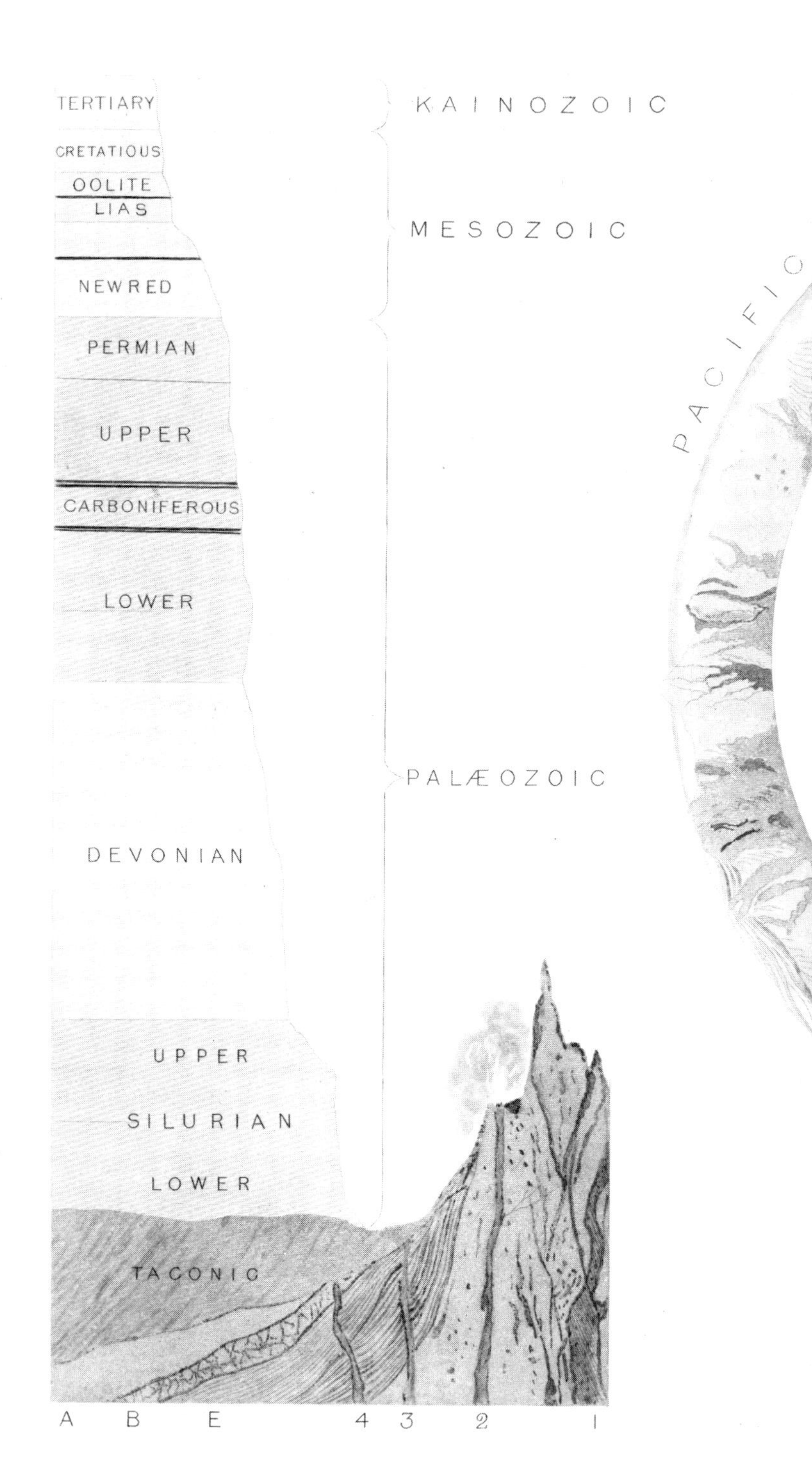

TERTIARY
CRETATIOUS
OOLITE
LIAS
NEWRED
PERMIAN
UPPER
CARBONIFEROUS
LOWER
DEVONIAN
UPPER
SILURIAN
LOWER
TACONIC
KAINOZOIC
MESOZOIC
PALÆOZOIC
PACIFIC
A B E 4 3 2 1

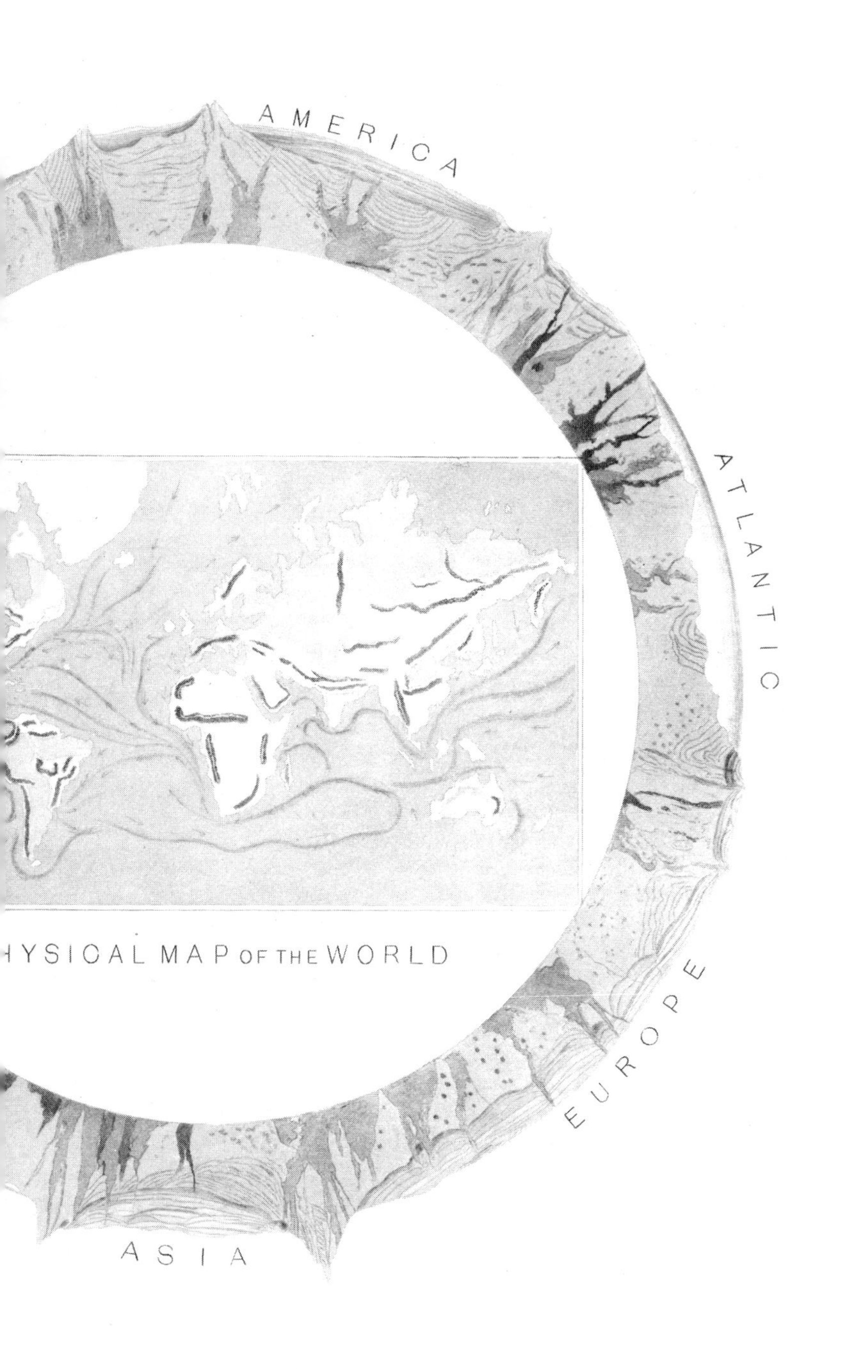

SECTION OF THE EARTH'S CRUST

AMERICAN GEOLOGY.

PART II.

THE TACONIC SYSTEM.

PRELIMINARY OBSERVATIONS.

§ 1. We are now prepared to engage in that part of American Geology which treats of the sediments. It comprehends the third division in the proposed classification, the *hydroplastic rocks*, or those which have been laid down in and moulded by water. They usually contain in themselves the evidence of their origin. Their component particles have been rounded by attrition. There are exceptions, however, to the rule, but even in those masses which are exceptions, they are so connected with those that are made up of sediments that we need not be led into error in their determination. The attestation of their origin by pebbles is usually confirmed by the presence of fossils. Fossils may be contained in the pyroplastic rocks in consequence of their having overflowed those places where they were collected upon the surface; but such instances are comparatively rare, and speak for themselves. Neither fossils nor pebbles are ever found in either division of the pyroplastic rocks.

The deposition of sediments has been in progress without interruption since water began to accumulate in seas and oceans. This accumulation has not progressed at a uniform rate, neither have the same areas been under the dominion of water during this entire period. Oscillations of land have often occurred, so that in fact the oceans may be said to have traveled sometimes to the east and sometimes to the west, but moving in obedience to those subterranean forces of which I have already spoken.

These movements are indicative of stages in geologic time. They seem to have often been paroxysmal, and to have been followed by periods of repose, during which sediments accumulated quietly upon the ocean's bottom.

The sequence in which the sediments have followed each other, has been an important geological problem. The best evidence of sequence or the order in time in which rocks have been deposited, is superposition, or the association of the fragments of the older with the newer rocks. Succession is often indicated when the overlying mass contains the materials of the older or inferior rock. Sedimentary rocks are the only ones which are stratified, stratification implying a succession or accumulation of materials under water, which are spread out in thin strata over the ocean bottom by the same agent. Different strata are indicated by the different materials received from the land. Sediments require to be studied with reference to their kinds, their source, their texture, the rapidity of their accumulation, and the time they were forming, and their order of succession.

The first two will give us information as to the direction in which the land lay which furnished the sediments; the third the probable distance they were transported; the fourth, with information respecting the force and rapidity of the transporting streams and currents, and the probable existence of floods alternating with drouths. Sediments are measuring lines, all things being equal; they reach into past time directly according to their thickness. The table of arrangment of rocks shows that the older sediments are thicker in their periods and stages than the new. Comparing the Palæozoic with the Kainozoic sediments, the disparity is immense. The divisional lines of the great triads of past time, as measused by the sediments, show a rapidly diminishing rate from the bottom to the top. The base is immensely spread out in the Taconic system, but the systems converge rapidly from the carboniferous, as if in the long preparatory stages for man's entrance upon the scene of earth's conflicts, nature had become impatient, hastened the later periods to their ends by crowding the greatest events into smaller spaces.

§ 2. But it is not with sediments only, that the geologist has now to deal. In the sediments are the vestiges of life. Since the sediments began to collect, we know that this new element was introduced upon the globe. But life does not seem to have begun with the sediments, for they appear to have been accumulating for immense periods before we find even traces of it in their beds. The precise time is not yet determined, but it appears now that it was clothed in a humble dress, that it began in the lowest rank both in vegetables and animals, and that they began simultaneously and as occupants only of the seas.

Sediments, are distinctly separated from the massive and laminated igneous rocks, and the line of demarkation is well defined. We have therefore a distinct sedimentary base.

§ 3. Life, beginning in the sediments, has also a base which is termed the *palæozoic base.* The use of this term seems to imply that there is also another base, but it means simply, the period when life first appeared upon the earth. Virtually, however, we may work instructively from a mesozoic or a kainozoic base; for in either case, it is possible to make comparisons of the organisms in their aggregate belonging to either of those divisions. It is well established by observation that the organisms preserved in each of these grand divisions differ on a whole from each other.

Sediments, taken in conjunction with life, are the elements of our knowledge of geologic time, or the events which occurred long prior to the historic era. Since the historic era, events strictly geological have been recorded with the passing events of the day. But the events prior to history are recorded only in the physical phenomena in the earth's crust. Their interpretation is the highest duty of the geologist. As an illustration of interpretation of phenomena we may cite the occurrence of volcanic dust as interpolated in beds among the sediments; or, the flexures and wrinkling of strata, the first showing the activity of volcanic forces in past time similar to that of our own, the latter also of earthquake movements, differing in no respect in kind from those of the present. The course of nature therefore, in the long run, has been uniform. It has been governed by the same laws in the remote, as in recent times. Life, it is well known, does not admit of wide deviations from certain normal conditions of the earth's atmosphere or its waters. The limits within which the integrity of vital forces can be maintained, are within narrow bounds. Hence, its frequent interruption. The types of animal and vegetable life are few, and however protracted time has been in the past, they have never been increased since the middle of the Silurian system was laid down.

Perturbations of the earth's crust are cotemporaneous with the changes in the organic world; they seem to mark the outgoing and incoming of the geologic periods. The systems are particularly indicated by physical changes. Hence the outgoing of organic forms are connected together as cause and effect with the perturbations of the earth's crust. We are not, however, to carry this doctrine too far; facts seem to show that many species survive a perturbation, and in historic times species have become extinct from causes which can not be traced to perturbations of the earth's crust.

The sediments with their organisms is a richer field for thought than the pyrocrystalline masses. The former, it has been aptly said, contain the medals of creation.

OF THE OLDEST SEDIMENTS. COLLECTIVELY THEY FORM THE TACONIC SYSTEM. ITS BASE, AND THE ORDER OF INVESTIGATION, ETC.

§ 4. My first business is to sketch a picture of the oldest of the sediments, as they are exhibited in a series which collectively constitute the Taconic system, and as it is developed in the Taconic ranges of Berkshire and the adjacent country immediately north and south. It is of the first importance to state the constitution of the masses, and to connect therewith a statement of their relations, as they appear, in the order of their sequence, taking occasion also to make such comparisons with other systems, as the nature of the facts and phenomena require.

The natural order, and the one which should be followed, is to begin at the base. We then follow up the series in the order of events. It is impossible to work down to a base; certainly such a course deprives us of a continuous narrative; it would be like reading American history backwards—beginning with Franklin Pierce and going back step by step, to George Washington and the events of the Revolution.

The Taconic system has a clear and well defined base, which is rarely obscured by passages into the primary schists, the pyroplastic rocks, sienites or granites. We have no intermediates, and no masses which may not be distinguished, either by their composition, or their relations to conglomerates and pebbly beds, the invariable characteristics of sediments. It appears therefore, that we are justified in the conclusion, that when the base of this system was laid down, water had become an established element upon the face of the globe, not subject to dissipation in vapors by excessive heat of its crust, for the pebbles in the rocks exhibit an attrition similar to what takes place upon our shores at this day. If my views are correct, and I have endeavored to sift them of error, we can go back no farther; we have no older sediments. When, in 1836, I determined that in New York, the Potsdam sandstone was the base of the Silurian system, it seemed that we had at that time, the base of the sediments; but, when, two years subsequently, I had observed the same base resting on sediments still older, as those along the eastern side of Champlain and elsewhere, it became evident that there was still a series older than the Silurian. The proof of this has been accumulating ever since; and the Taconic system is found to rest upon primary rocks without an exception; and it has now been observed through the whole length of the states, from N. E. to S. W. It is worthy of note, that through this whole extent, the base is continuous. The most northeasterly point at which I have observed this system, is at the Fox Islands, off the coast of Maine; but I have good reason to suspect its existence in Newfoundland. If so, it ranks among the most persistent geological formations of this country.

§ 6. The evidence of the existence of a system of rocks, beneath and older than the Silurian system in this country, rests on many well determined facts. These facts are not all of equal importance; but those which are not direct, serve to corroborate and sustain those which are. The facts which bear directly upon the evidence alluded to are superposition, succession, unconformability and the presence of fossils distinct from those of the Silurian system. Those which corroborate and sustain the independent existence of the system are, a thickness greatly superior to the Silurian system, an arrangement of its masses quite different from the latter, and the absence of fossils where they should exist, provided there is a correlation of the two series. The foregoing views will appear more conclusive when we take into account the fact that the lower Silurian rocks of this country, consist of well determined members through their whole extent; they are arranged in a determinate order, and hence are more easily recognized by their fossils, and also by their lithological characters. The lower limestones, it is true, are not eminently rich in fossils at every location, but they may be discovered by careful search, and moreover the true calciferous area, if wanting in fossils at any given place, is readily known by its relations. The present advanced state of the science requires of the geologist no better evidence of age, or a better foundation for establishing the age and infra position of the Taconic system, than the determination of the foregoing facts. This is especially true, when we have sifted our observations of certain errors which might occur in regard to superposition of an older rock upon a newer by excessive derangements. Fortunately, the most important points where superposition and succession occur, are those of only moderate derangements; indeed, the phenomena neither indicate excessive upheavals, nor downthrows, the sliding of the inferior mass upon an older, the passage of an anticlinal axis into a fault, nor the deceptive plications of contiguous strata; nor lastly, the folding beneath of a superior mass, or what we may possibly deem the mistaking of cleavage planes for those of deposition. When our observations are free from the possibility of error of the foregoing kinds, we are forced to maintain that superposition is due alone to succession; or we are forced to concede one of two things, either that there is a system of rocks older than the Silurian, or else that it is possible that to the Potsdam sandstone there succeeds conformable slates, limestones, conglomerates and sandstones whose joint thickness is between 25,000 and 30,000 feet, forming a prodigious appendage to a system whose base, all admit, is the Potsdam sandstone.

Lithological characters should not be neglected, though it is true, that in cabinet specimens no reliance can be placed upon these characters; yet in the field and when investigated in mass, they really become important aids in clearing up the difficulties which lie in our path.

§ 7. If it was my only aim and object to place the evidence I have of an independent system of rocks beneath and older than the Silurian, I should proceed to point out at once, that there is a slate beneath and older than the Potsdam sandstone, or in its absence, the Calciferous sandstone, and that a succession of rocks of great thickness lie in conformity to this underlying slate; and farther, that the members constituting this system, are never incorporated into the Silurian, or lie in conformity therewith; or stand in the relation of sequents to any of its members. This slate I should select because its position would enable me to exhibit its relations to the lower Silurian rocks, and this would be the more satisfactory and conclusive, inasmuch as it is maintained that the Taconic system is only the lower Silurian, embracing those members which lie between the top of the Hudson river group and the Potsdam sandstone. I shall not however pursue this plan, as I wish first to exhibit the sequence and relations of the members composing the Taconic system. In this country, seeing that the lower Silurian system is remarkably well developed and defined, it has appeared to me that American geologists should avail themselves of this fact and employ it as one of the instruments or means for the determination of the true palæozoic and sedimentary bases. It certainly gives us important advantages over European geologists. The bases of both systems are remarkably well defined, and the arrangements of the lower masses in both are so uniform that a comparison in detail is by no means difficult. This is true even in the disturbed districts. We may admit the existence of folded axes, or, that in the eastern district the masses are metamorphic, still the comparison of one with the other need not involve us in error.

ROCKS WHICH GENERALLY COMPOSE THE TACONIC SYSTEM.

§ 8. The sediments of all systems must necessarily consist of the same materials. Sandstones or the debris of the silicious rocks, limestones and slates with various intermixtures must make up, lithologically, the matter which compose them. Conglomerates and breccias are also constituents in a less amount and degree, yet not less important geologically, as they serve to mark more distinctly the physical changes which have taken place in the course of time, and during which the depositions have been going on.

The important point to be observed is the arrangement of the materials, although in the same system at different places there may not be a parallelism of deposits, yet there is a great similarity in districts which are widely separated. The Silurian system in this country is quite uniform as to the arrangements of its subordinate members — the northern limits of the system in New York is almost identical with the same part of it in Virginia and Tennessee; this is especially the case in the lower Silurian. The Potsdam sandstone, the calciferous, birds-eye, Chazy, Trenton, and the upper slaty and shaly masses, can not be distinguished from each other at these wide distances. I am not called upon to account for this remarkable fact. I have only to state it in this place. This constancy of mineral character becomes an available fact, where it is necessary to compare the corresponding parts of two adjacent systems. For example, a comparison of the lower members of the Taconic system with those of the Silurian in this country shows a decided difference in the mineral constitution. The first partakes of the primary character of the talcose and mica slates of the pyrocrystalline rocks; indeed, it is often difficult to distinguish the lower slates of the Taconic system from the schists which are intimately connected with the gneiss and hornblende. So close indeed is their resemblance that they were regarded by the old Wernerian geologists as primary rocks of the same age as hornblende and gneiss; and the same is true of the quartz rock occupying the same geological position. When, however, we examine the lower Silurian masses their origin is not doubtful; they all bear the impress of a sedimentary origin. This difference has usually been explained by metamorphism. This explanation, however, is not satisfactory, inasmuch as the base of the Silurian system in northern New York reposes on the pyrocrystalline rocks, and so far as we at this day are able to judge of cause and effect were as likely to have become metamorphic in this region as the lower Taconic rocks. It has appeared to me, therefore, that the difference in physical condition is due mainly to composition. The lower Taconic rocks are derived directly from the pyrocrystalline rocks, granite, gneiss, mica and talcose slates; and the latter being very uniform in composition through the whole length and breadth of the Appallachian system of mountains, the source from whence their materials were derived, they have of necessity a constancy of mineral constitution. So also the same accounts for the constancy of the mineral constitution of the lower Silurian, whose materials have been derived from the Taconic system in a parallel belt and of an equal extent; and this view is by no means theoretical. Years ago, I had obtained masses of a gray sedimentary limestone from the lowest part of the Silurian system at Chazy, which resembles most perfectly the limestone of the Taconic system in the neighborhood. The carboniferous system of Rhode Island is in part composed of talcose slates which contain the stems of lepidodendra; so in the masses of conglomerates of the same system at Wrentham, Mass., masses of talcose slates are by no means uncommon. All these variations of mineral constitution have been attributed, as I have already said, to metamorphic action. I can not but regard it, however, as erroneous in all the cases I have cited. While I recognize metamorphic action as important, I can regard it only as a local result, and limited in its effects. At most, such seems to be the ground upon which it is to be placed in this country; and hence, in comparing the lithological characters of the systems, it is more important to notice the mineral constitution of their masses. This mineral constitution, I maintain, is not a secondary affair. If, for example, we find magnesia as a constituent of the limestones or talcose slates its presence is to be regarded as having coexisted in the sediments, and not as material which has been introduced subsequently through the influence of chemical and physical forces.

§ 9. The rocks which compose the Taconic system are sandstones, often vitrified; slates, both green and black, the former varying somewhat in constitution, in some instances they are talcose or magnesian, in others argillaceous; and others still may more properly be denominated chloritic; and limestones purely

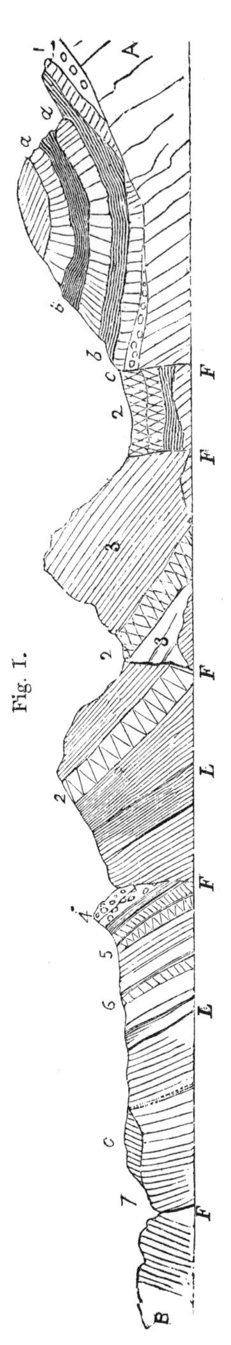

carbonates of lime. containing beds of dolomite, black slates colored with carbon, and decomposed sulphuret of iron, and calcareous sandstones. Subordinate to these rocks, I may also enumerate fine and coarse conglomerates, breccias, limonites, and schists of various colors. The system contains also veins of magnetic and specular iron, auriferous and cupriferous quartz veins, and in the metalliferous districts sienitic and compact trap dykes.

§ 10. The general order in which the rocks are arranged is exhibited in figure 1.

1, conglomerates and breccia at the the base of the system. The iron breccia and limonite veins occupy this position; so also quartz in the condition of an ordinary sandstone or an even bedded flagstone; and I may also add talcose or magnesian slates alternating both with the quartz rock, a, c, d, and the subsequent members; 2, limestones; 3, slate, which in its usual enormous development seems to constitute the principal feature of the system; all the other masses might be regarded as subordinate beds; 5, slates, which I have denominated Taconic slates, and which are usually of a dark color, and contain beds of sandstone, sometimes a purely silicious rock, at other times they are silico calcareous; 6, sparry limestones in the Taconic slates; *L*, limestone; *F*, fractures.

This section exhibits the order of arrangement, or the sequence of the members. In certain geographical districts it is in proximity to the primary schists and granites, as in North Carolina and Maine; in others, to the primary schists and the carboniferous, as in Rhode Island, and New York, Virginia, Pennsylvania and Tennessee, to the Silurian and carboniferous.

THE GENERAL GROUPING OF THE MEMBERS OF THE SYSTEM, OR THE DIVISION WHICH IS PROPOSED FOR THE TACONIC ROCKS.

§ 11. The Taconic system is susceptible of a division into two parts, the lower and upper—the line of demarkation is tolerably well defined. The first or lowest division terminates with the slate overlying the Stockbridge limestone, 2, 2. The entire mass of this division exhibits the primitive schistose aspect of the laminated pyrocrystalline rocks, and were it not for the presence of conglomerates in this formation, it would still be regarded as belonging to a period in the earth's history which preceded the deposition of sediments. As the phenomena exist, however, it must be regarded as the sedimentary base of all of the hydroplastic rocks. It is in the upper part of the second division that we first find fossils, the fucoids and graptolites of the Hoosick roofing slate and the adjacent beds of the same period. This fossiliferous part of the division still retains in part the primary aspect of the lower beds, but the colors of the slates are darker, some are purple and chocolate colored, and the sediment is much finer, and more homogeneous. In this part of the system, therefore, the rocks are coarse in texture at the base, and become finer and homogeneous in the ascending scale. Oak hill itself, with its adjacent mountains of protogine and gneiss are the most interesting localities for an exhibition of the arrangement of the lower Taconic rocks, in this county. It exhibits the junction with the protogine, the order of arrangement of the members nearly to the first limestone. This order is exhibited in figures 1 and 2. But there remain several facts connected with the localities which require a farther statement. In addition to the dip of the several masses exhibited upon the first section, there is a steep southern slope upon which the masses plunge rapidly, and beneath the narrow valley leading from Williamstown to North Adams, the formation exhibited upon the western slope of Oak hill is lost, in part, beneath the soil of the valley. The upper mass of quartz may be seen near the Adams road, not far from the bridge crossing the Hoosick.

The out crop of the quartz occurs again two miles south near a mill at the junction of the Hopper creek and Green river. A small part only of the mass is exposed dipping southeast and towards the high range of mountains known as Saddle mountain and Graylock.

The upper division begins with coarse slates and sandstones, and terminates in a fine black slate. These sandstones and slates are greenish, and rather chloritic than talcose; and the masses, as a whole, often resemble greenstone. This appearance is no doubt due to the presence of chlorite and perhaps the debris of hornblende. In New York these rocks are by no means altered rocks; there are no veins of greenstone or porphyry in connection with them. The upper part is much more protean than the lower. Dark colored slates predominate, but they contain a multitude of subordinate beds, as olive sandstones, intercalated with fine green slates; calcareous sandstones, which weather to a brown or drab, dark green flags with fucoids; sparry limestone; green and black slates; beds of quartz free from calcareous matter; beds of conglomerates and black shaly limestone, which occupy a superior position in the series.

These beds are also interlaminated with a thin bedded sandstone, which we may not be able to identify at distant points, or which may prove to be due to local variations, and not persistent at distant points.

MEMBERS OF THE LOWER TACONIC SYSTEM. THEIR CONSTANCY OF MINERAL COMPOSITION. THEIR THICKNESS. ABSENCE OF ORGANIC BODIES.

§ 12. The principal members of the lower Taconic system are few in number, for though the beds are numerous, it is not deemed necessary to multiply names of rocks where there are no fossils, and when the differences are due to position only. For example, it is difficult to distinguish the beds of slate which lie between the masses of quartz, from those which overlie the limestone. It seems to be one rock prolonged through a long period, the continuity being broken occasionally by a substitution of silicious and calcareous deposits.

As the masses referred to possess characters which are essentially the same from Maine to Georgia, I shall select for description a section which I have worked out in detail, and which is also accessible to those geologists who have sufficient interest in these rocks to examine and test for themselves the truth of my statements. A brief reference will also be made to the same rocks which are situated at distant points.

The second line of section which I shall now describe, extends also through the north part of Berkshire county, Mass., and immediately across Prospect hill, S. E. of Williams college.

§ 13. The western declivity of this mountain furnishes the necessary information relative to succession of the limestones and slates, which are superior to the rocks of Oak hill. This succession has been satisfactorily made out and is exhibited in section 2. It passes directly over the mountains a little to the north of Graylock, so as to cross the valley about two miles south of South Adams. The dip of the limestone, quartz, &c., near the mill referred to, is S. E., the limestone reappearing on the Adams side of the mountain in a reversed dip, forming therefore in the range of mountains through which it passes, a synclinal axis. The rock above the limestone is entirely slate and can scarcely be distinguished from primary talcose slate, and yet it is a sediment. The height of Graylock above the outcropping of the limestone is 2000 feet. The limestone is 500 feet thick, and the mass between the limestone and upper bed of quartz is at least 500 more.

On the east side of the valley, near South Adams, the rail road cutting exhibits a fine section of slate; it dips westward steeply, and behind it towards Hoosick mountain, which is composed of gneiss, the limestone and quartz reappear in their true relations. These rocks therefore do not dip towards the primary range, but westward.

§ 14. To enter now upon an enumeration and description of

the members of the lower division as they occur in an ascending order, I proceed to speak of them as developed in Williamstown and Adams, and to which an allusion has been made.

1. The lower rock or base of the Taconic system at these places is a quartzite. The propriety of the name rests on the fact, that quartz in some form is the predominant element. The inferior mass is sometimes a slate in which talc is abundant, and which generally contains pebbles. Lithologically, it is a silicious talcose slate, closely resembling the talc-schists of the primary rocks of which hornblende is one of the associates. At or near the base we find beds of conglomerates, which are usually made up of rounded quartz pebbles, as well as angular stones imbedded in a talcose paste. Another brecciated conglomerate consists of a paste of limonite in which both rounded and angular masses are enclosed. Proceeding from the base upwards, the masses consist of quartz of various colors and degrees of fineness, alternating with talcose slate. Associated with them in thick beds, is a quartz charged with feldspar, which often looks like a porphyry. The feldspar decomposes and leaves small ragged cavities, when the mass becomes a *burrhstone*, and is often used as a millstone. The quartz rock is granular, often friable, vitreous, and compact; usually brown, but frequently gray or grayish brown, or white; sometimes snow white, and an excellent material for glass. The different varieties appear to occupy a given place in the series, but the vitrified kinds do not lie at the bottom, but frequently occur at the top; and so, of the distinct sandstones or friable ones, the beds are more frequently at the bottom, or near it. The grains have the form of the common sandstone of the Silurian system. Hence the vitrified masses became so by circumstances which attended their deposition, rather than by a subsequent vitrification by heat. So much do the bottom rocks appear like primary rocks, that no one would suspect they were sediments, were it not that they are accompanied by conglomerates. The conglomerates are sometimes obscure, in consequence of a thin investment of talc, which often appears

pressed into the pebbles. Among the pebbles, angular masses of quartz and hard green slates are often found so closely resembling the accompanying talc-slate, that it might be inferred, that the slate had been broken up and incorporated with the mass, but it is probable that they belonged really to the primary schists which accompany hornblende and gneiss.

§ 15. That the character of these lowest sediments may be clearly distinguished from the lower Silurian, I propose to speak of them in greater detail: Fig. 1.—1. The lowest mass and which reposes upon protogine about one mile east of the crest of Oak hill, is a porous silicious slate, composed of talc, and some mica, quartz and feldspar. The feldspar and quartz is angular, sometimes the particles are rounded, but in either case the rock has an open structure arising from the decomposition of feldspar, and the small cavities are rough and unequal. The porosity of the mass fits it for a millstone. It is a thin mass at this place. At other places its pebbly character is remarkably well developed. It lies directly upon the granite here and its materials clearly indicate that it was derived from this rock. Its thickness is variable, not exceeding thirty feet at this place.

The mass reposing upon this millstone is, 1, a coarse slate made up of talc and quartz, it is a talcose slate. Its stratification is uneven from the presence of coarse masses of quartz. We meet occasionally with needleform schorl. It is a much more crystalline mass than the slates above it. It is seventy feet thick. 3. The next rock, c, is a sandstone. It is made up of white quartz grains, and forms a flagging stone susceptible of division into very thin layers. It seems to be purely a sandstone unchanged by metamorphic action. It is 100 feet thick. thick. It dips southwestwardly at an angle of 15°.

The fourth rock, b, is talcose slate. It is gray, soft, and even bedded. It is comformable to the masses below it. It is forty feet thick. 5. A mass of quartz, d, succeeds this fine slate. It is brown and rather massive and jointed. It is fine grained, and destitute of pebbles. It is 400 feet thick. It extends nearly to the crest of Oak hill when it slopes to the south, but there still overlies it a bed of quartz and a silicious talcose slate which seems to be the remains of a much thicker mass; the latter is about thirty feet thick, and lies obliquely across the axis of the hill. The strike of the rocks therefore is not coincident with the ridge or crest of this chain of hills, the former bearing a few degrees more to the westward.

The masses which I have described crop out upon the east side of the ridge and dip S. W. at a moderate angle. They reappear on the west side again dipping eastward, forming in consequence of their opposing dips a synclinal axis. Oak hill is 1700 feet above the valley of the Hoosick, which flows at its western base, but the rocks are exposed only at or near its crest on the west side, while on the east side the succession is perfectly clear by an exposure of all the masses from the protogine to the top of the mountain.

§ 16. The sequence being made out in the mode I have stated, it becomes necessary now to ascertain what rocks succeed those which have been described. This determination may be satisfactorily made out by tracing these same masses along the southern slope of Oak hill into the valley already referred to as leading from Williamstown to North Adams. We learn from the direction they take that they plunge beneath Saddle mountain, and by tracing some of the members already spoken of along the western flank of this range, we at length find the quartz outcropping in a small creek which comes down from the hopper of Graylock the summit of Saddle mountain, and the highest point of land in Massachusetts. This point is 3600 feet above tide, and 2800 above the Hoosick. It forms a heavy mountain mass between Williamstown on the west and South Adams on the east. By ascending the western slope of Saddle mountain to Graylock, from the quartz which outcrops at its base, we obtain the succession of the masses which succeed the formation of Oak hill. The rock overlying the quartz is again talcose slate, silicious at its base, but purely a talcose slate as a mass and which requires no farther description. It is between

400 and 500 feet thick and extends up to the limestone, which constitutes the seventh member of the lower Taconic system.

This limestone, in a former treatise upon the Taconic system, I denominated the Stockbridge limestone, as it is extensively known in the commercial world as a marble and a valuable building material. It is however less constant in its composition than we should expect judging from the pieces of marble exhibited in market, or at New Ashford. It is in the first place frequently a dolomite, and in this composition is flexible when first removed from the quarries. Its layers along the plane of bedding are profusely sprinkled with green talc, and though in Berkshire county it seems to be confined to one mass, yet at other places not far distant it is interlaminated with slate.

The colors are white, gray and occasionally very dark. It is reddish at Williamstown and is intimately blended with silex. It is also seamy or sparry, containing calcareous spar, magnesian spar blended with a variable quantity of carbonate of iron.

It crops out on the western slope of Saddle mountain dipping south of east, but on the east side at South Adams it crops out at its eastern base, dipping west. Its limits and position are clearly defined at Williamstown and Adams, and indeed for the whole of western Massachusetts and eastern New York, as it is found in the same relation throughout this entire region. It is 500 feet thick as exposed on the western slope of Saddle mountain.

§ 17. The eighth member of the lower Taconic system which succeeds the limestone already described is a talcose slate. From the termination of the limestone to the top of Graylock, the talcose slate is uninterrupted. The thickness of slate above the limestone is about two thousand feet. These beds of slate are similar and uniform in their composition and structure not only in Berkshire county, but in Virginia, North Carolina and Tennessee; the beds at the south can not be distinguished from those at the north.

By referring now to fig. 2, the reader will see at a glance the relative position of the rocks which I have described.

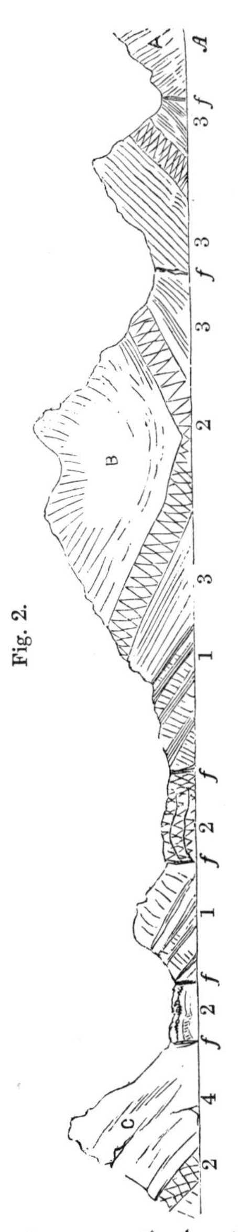

Fig. 2.

1, 1, Quartz with its interlaminated slates, at the foot of Saddle mountain; 2, 2, limestone outcropping upon both sides of this mountain; 3, 3, slates beneath and above the limestone. I have also shown in the section the relative position of the mountain ranges which traverse Berkshire county and eastern New York. A, Hoosick mountain composed of gneiss and mica slate, hornblende, etc.; B, Saddle mountain; C, the main belt of the Taconic range along which runs the boundary line between Massachusetts and New York. This range, C, is composed of the same talcose slate as B. It is another uplift of the same rocks. In the valley between, Stone hill forms a prominent point and shows a repetition of the rocks of Oak hill. It is 400 feet high and is bounded on both sides by the limestone No. 2. This limestone is crushed and contorted.

The prominence represented between A and B is a sharp ridge in South Adams, on the west side of which the rail road cutting exposes the slate dipping steeply to the west.

§ 18. *Iron breccia of the Taconic rocks.* This singular mass should not be passed by without a brief notice. It is composed of angular quartz cemented together by limonite, or brown hematite. It does not form as I have ever been able to find a continuous stratum, but it is associated with this part of the series. It occurs in Vermont, Massachusetts, Pennsylvania, Virginia, North Carolina and Tennessee, and always in the same relation. At North Adams it was worked for iron at one time, a thick seam of it being discovered in place near the junction of the slate and quartz. It is always too silicious however to be profitably worked. Its origin is unquestionably igneous, as the phenomena seem to indicate at the locality just referred to, and it also seems highly probable that the immense beds of limonite which extend through the states I have named above have been derived from the originally injected masses of iron breccia. However this may be, the mass is highly interesting. It is easily recognized as it is composed of quartz and iron. The quartz is sometimes in rounded masses as in Hillsborough, N. C, and hence is a conglomerate. Ferruginous quartz also occupies the same geological position and is as widely distributed. It is common in Berkshire, Mass., Buncombe county, N. C., and near Abingdon, Va.

§ 19. The sequence of the lower Taconic rocks which has been stated and illustrated in the foregoing pages is essentially the same from Maine to Georgia. I shall illustrate this position from the careful observations which I have made in this and in former years. The order of arrangement, therefore, which has already been placed before the student, must be regarded in no other light than as confirmatory of the views I have adopted in this and my former treatise upon this system.

§ 20. *Taconic rocks in Maine.* The Taconic system is surrounded by primary schists, granite, &c. I have been unable to discover any masses which could be referred to the Silurian system. The earliest information respecting this system in Maine was communicated by Dr. Jackson. The discovery of fossils, which were regarded as vegetable impressions at the time, were subsequently referred to the class of Nerecites. At this time they may be placed among the foot prints of moluscs. The slates which contain them are at Waterville, on the banks of the Kennebeck. They are soft, green talcose or magnesian slates, placed nearly in a vertical position and whose trend is N. 10° E. They are connected probably with the fine roofing slate of the Piscataqua river. They seem, however, to be isolated and separated from the lowest members of the Taconic system. The slates extend over a width of country, for about 15 miles. It is at Camden, however, that we observe a section of the rocks which coincides with those of Berkshire county. The intervening country is partly covered with drift, over which granite, gneiss and mica slate prevail.

One of the most interesting rocks of Camden forms a very conspicuous mass in the neighborhood. It forms an eminence 700 or 800 feet high. It is the Meguntieook mountain, and is a mass of conglomerates resting unconformably upon mica slate. It is isolated, and the members once connected with it have been swept away by diluvial action. The conglomerates are at least 400 feet thick. A large proportion of the quartz is still angular. A succession of the lower rocks of the system is exposed on Goose river and harbor in the vicinity.

Figure 3 shows the succession referred to, enumerating them in the descending order.

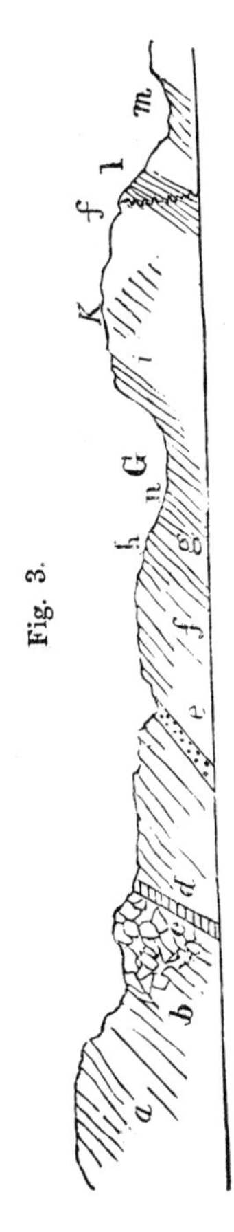

Fig. 3.

a, wrinkled magnesian slate; *b*, limestone; *c*, trap dyke; *d* and *f*, slate more or less silicious; *e*, granite vein; *h*, hard quartz rock; *i*, slaty and contorted quartz; *l*, slate containing imperfect macles; *n*, second mass of granular quartz; *m*, magnesian slate. At *F* there is probably a fracture and hence the foregoing masses may be repeated; *G*, Goose river.

The rocks at Camden are identical with those of Berkshire. Their order of arrangement the same. All the differences arise from their having been changed by their vicinity to trap and granite, as in the slates macles are partially developed. This section has one advantage over the Berkshire section, as it evidently gives us the rocks according to their succession. Their aggregate thickness I estimate at 2,000 feet. Conglomerates occur in the mass at K. The Taconic rocks extend from Camden seaward. They appear and form the Fox islands 12 miles from Camden; where they are much changed by igneous rocks at certain points, especially where they are traversed by dykes.

At Thomaston, seven miles distant, the slates and limestone similar to those of Camden are the prevalent rocks. The limestone is traversed by a huge trap dyke.

§ 21. In Rhode Island near Smithfield, quartz, slate and limestone are well known.

The succession here is as follows, and is represented by fig. 4. The rocks lie in a trough bounded on one side by granite and on the other by a conglomerate of the carboniferous system.

Fig. 4.

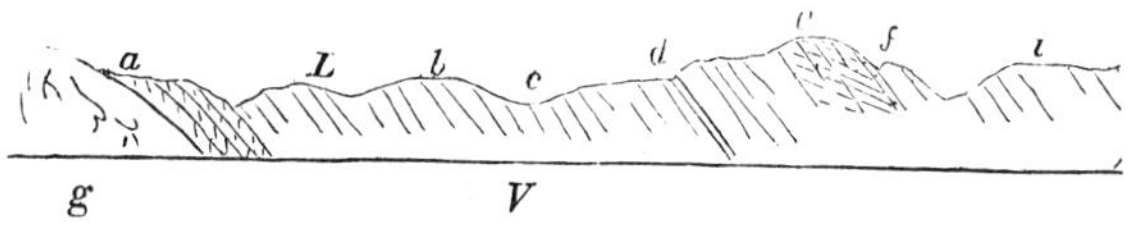

g, granite; *a*, altered magnesian slate; *L*, *c*, *e*, limestone *b* and *d*, dykes, in the form of hornblende; *f*, slate partially changed to serpentine; *V*, valley of the Blackstone; *i*, granular quartz interlaminated with talcose slate; on the right of the section, the coal conglomerates occur in juxtaposition.

In this section the older rocks seem to rest on the newer. The section is made out so far that we can only speak of the masses which represent it. The succession is obscured by drift and intruded rocks. In general, I may observe that the quartz, limestone and slate resemble the rocks of Berkshire.

It is important that the student should be acquainted with certain phenomena respecting the members of the Taconic system, which obscure their relations at many places. One of these phenomena is produced by the intrusion of pyrocrystalline rocks, granite or sienite. These intruded masses separate the members of the system from each other, and at the same time change the direction of dip. This fact is illustrated in a belt of country near Fisk kill at the Rocky Glen factory.

Fig. 5.

Thus *a*, magnesian slate; *c*, limestone; *b*, e, granite, which separates these two members from each by a rocky ridge.

It happens, however, that ridges of mica slate are interposed in the same manner. Probably the deposit was originally thin, and hence has been worn away by diluvial action. Some five or six miles sometimes intervene between two members of the system, as in Cherokee county, N. C., where a mica slate like that of New England with staurotides intervenes between the quartz and talcose or magnesian slate. This subject may be reverted to again.

THE TACONIC SYSTEM IN THE SOUTHERN STATES.

§ 22. The lower Taconic system is equally well developed in the southern as in the New England states. From the northern part of New England it is prolonged southwards, and upon the line of prolongation it continues uninterruptedly for more than one thousand miles. For the purpose of illustrating its development in the southern states, I have selected the region of the Warm springs, in Buncombe county, North Carolina. The series at this place with their sequence is illustrated by fig. 6.

It is situated in the midst of a cluster of rough mountains, along the bases of which flows the French Broad river. In some respects it is admirably adapted to geological investigations; in others it has its objections, the surface or slopes of the mountains are too rough to be traversed where great accuracy is required.

The succession of the rocks however is determined without much labor, especially the inferior masses, those which are under consideration. So also their relations to the primary rocks upon which they repose. The latter I regard as an important consideration at this time, inasmuch as it seems necessary to determine them in order to remove certain objections to my views which have been started since the publication of my report upon this system.

We find at this locality an anticlinal which clearly separates the pyrocrystalline rocks from the base of the Taconic system. It is not a local disturbance nor confind to an area of a few hundred yards. The older masses, consisting of gneiss, mica slate and hornblende, dip generally S. E. from Ashville to the Warm spring, or indeed from the Blue ridge itself. The Taconic rocks on the contrary dip westward for fifteen to twenty miles, though in the Paint mountain range there are numerous changes of dip. In fine, the locality at the Warm springs is a good exhibition of the development of the lower Taconic rocks in the southern states.

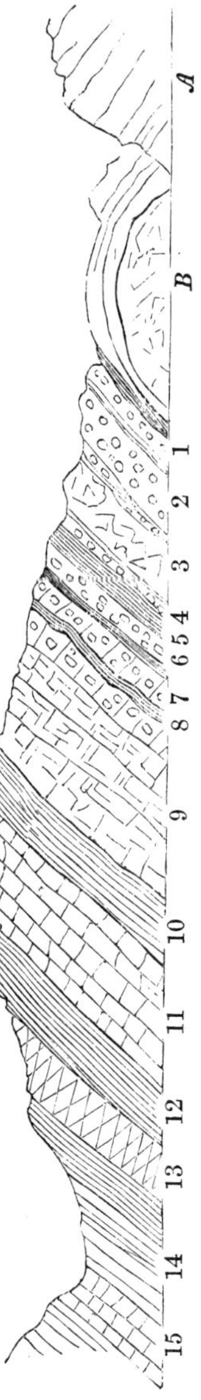

Fig. 6.

B, Sienitic granite, which in consequence of its protrusion at a period subsequent to the deposition of the lower rocks of the system, throws them into an arch.

1, talcose slates and slates with pebbles; 2, seamy ferruginous sandstone, 115 feet thick; 5, slaty conglomerate and slaty sandstone; 6, vitrified quartz of a brownish color, 200 feet. It contains pebbles and frequently fragments of slate. 7, brown silicious slate passing into quartz with a few pebbles, 85 feet; 8, granular quartz; 9, talcose slate, 200 feet; 10, quartz, the two beds with the slate included is 1500 feet; 11, slate, 250; 12, gray limestone, 500 feet; 13, gray fine grained talcose slate, over 150 feet thick; the mass is too much concealed to admit of its exact measurement. Adjacent to this mass I find a feldspathic slaty sandstone of an undetermined thickness, and its relations are obscurely developed.

The foregoing statement respecting the thickness of the masses at the Warm springs, is as near their actual thickness as the exceeding rough and steep surface of the mountain will permit. The united thickness of all the masses is therefore about 3000 feet. The sequence of the rocks is exact. The dip of the superior masses is west, that is from the axis of the primary schists A, adjoining them on the east; and I may here remark, that the line of demarkation between the former and latter rocks is well defined. The dip of the inferior rocks is north of west according to their distance from the axis of elevation.

The rocks at the Warm springs when compared with those of Berkshire county, embrace a larger development of quartz than of slates, and the beds of conglomerate are more remarkable, extending far up in the series; but the slates and quartz, lithologically, are undistinguishable from those of Berkshire. The masses are widely separated and are unquestionably correlative; besides, we have the evidence of continuity of *rock*, that same clear well defined separation from the rocks of the Blue ridge that we find both at Williamstown and the Warm springs of Buncombe. When we go farther and compare the subordinate masses we find that the same resemblance and identity holds good; for example, the limonites and manganese, the ferruginous quartz, both yellow and red, chert, agate, etc., are common to the whole belt of which I am speaking.

§ 24 In Cherokee county, N. C., all the members are present which I have given as belonging to the formation at the Warm springs, but the system is separated a few miles by the interposition of a ridge of primary schists with staurotide. Conglomerates occur three miles northwest of Murphy, Cherokee county, which dip eastward, to which a series of slates and quartz rock and limestone succeed, accompanied with beds of limonite, as in Berkshire, Mass. To the westward of the conglomerate, the ridge of primary schists come in, after which the dark colored slates and brecciated conglomerates and sandstones, &c., occur as in Columbia, Rensselaer and Washington counties, in New York, and as they recur at the Paint rock in Tennessee, six to ten miles west of the Warm spring in Buncombe county, N. C. Although there is a separation of the older Taconic rocks from the newer by the primary rocks to which I have alluded, still, there is no question as to the relationship which exists between them. I find a similar separation in New York by the intrusion of granite. The intrusion of granite is a phenomenon which is not likely to lead to error. In the recurrence, however, of the primary rocks, particularly the fine mica and talcose slates, the geologist might regard them as interlaminated masses, but a careful examination of relations of these rocks to each other will result in the conviction that the primary rocks are underlying and older masses, and have no connection with the sedimentary rocks which they geographically separate.

§ 25. The cluster of mountains which the French Broad river has cut through at the Warm springs is prolonged northwards and southwards. The range is necessarily crossed in going from Ash county, N. C., to Abingdon, Va. Stone mountain which rises near this route is quartz, the base of the system. It is succeeded by slates and limestones, and conglomerates and sandstones, in the order which I have already stated in sufficient detail. The limestone, however, is not a marble as in Massachusetts and Vermont, except at a locality about half way between Taylorsville, in Tenn., and Abingdon, Va. It is here a white marble, and is evidently an altered mass. The change is quite local and has taken place on a line of disturbance, and where the rocks are contorted and wrinkled. The route follows a rapid river, known as the Laurel. This range has received different names at different places. Between Macon county, N. C., and Tennessee, it is called Iron or Smoky mountain; between Cherokee county and Tennessee, it is known as the Unaka mountain; between Yancey county and the same state it is Bald mountain. From the northwest corner of North Carolina to Cherokee, quartz rock is the predominant rock of the range. It is traversed in the line of North Carolina and Tennessee by the Watauga, the Nolachuky, French Broad, Pigeon, Tennessee and Hiwassee. All of these rivers cross the line of base of this system of rocks and furnish fine sections and exposures of it. The conglomerates and quartz, however, in Cherokee county, are upon a belt eastward of the mountain range, while further north of this hard rocks form their crests and the highest points upon their ridges.

§ 26. *Sections of the lower Taconic rocks in Virginia.* — Another section a little further north is opened along the course of New river, which rises in North Carolina and flows northwestward into Virginia. I examined this section on a route leading from Hillsville to Wytheville, Va. I crossed the Blue ridge at the Fancy gap. The quartz rock and its slates and conglomerates will be crossed about fifteen miles east of Wytheville. It is sufficient to remark in this place that the rocks and their sequence differ in no respect essentially from those of the Warm springs in N. C. At an old iron furnace near the New river, the occurrence of the limestone is more abrupt than usual, the slate which intervenes between the quartz and limestone at other places is wanting at this, and the limestone, in consequence of the inclination of the quartz, seems at the surface to dip beneath it. It is not, however, an underlying mass.

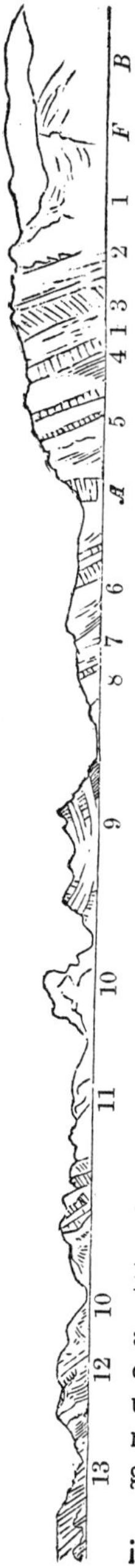

The limestone is gray, and only semicrystalline. It is overlaid by a dark colored slate. On this line of section, the rocks which succeed still further on the route to Wytheville, will be given in detail in the proper place. Thus far, the section corresponds with those I have already given.

§ 27. The next section is represented by the annexed cut, fig. 7. It leads from Waynesboro to Staunton, Va. Beginning at the western base of the Blue ridge, the series consists of, 1, Talcose slates accompanied with seams of iron breccia, seventy-five feet; 2, green talcose slate with breccia thirty-four feet; 3, quartz and iron breccia, twenty-five feet; 4, green thick bedded slate, seventy-five feet; 5, thin bedded quartz and iron breccia, alternating also with slate and quartz in beds presenting a ribbon-like appearance, 100 feet; 6, talcose slate, 100 feet; 7, quartz and iron breccia, 400 feet; 8, reddish and purplish slates and two or three beds of quartz; 9, thin bedded quartz; 10, of white and gray vitreous quartz, 500 feet; 11, talcose slate or light colored magnesian slate, too much concealed under debris to admit of a determination of thickness; 12, gray limestone, 300 feet; 13, black and gray slate with beds of shaly limestone.

In this section the recurrence of the iron breccia is the most remarkable feature. There is less quartz than at the Warm springs or at the Stone mountain. Still there is a great similarity in the group with those at distant points along this remarkable range.

§ 28. It remains now to describe the Harper's ferry section, which runs along the eastern slope of the Blue ridge with a southeasterly dip instead of the western slope. The Potomac crosses the base of the Blue ridge at this place. The rocks are finely exposed and the succession can not be questioned. Neither can the relations of the sedimentary rocks be misunderstood. The steep mountain which rises from the river on its north side, exhibits very clearly at a distance the line of demarkation between the primary and Taconic rocks; and on a near approach to this line, we find it is produced by the difference of the rocks. The primary and underlying rock is a wrinkled talcose slate, similar to the New England slates and Green mountain slates of the same name, and which are associated with gneiss and hornblende. 2, a brecciated conglomerate, eight feet thick. This rock makes the narrow belt on the south face of the mountain, and causes the distinct line of demarkation between the older and newer rocks. 3, greenish and purplish slates containing brecciated quartz, 500 feet; 4, slates, three to 400 feet; 5, purplish quartz, two feet; 6 slate, two feet; 7, breccia, four feet; 8, clear and vitreous quartz, seventy feet; 9, purplish slates, traversed by numerous seams of milky quartz, sixty feet; 10, quartz, thin bedded below, and thick above, 200 feet; 11, slaty quartz; 12, hard contorted quartz. Slates with limestone succeed the foregoing, the latter crops out from beneath the Trias, at a point upon the Potomac, from which were taken the brecciated pillars which ornament the halls of the capitol at Washington. The rocks eastward of Harper's ferry, however, are too much concealed on the line of rail road to enable me to give their arrangement in detail, still, enough is exposed to warrant me in speaking confidently of their sequence which as I have already intimated, does not diffe r materialy from those of the localities which I have already given. In the minor details there are certain unimportant differences particularly in regard to the thickness and position of the brecciated conglomerates and a few intercalated beds of slate and quartz. The former are noticed for the purpose of proving the sedimentary origin of these rocks, an origin which many geologists would formerly have questioned, and even at this day were it not for their presence.

The quartz and slates still continue, and these are succeeded by a gray limestone, which crops out from beneath the Triassic, or as some regard it, the Liassic system east of the locality which furnish the Potomac marble.

§ 29. From Harper's ferry the series continues through Pennsylvania into New York, and then into Massachusetts and Vermont, retaining the same relation to the primary rocks throughout its long extended line. If we trace the relations of the quartz rock to the Blue ridge and Green mountains of Vermont, we find it varies exceedingly along this great range. In North Carolina, the quartz rock at the base of the Taconic system forms independent ranges of mountains sixty or seventy miles to the west of it. As the quartz extends northeastward, the Blue ridge and the quartz approach each other. At Waynesboro the quartz reposes upon its western face. At Harper's ferry the quartz reposes upon its eastern face, having crossed the Blue ridge in its northern prolongation. The rocks which compose the Blue ridge in North Carolina and the southern part of Virginia, diverge from it south of Charlotteville and strike towards Baltimore. If the Green mountains and Hoosick mountain of Massachusetts, are regarded as a prolongation of the Blue ridge, we again find the quartz and slates at the base of the system reposing upon its western flank, though here these rocks form of themselves a distinct and independent range, which has long been known as the Taconic* mountains. I have already spoken of the wide distinction there is between the system of rocks composing the Green mountains and those of

* The word has been spelled Taughconnac and Taghconic, but those modes of spelling have given way to Taconic.

the Taconic system. When we trace however the relations of these masses throughout the entire extent of the country, the importance of regarding this distinction as established can scarcely be doubted. If we assume that the Green mountains are composed of the altered Hudson river slates, the same assumption must be made respecting the whole of the Blue ridge. The only belt where the Blue ridge is composed of rocks which can be regarded as metamorphic, is at the crossing near Waynesboro. At Charlotteville and westward towards the Blue ridge, gneiss, mica slate, hornblende and sienites, prevail. But near the ridge, a sedimentary rock which occurs extensively in North Carolina, comes in; and I found in crossing the ridge, that the green slaty rocks were charged with epidote, with scarcely a mass of gneiss or hornblende which could be recognized. I can only say, that the Blue ridge for a short distance exhibits a different composition, while at the same time the Taconic system, which appears on its western flank, is quite as clear and distinct as at any other point on the line of this great belt.

THE TACONIC SYSTEM ON THE EAST SIDE OF THE BLUE RIDGE.

§ 30. The account which I have given of the lower Taconic rocks, relates to the series which flank the Blue ridge on the west. I shall now describe a parallel series which skirt the same range on the east, and which though not equally continuous with the former is still the predominant system over a wide belt of country, extending in a northeasterly and southwesterly course from South Carolina to Maine. The Taconic rocks which I propose to describe in this connection, are destitute of that regularity and symmetry which mark those upon the west side of the mountain. The series also is incomplete, I find the upper members are wanting, and in placing this series in the Taconic system, I am guided by lithological characters.

§ 31 *Members of the series.* These embrace conglomerates,

quartz rock and quartzite,* talcose, chloritic and argillaceous slates.

The mode in which the foregoing rocks are arranged, will be understood by describing them as they occur upon a transverse section. If we descend the eastern slope of the Blue ridge in North Carolina, in the direction of Wilkes county, the series which represents the geology of the country traversed consists mainly of mica slate and gneiss and intruded granitic masses. I believe they resemble the New England rocks of the same name as perfectly as possible. These continue upon its flanks until after we pass a low range called the little Brushy mountain, which traverses the country about ten miles east of Wilkesboro, in a N. E. and S. W. direction. This formation continues to a point about one and a half or two miles east of Lexington, Davidson county, N. C. At this point and in a line ranging with it nearly N. E. and S. W. the system changes, the primary giving place to a series of sedimentary masses which continue to a line about ten miles west of Raleigh, in Wake county; it comprehends a belt of country about seventy miles wide. This belt is prolonged, in the direction I have stated, into Virginia and South Carolina. It is necessary, however, to state, that the lines of demarkation of its borders are not equally well defined at all points, but they are usually recognizable by making offsets at short distances to the right or left.

The prevailing dip of the rocks of the Blue ridge is S. E. When, however, we have passed from the pyrocrystalline rocks to the Taconic system, the dips are changed, first to the S. W. and finally to N. W., which is the prevailing dip. The angle dip varies from 40° to vertical. The first mass which indicates a change in the series of rocks, occurs a mile or two east of Lexington. It is a coarse talcose mass. Farther on it becomes

* The term quartzite I use to designate a kind of silicious mass which closely resembles chert or hornstone associated with many of our limestones.

a fine talcose slate which soon becomes silicious and forms a tough mass, which I shall denominate a coarse talcose quartzite. Its color is light green, it is often obscurely porphyritic, and though jointed, it is one of the most tough rocks which is ever encountered by the geologist or miner. When moderately silicious, it decomposes readily and forms a deep red soil. As it is nearly vertical and as it weathers unequally, the outcrop presents a singular succession of sharp edged masses, which are rounded in the direction of the strike. It presents an interrupted succession of hatchet shaped eminences. These are sometimes seven or eight feet high and ten feet at the base, and it sometimes also happens that two of these projecting masses stand within a few feet of each other, and hence are conveniently roofed over for a small outhouse, the rocks forming the two parallel walls. These projections are so numerous on the Three Hat mountain, some eight miles east of Lexington, that its surface can not be cultivated. The rock between Lexington and Spencer postoffice is mostly quartzite, with bands of flinty slate which breaks with a smooth conchoidal fracture, and a soft even bedded talcose and chloritic slates. It is difficult to assign a satisfactory reason for the occurrence of the quartzites in the midst of the unaltered slates. The change does not appear to be due to the proximity of trap or igneous masses. I have therefore for want of better reason regarded these cherty masses as products of chemical forces acting at the time of their deposition. The quartzite of North Carolina is so peculiar that I deem it useful to describe it still more particularly. In the extreme variety it is a pure bluish hornstone, and breaks with a flat conchoidal fracture. In this form, it is rather easily broken, but when less pure, it is tough; the fracture is then uneven as if the masses were torn apart. It is translucent on the edges, which are sharp like flint. When struck it is often sonorous like cast iron. It is not perfectly homogeneous as a mass, as small crystals of feldspar and sulphuret of iron are often disseminated through it. In disintegrating, the outside weathers to a drab. In Chatham,

Randolph and Davidson counties, it never forms ledges of jointed rocks, but crops out in the manner I have just represented. Its strike is N. 25° E. while the ridges which it traverses, and which are narrow, run N. 10° E. The ridges are quartzite, the valleys are usually underlaid with a soft slate. The quartzite in many instances is massive, but passes into jointed silicious slates. The latter are frequently associated with novaculite and other fine and coarse grits. Numerous varieties of quartzite are met with, as the blue and purple, light gray and green of many shades, and composed of impalpable individuals, also yellowish and yellowish brown, and traversed by seams of quartz; or deep green and banded varieties, or coarsely agatized ones. The quartzite is constantly undergoing decomposition. The atmospheric influence penetrates deeply into the rock; and its extent is marked by an opaque white friable border, &c. &c. Indeed all the varieties of quartzite disintegrate, decompose, some slowly and others rapidly, the silicious and ferruginous varieties forming a deep red, and the apparently pure silicious ones, a pale drab colored soil. These belts of quartzite are frequently half a mile wide, in which a few beds only of slate occur. The most interesting localities are in Davidson and Chatham counties. They are not confined however to these counties.

Randolph county, which succeeds Davidson on the east, furnishes a large amount of quartzite in its rocks. All the ridges may be set down as formed of this kind of rock, excepting those which belong to a narrow belt of sienite which comes in about six miles west of Ashboro. Fine and coarseish grits resembling certain varieties of novaculite are frequently met with in and about Ashboro. But at Franklinville, eight miles east, the green chloritic slates predominate with only a few beds of quartzite. The soft green slates are traversed by veins of milky quartz, but they are rarely auriferous. A few miles further east the slates are replaced by a thick bedded chloritic sandstone which bears a trappean appearance, and as it weathers readily the detached masses become round. These thick bedded

5

rocks are again succeeded by thin bedded, which contain pebbles of quartz. Such masses are repeated frequently in the distance of twenty miles. Deep green slates, rather thick bedded chloritic sandstones and beds of brecciated conglomerates belong to the same belt. Twenty miles west of Pittsboro or about twelve miles east of Franklinville a belt of this brecciated rock may be observed near by and crossing the road. At Pittsboro, red and purple slates with breccia prevail, particularly near the village. At Jones' falls, on the Deep river, the quartzite and brecciated quartzite and porphyries form a belt half of a mile wide and extend some five or six miles in a northeast direction. The porphyries are no doubt products formed by the adjacent trap of the neigborhood, as the Taconic rocks emerge here from beneath the sandstone of the Trassic or Permian system.

The belt of Taconic rocks which lie between Lexington on the West and Deep river at Jones' falls on the east, passes up through the east part of Guilford county, form most of Alamance and Orange counties, and so onward through Granville into Virginia. At Hillsboro, Orange county, a range of hills come down from Granville county. These hills are frequently composed of quartz and iron breccia; and in the northwest part of Granville county, an extensive range of slate conglomerate occurs, which has been traced to the Dan river in Virginia, and south some ten miles. This bed of conglomerates has a very steep dip from 60° to 65° and occupies a belt at at least one fourth of a mile wide. From the northwest corner of Granville county, the Taconic rocks extend nearly to Roxboro, when they are discontinued and replaced by a tabular granite.

This belt in North and South Carolina consists of quartz which is developed only upon a limited scale, but which is accompanied by the iron breccia as in Berkshire county, Mass., talcose, chloritic and argillaceous slates. The first is undistinguishable from those of Vermont and Massachusetts, which occupy the same geological position; slate conglomerates as in Granville county, and extending into Virginia; quartzite which is far more abundant among the slates than in Massachusetts.

Beds and ranges of argillaceous slates of various colors seem to be distributed without much order; or in other words it is impossible to trace them far in the direction of their strike. Another range of the Taconic slates passes through Johnson county. It is exposed near Smithfield. In range of this belt, a similar rock crops out near Gaston, and appears three or four times on the road to Weldon. The Johnson slate is separated from the slates and conglomerates of Orange and Granville, by a strong belt of primary rocks, among which granite is the most conspicuous member. It overlies, I believe, the hornblende and gneiss and mica slate rocks, and in this respect is similar to the Maine granites; it appears to have been ejected through fissures and to have overflowed the primary schists. They are exposed as underlying masses at Warrenton in Warren county.

§ 32. It can not have escaped the reader that the limestones which are so important and prominent in the series on the western side of the Blue ridge and Hoosick and Green mountains is probably wanting in the series I have just described. In North Carolina I have not discovered it, and I have made only imperfect examinations in South Carolina on the line in which this series is prolonged. In Virginia there is a range of blue limestone at Gordonsville, but it scarcely passes further south. It appears to belong to the series I have just noticed in North Carolina.

At Gordonsville there are soft talcose slates reddish white, and in connection with it, or nearby, there are beds of slaty blue sparry limestone, dipping southeast. Westward of Gordonsville the rock is argillaceous. In this connection the purple, red and green roofing slates come in. They are similar to the slates of Maine, which contain the nereites. They are the prolonged beds of roofing slate of Chatham county, N. C. With the latter, thick seamy beds of conglomerates occur. The roofing slates, west of Gordonsville, Albemarle county, are succeeded by thick, heavy beds of green chloritic slates, and alternate with them; they are seven miles east of Charlottesville. These beds, as we approach Charlottesville become more and more like the corresponding formation at Franklinville, Randolph county, N. C.

One mile east of Charlottesville, near the bridge, the slates and thick bedded chloritic masses have been altered by heat, having become hard and epidotic, accompanied with a few seams of asbestus So also in a cut for the rail road near by, I found thin seams of specular iron. The rock is also cut up by seams of milky quartz. When the rock is unaltered, it is soft, green and resembles the rocks of Randolph and Chatham county, N. C., which are associated with conglomerates. They also decompose and form a deep red soil. These Taconic rocks are interrupted at Charlottesville, at the university, by sienites which are protruded among the slates near their junction with the primary rocks which now come in and succeeds them on the west. I saw indications of conglomerates in the vicinity, but none in place; but the whole formation between Gordonsville and Charlottesville is almost identical with that of Randolph and Chatham counties, in N. C.

§ 32. It is difficult to describe in a satisfactory manner the unfinished series of the Taconic rocks in the southern states. They are widely spread out, but as they have disintegrated and the debris remains in place, it is impossible to make out the succession as it really is. It has therefore been described as a wide belt consisting of a few members only, and as extending northeastwardly and southwestwardly, and dipping generally to the northwest at a steep angle. That these rocks are all sediments seems to be established by the relations they hold to conglomerates and brecciated conglomerates. This series represents the lower Taconic rocks if lithological characters may be relied upon. No carboniferous or black slates appear in the series in South or North Carolina, or in Virginia.

§ 33. Another remark seems to be called for, viz: that the North Carolina series belongs to those isolated fields of Taconic rocks which occur in Rhode Island and Maine. They occupy the same relations to the principal mountain chain which traverse the country from the northeast to the southwest.

I have stated that this belt of rocks is prolonged from North Carolina into Virginia. An interesting locality within the state last named, for illustrating the changes which rocks have sometimes undergone through the agency of heat, may be witnessed at the rail road crossing leading from Charlottesville to Staunton. On the east side, and upon the flanks of the Blue ridge the chloritic slates seem to have become massive, accompanied also with the development of epidote in nests and geodes; and indeed masses seem to be semi-epidotic throughout. But I did not observe at this crossing, the primary schists, gneiss, mica and talcose slates, or well developed hornblende rocks. If I was disposed to assume that the rocks of the Blue ridge were only the altered rocks of the lower Silurian or Hudson river, for example, I would refer the reader to this locality for proof. But the character of the Blue ridge and that of the Green mountains of Vermont, can not be changed by such local phenomena; and it seems to me a waste of words to state the facts which refute an assumption so palpably erroneous as that respecting the rocks of the Green mountain range. I refer to the assumption that the Green mountains are composed of the altered masses, which are known as the Hudson river rocks. But I may present here another view which discredits this assumption, it is that which recognizes in the Taconic system the source from which its materials have been derived; such is the similarity of the coarse particles of the quartz, and its conglomerate, gneiss and mica slate, that we are not left in doubt a moment respecting their origin; we may trace the materials to the parent rock, and this parent rock is a constituent part and member of the Green mountain range, a preexisting series which have furnished the debris from which the Taconic system has been constructed. From this fact it follows, that in whatever light we may view the Taconic system, either as an independent system or as an altered series of the age of the Hudson river group, the Green mountain rocks themselves, can not be regarded as of the age of the latter. Neither is there any evidence, any monuments, or phenomena which will justify the assumption, that the gneiss, mica and talcose slates and hornblende of the Green and Hoosick mountains were ever sediments at all. Although I have not stated all the facts respecting the lower series of Taconic rocks in the foregoing pages, still the limits within which I propose to restrict myself will not allow of further details. I have shown in numerous sections that the sequence of the members of this part of the system differ materialy from the lower Silurian, and that the difference can not have arisen from a local accident or occurrence as we are accustomed to regard a few local changes, but must be due to general facts, such as prevail from Maine to Georgia; and as the lower Silurian lies along a belt nearly parallel with the Taconic system, and still the members of this system differ so much from each other, I feel justified in the opinion that these systems are unlike and differ in age, and can not be brought into correlation. It is not metamorphism which causes this difference; metamorphism has nothing to do with sequence. It is not by intercalated beds that the Taconic system exceeds in thickness the Silurian. It is not to metamorphism that we are to attribute the general absence of fossils, for a large proportion of the rocks are not metamorphic. Those who maintain then that the Taconic system is identical with the Silurian, are bound to show how these important differences are to be reconciled. They are not due to an inverted axis and if not due to one or the other of the causes I have named, it follows that the distinctions are essential and not accidental, and we are required to consider the systems themselves as different systems.*

---

* Since the foregoing was written I have observed that Mr. Hall assumes that the slates of Maine in which the so called nereites occur, are either Carboniferous or Devonian.* As in the case of the assumption respecting the rocks of the Green mountains of Vermont, so in this it seems a waste of time to add facts to disprove a mere assumption. In this case, however, it may be well to inquire what fossils of this locality have ever been found of this age, and as the doctrine of metamorphism can not be adduced to sustain the assumption,

* Silliman's Journal, p. 434, vol. xix.

§ 34. *Roofing Slates.*—Roofing slates, if I rightly interpret the phenomena of the districts where they occur, are found both beneath and above the limestone which is marked 2 in sections 1 and 2. They can not however be restricted to these positions. But the beds which I propose to describe now, occupy those places.

The beds below this limestone, however, scarcely differ from those above. Of this I have satisfied myself after an examination of the slates taken from certain beds in Columbia county, N. Y., about three miles south of East Canaan. Their characters scarcely differ from those of Hoosick in Rensselaer county, belonging to beds which I believe are above this limestone. The Columbia slates are blue, green, or greenish and purple. They split evenly and appear hard and sound, and sufficiently firm to resist the action of the weather.

The beds may be observed in the road between East Canaan and the state line. The quarries are not opened on this road, but just beneath the limestone, fine blue fissile slate occurs in thick beds. I regard these beds as older than those which have been opened at Hoosick. There can be no doubt of the relations which the Columbia slates bear to the limestone I have referred to, but I am unable to state the thickness of these beds in consequence of their concealment under drift.

From the view which I have taken of the relations of the lower Taconic rocks, it follows that the slates beneath the limestones have increased in thickness in their westward extension. These are also finer and much more even bedded, and contain but little quartz either in dissemination, in grains,

---

and as there are no fossils belonging to either of the systems, it must be erroneous. The most delicate imprints are preserved in the slates of Waterville, and here the coarse and prominent fossils of the Devonian or Carboniferous should be found, besides, there is not the least resemblance to the Devonian in mineralogical characters; but the rocks of this system of slates resemble the rocks which occupy the same position in Rensselaer and Washington counties, in New York, as well as the slates which extend through the southern states.

or bunches. Sulphuret of iron in fine particles is often present, and destroys the slate sooner or later by its decomposition.

§ 35. The beds of roofing slate and which are situated above the limestones, are the best known at the Hoosick quarries. They have been extensively opened at Hoosick Four Corners and at North Hoosick. The slates are blue, fine grained and even bedded. At North Hoosick, slates of a suitable thickness for roofing have been obtained five feet long and two feet wide. One surface of the slab was as perfect as could have been made with a plane.

The dip of the rock is east 20° south. The angle of dip 45°. There are local variations, both as to the direction and angle of dip.

The most interesting fact respecting the Hoosick slate is, that they contain fossils, and so far as discoveries have as yet been made, they are the lowest and oldest rocks in which fossils occur in this system.

A difference of opinion existed at one time, respecting the class to which they belonged; from their close resemblance in texture to fucoids in the flags and slates of Washington county, they were regarded as vegetables. I still entertain doubts upon the question, but as it seems to be generally conceded that they are animals, belonging to the family of graptolites, I am disposed to concur in that opinion.

As in the lower roofing slates so in these it is difficult to determine their thickness. It is evident that a dislocation exists at Hoosick falls, three miles northwest of the quarries. This dislocation pursues a route which leaves the quarries, about half a mile in a straight line, to the eastward. The first quarries which were opened, were near this line of dislocation at a place on the west side of the Hoosick at the bridge. Westward, the slate forms the base of the hill. Going east however from the old quarry at the bridge of which I have spoken, the series of beds is unbroken. A good section may be obtained by following up a small stream about half a mile southwest from Hoosick falls which runs down the slope of the hill, where the quarries are opened. This stream lays the rock bare at many places, disclosing a uniform dip and a regular succession of beds. The same mass of slate extends at least a mile in a direct line east from the quarries, passing over a wide ridge, at the base of which it is probable there is another dislocation. The next ridge east is still slate. But if the estimate of thickness is based upon that part of the series which forms the ridge spoken of, the entire mass is 5000 feet thick. The lower part of this mass is silicious, and coarse, and resembles the rock which overlies the limestone near the state line. It soon becomes softer, finer grained, in the ascending series of beds. But as a whole it is a roofing slate and I have not been able to discover those beds of sandstone which resemble those so common in the higher part of the series. The slates therefore form a mass which are properly worthy of a distinct notice; and, as they contain the oldest fossil known, they become still more interesting as a series of beds.

The graptolites are confined to a limited space. In this respect they agree with all fossils belonging to the oldest series of rocks.

There are probably two species, but as the impressions they have left are obscure, I have figured but one, selecting that whose characters are preserved the best.

§ 36. Roofing slates extend southwards through Pennsylvania, Virginia and North Carolina. Those of Albemarle county in Va., are fissile and generally blue, but purple and red beds occur. In Chatham county in N. C., upon Rocky river, light green slates are abundant which split in very thin lamina on being heated. Six miles north of the Gulf, Chatham county, fine purple slates have been discovered, but as there is a covering of soil it is difficult to determine whether they are of any value commercially. Geologically they show the persistence of the rock. It should be stated that those of Albemarle county, Va., and Chatham, N. C., are upon the east side of the Blue ridge. They probably belong to the oldest beds, or those which lie beneath the limestone. In North Carolina, I have been unable to determine their relations to the limestone, as it does not exist on the east side of the Blue ridge. They occupy a position which proves their sedimentary origin, inasmuch as they are not removed very far from beds of conglomerate.

ABSENCE OF CARBON AND ABSENCE OF FOSSILS IN THE LOWER TACONIC ROCKS, ETC.

§ 37. The absence of fossils in the lower Taconic rocks excepting in the Hoosick slates, has been attributed to the changes which they have undergone by the agency of heat. This doctrine, however, should not be adopted without an examination of its claims to our belief.

If the relations of limestone of Berkshire are carefully examined it will be found that they were deposited upon sediments beneath some 1000 feet thick. The question then comes up, is it probable that sufficient caloric could have penetrated through this thick mass of different rocks varying considerably in their conducting power to have given the particles of carbonate of lime that mobility so necessary to their crystallization—for it must be conceded that it is only through this thick series of rocks that heat could have affected the limestone, since porphyries, trap, or pyroplastic rocks are wanting through the Taconic range. Not only are the traps rare in this range but dykes of injected rocks are rare in a large part of the Hoosick mountain range.

If we turn our attention to the slates, they appear to be formed of materials scarcely differing in the condition they were before consolidation had taken place. If the breccias and conglomerates are examined their fragments are still like those of the parent rock.

When the color and grain of the Berkshire marble is compared with that of the Trenton limestone in the southwestern part of Virginia, the color as white and the crystallization is nearly as perfect in the latter as the former, and yet the fossils are preserved in the latter with all the distinctness and perfection that they are at Trenton falls in N. Y. Crystallization does not obliterate fossils.

I infer from the foregoing facts that there is little probability that animals and plants existed at the time the Berkshire limestone was deposited.

At this period there seems to be an absence of carbon. The dark color of the limestones of Berkshire is due to the slates or else to sulphuret of iron, and besides the uncrystallized gray limestones of southwestern Virginia and Tennessee are as destitute of fossils as the Berkshire marbles.

Again I remark that it follows, that the absence of fossils is due to the period and not the condition of the rocks; and hence too, it follows on palæontological grounds the Taconic system is distinct from the Silurian. It may be laid down as a principle in geology that in all countries the formations are to be worked out and determined on evidences furnished by each country respectively. American geologists will not try their formations exclusively by those of England or Russia.

ORIGINAL POSITION AND VARIABLE THICKNESS AND EXTENT OF THE MEMBERS COMPOSING THE TACONIC SYSTEM. LONG PERIOD OF REPOSE DURING WHICH THE MEMBERS WERE DEPOSITED, ETC.

§ 38. That the present condition and relations of the members of the Taconic system may be understood and made intelligible to the student, I have presented in a diagram the original position of the members of the series as they were probably deposited. It is designed to illustrate also the facts respecting the variable thickness of the members of the series at different places and their discontinuance in consequence of thinning out, and the limited area over which some of them were expanded, and the probable fact that the plane of deposit was subject to oscillations, or an upward movement in consequence of which the superior rocks would not be extended far enough towards the base of the system as to cover or overlie the oldest members of the series, their outer margins would be confined within the outer rim of the basin.

Of all the members of the series, it is probable the conglomerates and quartz rocks are the most limited, or confined to the smallest, and the slates and limestones are the most expanded or spread out over the largest areas.

In the Taconic system we have the usual evidence of a long period of repose during which the rocks were formed, for although these may have been feeble upward movements as stated in the foregoing paragraph, still the evidence seems to be that disturbances of the soil were scarcely greater than during the period when the Silurian rocks were deposited.

Towards the close however of the Taconic period, we have abundant evidence of a change in the condition of the surface, for we find brecciated conglomerates among the upper members of the series, and besides at many points where the junction of the lower Silurian and Taconic rocks are exposed, we may observe that the beds of the latter are inclined, or had undergone a change of position prior to the deposition of the former. When they received their present steep southeasterly dips is not easily determined, but that it was just prior to and during the lower Silurian period is probable from the fact that the outliers of this system were not expanded over the lower beds of the Taconic system in consequence as it would seem of the elevation which the surface had undergone. In confirmation of these views I shall show that this period was one which was marked by numerous fractures and the intrusions of trap and metallic veins.

But leaving for the present the consideration respecting the period when the rocks of the Taconic system received their present dip and position, I may observe that as a general fact they dip towards the southeast at an angle varying from 20° to 45°, and even frequently they are much steeper. This inclination furnishes so few exceptions that it may be regarded as a general fact. In consequence of this fact we are obliged to offer some explanation which shall account satisfactorily for the apparent position by which the newer rocks appear to be older than those which lie at the base of the system.

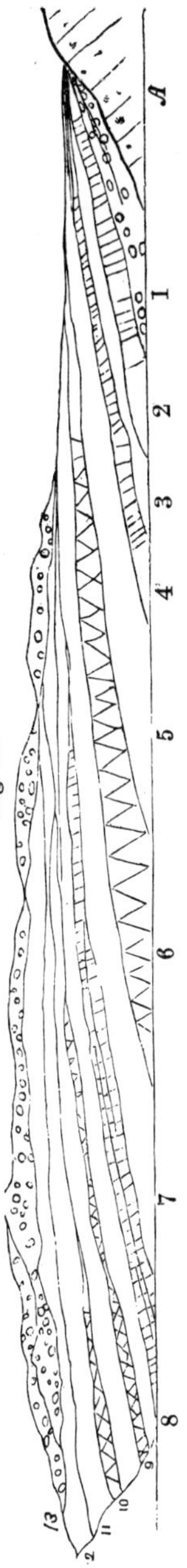

Fig. 8.

This explanation it appears to me is found in the numerous dislocations which are known to exist in the belt of country occupied by these rocks, and whose geographical position in New York, Massachusetts and Vermont, is upon the western slope of the Green mountains and which is bordered on the west by the Hudson river and Lake Champlain.* If we take our departure from the Hudson river and travel east or rather east 15° to 20° south we pass over the series at right angles to the direction of their dip, and if we take no notice of the dislocations which intervene between the river and the mountain range, our inference would be that we were passing from the older to the newer.

If we take notice however of the dislocations we shall find the series broken up into segments, each successive segment embracing a series older and below the one immediately west on the line of our section. In some instances it is true there is a repetition of rocks, and probably each segment always embraces members or parts of them which appear there in consequence of a repetition in two or more adjacent segments.

On reference first to figure 8, which represents the original position of the rocks of the Taconic system, we can readily conceive that a series of dislocations occurring at 2, 5, 7, and 8, would bring up all the members of the system. The dislocations through 6, 7, 8, would place the members of two segments in a position which would cause the newest member 13, of segment 8, to dip beneath segments 6, and 7, embracing the limestone and roofing slates already described.

---

* I am obliged to insert in this note, that I do not assume that the Hudson river or Lake Champlain are necessarily boundaries

This fact, however, may be better understood by reference to figure 1, where the present relations of these members 3, 2 and 4 are given. If this view is admitted, it exhibits a true succession when the series is taken in segments. Other views have been offered in explanation of the phenomena, and as I can not perceive in the geological statics of the region reasons for adopting a different interpretation of facts, I prefer the one I have given, as it appears the most simple. It is important to observe in this connection that the force which breaks the continuity of the strata, exerts its maximum power nearest the mountain chain; and hence, it may be regarded as proceeding outward from it; and hence, too, the frequeney of fractures the nearer the rocks are to the mountain chain. This statement appears to be sustained by the Williamstown section, figure 1, where, in the space of two miles, no less than 5 distinct and well defined fractures or dislocations may be observed.

So in section 2, at the eastern base of Saddle mountain, and at its western base, this high range is interposed between two valleys; and about one mile from the dislocation, Stone hill breaks in upon the continuity of the valley, with a fracture on each side, which runs also along the base of the Taconic range which lies between Williamstown and Petersburgh. There is another class of phenomena which obscures the true explanation in cases similar to the one I have had under examination. There can be no doubt that hundreds of feet of solid rock have been removed by denudation, and certain rocks at exposed points have been entirely removed. Thus from Oak hill, in Williamstown, Berkshire county, the series is wanting in two or three members whose aggregate thickness is at least 2,000 feet; the quartz near the summit retains in perfect freshness diluvial grooves. It will be recollected that quartz which corresponds to this upper mass of Oak hill, crops out at the foot of Saddle moun-

---

to the Taconic system. But that it disappears nearly on this line, is a fact. It is here that it meets with the thick and heavy beds of the lower Silurian, beneath which it is concealed.

tain, and heavy beds of slate and limestone overlie it. There can be no doubt, that those overlying slates and limestone overlaid also the quartz of Oak hill. If so, these masses have been removed. Members of the series, therefore, may be wanting at many places, and hence the true succession may not be made out.

§ 39 An inspection of figure (last referred to,) shows that the entire thickness of a series can not be made out at any given place, as some of the lower masses thin out before the upper are deposited above them. Hence we are unable to determine either the true succession or the upper members of the series at or near the base of the system. This is the case with the Taconic system, the members do not extend eastward to Hoosick mountain, and probably never extended so far eastward, and probably too the lower members which lie against Hoosick, or the Green mountain, do not extend as far as the valley of the Hudson.

The greatest obscurity in the succession of the rocks arises from dislocation and diluvial action combined, and this obscurity is the greatest at the junction of two rocks, whose lithological characters are alike. Thus the Taconic and Hudson river slates, meeting as they do, in the valley of the Hudson, perplexes almost all observers. The difficulty is increased in consequence of the mechanical effects which the dislocation has produced upon the adjacent rocks. All vestiges of fossils in the soft slates are destroyed by the crushing which the rocks have undergone, and hence it puts it out of our power to make out the line of demarkation at certain points between the two systems. The Taconic rocks in consequence of this disturbance are elevated higher along the margin of the Hudson, than the slates of the Hudson river group. Many geologists, in consequence of this fact, maintain that the former are the equivalents of this part of the Silurian system. I shall give most decisive evidence that this view is erroneous. I have offered the foregoing explanation respecting the phenomena, which are calculated to obscure the relations which subsist between the members of the Taconic system. It seemed proper to do this before I described the upper members of the system.

SUMMARY OF FACTS RESPECTING THE LOWER TACONIC ROCKS.

§ 40. 1. I have shown that the lower Taconic rocks, though once regarded as members of the primary series, are truly sediments, the beds of conglomerates giving and furnishing ample proof of the doctrine.

2. That the sediments are derived from the pyrocrystalline rocks, at least in part, from the debris of those which compose the Green mountains and the Blue ridge.

3. That they extend from Maine to Georgia, flanking continuously the ranges I have spoken of on the west side, or else forming distinct ranges by themselves. On the east side, the development of the members is ample, yet not so perfect as upon the west side, and moreover differ from them in certain lithological characters which show certain differences in the circumstances which attended their deposition. Slates predominate on the east side, and masses of chert are extremely common.

4. The relations and composition of the members prove that they are distinct from the Green mountain rocks on the one hand, and from the lower Silurian on the other. The quartz rock of Berkshire, of Buncombe and of Cherokee county, is not the Potsdam sandstone, nor are the marbles and limestones of Berkshire the lower limestones of the Silurian system, neither are the slates, either talcose, magnesian or roofing slates, shales and sandstones the equivalents, or corelatives of any rocks which are known to belong to that system. It has been shown that there is no similarity in kind, there is no similarity in sequence, and there is none in fossils to the lower Silurian system. None of the differences alluded to, can be explained by metamorphism, or inverted axes, or an inverted axis passing into a fault.

The result of all these showings is, that the Taconic system in its lower members, is independent of, and distinct from the Silurian.

5. It will be perceived that the foregoing conclusions are founded mainly on stratigraphical evidence. This evidence establishes a physical group which can not be brought in co-ordination with the lower Silurian group to which it has been compared. This group is mostly anterior to the organic period.

THE UPPER DIVISION OF THE TACONIC ROCKS. GENERAL CHARACTERISTICS OF THIS DIVISION. GEOGRAPHICAL POSITION. SUCCESSION OF THE MEMBERS. THEY FORM A SINGLE GROUP, ETC.

§ 40. The direct succession of the lower division having been determined, and their members having also been proved to belong to, and form by themselves a distinct physical group, I shall now proceed to state the general characteristics of the upper division. I have already shown that the lower rocks are represented by slates, and that they predominate so much over the quartz and limestones, that it might be regarded as a slate system. In the upper division, slates also appear in great force and are found in all the subordinate groups; yet, it can not be regarded as the predominant rock. The pure argillaceous slates, for example, never form thick beds without soon becoming silicious or calcareous, and perhaps both. Silicious slates and silicious bands and beds are every where interlaminated with those which are argillaceous; and besides the latter, the magnesian series seem to be wholly wanting, and are not represented at all.

The series begin on the west, where we find in the lowest beds black slates and black calcareous beds, as at Easton, Washington county, and St. Albans, Vt. The black slate is succeeded by silicious slates, interlaminated by thin beds of fine grained sandstone.

Ascending still higher, slates and sandstones continue in which also we find thin bedded blue limestones without fossils, succeeded by thicker beds of sandstone. Among the sandstones, those which contain carbonate of lime are quite prominent, and may be found in at least three beds; the thickest of which is over one hundred feet; gray sandstones, interlaminated with slate; thin bedded flags, with fucoids; cherty beds, green slates, one or more beds of vitreous quartz, coarse green slates, shales and grits; roofing slates, blue, green, purple and red; coarse sandstone and coarse shale passing into sandstone and conglomerates, and brecciated conglomerates. The latter terminate the series eastward, and geographically near the Hoosick roofing slates. In the foregoing brief enumeration in the ascending order, the rocks follow each other in a conformable position, and beginning with thin black slates, end in thick bedded sandstone and conglomerates. We find no where the marbles of the Taconic range or their representatives; the quartz rock, or its representatives; the magnesian or talcose rocks, or their representatives; and moreover, the Columbia and Rensselaer roofing slates are absent. Hence it follows that the groups must be lithologically, and when we take into the account the order of superposition, they must also be, in all respects, a different physical group. But if the dip and conformability governs the succession, we are obliged to regard the series as enormously thick, which with those who have not made the necessary examinations to form an opinion would be regarded as an objectionable feature. But the dip and conformability, however, should govern the succession, provided proper vigilance is used to avoid repetitions of the masses in the different segments of a section.

There are two modes of guarding against error of this kind. The first is, to become so well acquainted with the members of the series, that they can be recognized. The second is, to examine not only the cross section on the line of traverse, but to examine every mass which crops out in the direction of the strike where it may be concealed on the line of traverse, so that when the rocks are concealed in a cross section, they should be found and determined on the lines of strike: in other words the succession should be made out by an inspection of all

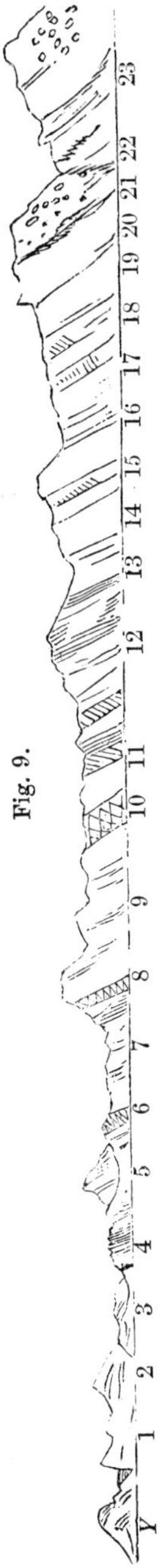

Fig. 9.

A 1 2 3 4 5 6 7 8 9 10 11 12 13 14 15 16 17 18 19 20 21 22 23

its members, some where near the line selected for examination. I can only say, that I have availed myself of all the means in my power to insure accuracy, and after examining many sectional lines, I am obliged to accept of the conclusion, that the series are enormously thick, and that it becomes necessary to regard the western rocks, as the oldest, or at the bottom, and, in consequence of this decision, the succession begins with the soft black slates and terminates eastward in the conglomerates and brecciated conglomerates in the neighborhood of the Taconic range.

§ 41. I propose to describe in the next place, in detail, the series as they are known in Rensselaer and Washington counties, N. Y., inasmuch as here they may be regarded as representatives of the division, and as types for comparison when it becomes necessary to speak of the same series in their prolongation to the southwest.

I have selected for my first section, the country lying between Comstock's landing, Washington county, and Middle Granville, a distance of ten miles in an easterly direction. This section is important as it passes over in the first place the primary and lower Silurian rocks. It is instructive, because it exhibits the relations between the latter and the Taconic system, and it gives us a very clear succession of the upper rocks of the system. The sequence of the series is given in tbe annexed woodcut, No. 9.

A, gneiss, Comstock's landing; 1, potsdam sandstone with the overlying calciferous sandstone; 2, calciferous sandstone; 3, chazy limestone; 4, slates interlaminated with fine grits of the Taconic system; 5, overlaying mass of chazy limestone; 6, thin bedded sandstone embracing four or five thick beds of pallure; 7, uneven bedded slates or shales; 8, thin bedded sparry limestone; 9, bluish slaty grits; 10, calcareous coarse sandstone; 11, gray sandstone; 12, flags containing fucoids alternating with thin bedded slates; 13, beds of cherty sandstone; 14, blue slates; 15, thick bedded sparry lime; 16 and 17, roofing slates, containing a bed of sparry limestone forty inches thick. These slates are blue, purplish and red; 18, hard thick and thin bedded coarse grits; 19, slates with beds of grits; 20, red slate; 21, slates and coarse grits alternating; 22, sandstone passing into breccia and embracing thin bedded slates; A, dislocation Pond mountain, and 23, thick bedded sandstone, breccia interlaminated slates. These upper beds are found at Bird mountain, Grafton and Alps, Rensselaer county, N. Y.

The series just enumerated and briefly described, begins at a point near North Granville which is now uncovered by removal of the lower Silurian limestones. The latter extend eastward on the sectional line about three miles and a half from Comstock's landing though not continuously. The Taconic rocks are exposed in a gorge through which the Mettowee flows, and in which the calciferous is exposed in one place abutting against a ledge of hard Taconic shales. Farther east, perhaps a quarter of a mile, the chazy limestones containing Maclurea magna, may be observed resting upon the slates, both of these rocks, the calciferous and chazy limestone, extend beyond the potsdam sandstone.

The fourth rock in the ascending order consists of a series of black slates and fine grits. The dip is east, varying from east to east 10° south. Angle of dip from 30° to 57°. A part of the series which succeed the foregoing, and which extend to the thin bedded chazy limestone, consist of thin bedded sandstone alternating with green thin bedded slates, embracing a few thick bedded sandstones. I am unable to state their thickness accurately, but have estimated it at 1500 feet. The thin bedded blue limestone which is destitute of fossils, is only 12 feet thick. 10, The calcareous sandstone, this is separated from the limestone by about 100 feet of bluish slaty grits No. 9. The calcareous sandstone is a coarse grained rock containing from fifteen to thirty per cent of carbonate of lime, and six per cent of oxide of iron. The carbonate of lime is usually dissolved out of that part of the rock near the surface, and hence it becomes brown and porous and frequently a light spongy mass of a drab color, sometimes it is a coarse brown friable grit of slightly coherent grains of quartz.

The different beds of this rock traverse the whole county. Their thickness is 12, 15 and even 100 feet. They are situated near the beds of limestone, being separated by slate and slaty sandstone, but I have never observed that the calcareous sandstone ever passes into the sparry limestone. The rock which succeeds the calcareous sandstone is a gray or greenish gray sandstone. It is finer than the former and does not weather to a drab, or disintegrate so much. 12, Flags containing an abundance of fucoids. These masses are really thick bedded slates splitting into layers from one and a half to three inches thick These alternate with thin bedded slate. Above the latter there comes in a succession of cherty beds, vitrified masses, mostly thin bedded and of green or greenish color. The rock breaks with a conchoidal fracture, and much of it is a perfect hornstone, and is about 100 feet thick. 14, Blue slates which contain a bed of quartz 30 feet thick, above the quartz rock the slate becomes coarse and is really a gray or greenish gray sandstone. 15, Sparry limestone. 16 and 17, Roofing slates; the lower beds are uneven bedded, but pass into even bedded roofing slates which are divided by a singular bed of sparry limestone, forty inches thick. The slates are very firm and free from sulphuret of iron. They are blue. purple and red, and will receive a high polish.

In addition to the common material for roofing, these beds furnish fine even beds of flagging, two and two and a half inches thick, which appear perfectly solid, and in which it is impossible to discover the slaty cleavage which in these beds coincide with the planes of bedding.

These beds occupy, it will be perceived, a position considerably higher than the Hoosick roofing slate, and it will be seen also that their relations differ materially from them. They contain branching fucoids, but no graptolites have been observed among them. Over the slates I find a gray coarse grit, or rough sandstone somewhat similar in appearance to the calcareous sandstone or grit already described. The beds are thick and uneven, and do not readily split. Between these sandstones and the bright red slates, there intervenes a thick mass of slates containing harsh greenish grits. The red slate is the most remarkable rock of the series, being perfectly well defined and extending through the whole country; they cross the road at Granville corners and pass Middle Granville about a half a mile to the east. The beds of slate is 150 yards across, and have an angle of dip about 45°. They are titanium red, and stand the weather better than any variety of slate in market. Grits and slates into which beds of sandstone are introduced succeed the red slate, and as the coarseness increases, the sandstones become predominant, and finally, we find conglomerates and brecciated conglomerates. They are exceeding hard and tough, and are only slightly subject to disintegration. They are usually greenish and often look like porphyry. Chlorite forms a part of the paste in which the pebbles are inclosed. This series terminates with the brecciated conglomerates. This line of section extends twelve miles. The average dip for the whole distance is about 40°, and as the succession is not repeated, or if so it is not recognized, the thickness is not less than 25,000 feet.

In comparing the series with the Hudson river, we can not fail to observe many striking differences. The greatest thickness of the Hudson river slates and sandstone, in Albany county, is 700 feet, and although Prof. H. D. Rogers gives 6,000 feet as the maximum in Pennsylvania, still in Virginia it is not over 700 feet on the western slope of Walker's mountain. If we begin an examination at a high ridge, three-fourths of a mile southwest from North Granville, and then extend our course east to the Pond mountains, we keep in sight the several rocks of the section, or we need not lose sight of them. At the western base of the Pond mountain, four miles east of Middle Granville, there is probably a dislocation. What I mean to assert is, that the evidence renders it probable that the succession is without a dislocation, or repetition. We find continually new members of the series in the ascending order. If this is true, and even admitting faults and dislocations, the series forms a physical group unlike either the lower Taconic rocks, or the Hudson river. It would be a perfect geological anomaly if we succeeded in making out both divisions of the Taconic system equivalents of the Hudson river group. The upper and lower divisions, however, are both claimed, but the sections for their illustration show that neither can be regarded as lower Silurian; for neither one nor the other can be brought as physical groups into coordination. The assumption becomes ridiculous when it assumes that both, which are so unlike each other in their general and special characteristics, are metamorphic Hudson river slates and sandstones. As the difference existing between the upper and lower Taconic rocks has been attributed to a higher degree of metamorphism only, it is fitting that this view should be met with a few remarks in this place. Metamorphism occurring in the sedimentary rocks is usually produced by the presence of pyrocrystalline, or pyroplastic rocks intruded among them. But from North or Middle Granville to the highlands of the Hudson I know of no trap, porphyry or granite. There is therefore a cause wanting to produce the condition claimed; a condition which has really no existence* unless it is found in the marbles of Berkshire, Rutland, etc. It is scarcely necessary, however, to dwell upon this assumption; for even if the fact respecting metamorphism was true to the whole extent claimed, it could not become a material argument in proving the Taconic rocks to belong to the lower Silurian

---

NOTE.—On opening quarries for slate, flags, etc., it sometimes happens that very limited dislocations occur, involving a derangement of some eight or ten feet, but by dislocations and inversions such as involve the repetitions of an entire series of beds, many hundred or perhaps a thousand feet thick, I am confident do not occur upon any line section which I shall give.

* In the case of marble the assumption is uncalled for, inasmuch as the limestones which belong to the primary series are white and crystalline; the debris from a limestone in this condition, would, when consolidated again, be as white and crystalline as the parent rock.

period. Metamorphism only changes the physical condition of rocks, and the causes which are intimately connected with it, may so far change the conditions of masses locally, as to obscure their characteristics, but the succession remains essentially the same, and in this respect it is impossible that metamorphism could have been an agent in effecting the differences which exist.

§ 42. *Topography of the western part of the belt through which the section runs.*—Before I attempt to describe the parallel series south of Granville, it will be instructive to bring before the student certain facts respecting the topography and geological structure of the belt which skirts the valley of the Hudson. If we ascend the highest points of land in the western margin of this valley, we shall see northward a series of knobs of a moderate height rising along this margin as far as the eye can see, and if we go up the valley we shall pass these knobs, which we shall find occurring every two or three miles and which are separated from each other by oblique valleys, which connect the bottom of the Hudson valley with the table land immediately east. This range of knobs is an important landmark, as it lies along the great Hudson river fault, and indicates its eastern side. This line of fracture and dislocation runs N. 37° E. East of this range of isolated hills, which rarely exceed 500 feet in height, the country rises in ranges of hills and low mountains, which run parallel with the line of fault; and where the ridges approach the Green mountains we find that they, too, are situated upon lines of fractures parallel with the first.

But to return to the consideration of the first line of hills or knobs. I wish to state, in the first place, that they are usually capped with the lower Silurian rocks similar to that of Cantonment hill; thus Bald Mountain, Mount Tobey and Willard's mountain, with many others, are lower Silurian at their summits. Now, when we look beneath the lower Silurian beds we invariably find them reposing upon the Taconic system. It is true we are not always able to see the line of junction; but one observation which gives us the result I have stated is enough to settle the fact as it regards all of them. But it is true that in almost every instance the junction can be seen, or else the relations are so clearly indicated by phenomena, in each respective case, that we are scarcely left in doubt respecting the supposition I have just affirmed. I do not wish to follow up this point any farther at this time, and I have only to state that my lines of section are selected with reference to these knobs. The one already described and illustrated with a wood cut passes through a hill three-fourths of a mile south-west from North Granville across this line of fracture where the thin bedded blue limestone occurs. The Silurian system does not reach this point, but remains in the valley west, occupying here the same relation to the Taconic system that it does in the knobs I have spoken of.

The next sectional line I shall describe runs through the trilobite locality four or five miles north of Bald mountain. The slate in which I found the trilobites is in the road near the house of Mr. Reynolds. The rock is a dark green slate, the surfaces of which are glazed, or often appear as if they had been covered with a black varnish. When split through planes which are only imperfectly fissile, the fresh surfaces are ferruginous. It is here that the Eliptocephalus and Atops were found, both of which have proved to be not only new species, but to belong to new genera, unless indeed the first is a Paradoxides with which it has certain characters in common.

The slates dip east 10° or 15° south at a steep angle; westward the surface rises into a ridge, and the rock becomes more slaty and takes in calcareous beds. Upon the top of the ridge one-fourth of a mile west, the calciferous sandstone occupies the ridge, and at its western base are the slates conformable to the trilobite beds. Eastward from the trilobite beds, the same rock continues 100 or 150 feet, when it becomes a hard quartz rock, some ten or fifteen feet thick, which passes into a black shining slate, which in its turn changes eastward into coarse grits. A meadow some eighty rods wide intervenes between the road and a ridge, three-fourths of a mile east; when

8

upon its western slope we found fine green slate cropping out at its base, which in ascending becomes coarse and ferruginous, and layers of a quartzose sandstone and thin bedded grits and slates. At the top again, or just over the eastern slope thin bedded quartz filled with particles of oxide of iron forms a mass twenty or twenty-five feet thick.

This locality with its grits, slates, quartz, &c., has been regarded as the equivalent also of the Hudson river group, the presence of trilobites is taken in proof of the assumption, both of which, however, have been shown to be new, and are not known to occur in beds with the other characteristic fossils of this group. Besides, between the trilobite bed, in the calcareous slates, which would be the equivalent of the Trenton limestone, fossils have not been found at all. So constant are Trenton fossils in Trenton slates and limestones, that there is no known exception in New York, Pennsylvania, Virginia and Tennessee, of their absence in the beds of this period. Besides, the beds above the trilobite beds, though there are very good exposures of the masses for half a mile, do not contain fossiliferous bands as in the true geological Hudson river group. It would be regarded as a very singular anomaly, if these rocks should turn out to be the Hudson river group and that all their fossils should disappear the moment they were raised to the surface on the east side of the Hudson valley, when at Snake mountain nearly opposite and a few miles only distant, the Hudson river group is finely developed and its members highly charged with fossils. Again, the group is physically different and its characteristics stand out strongly in contrast.

Dr. Fitch's Section through Washington county.—Dr. Fitch of Salem has published* a section through Bald mountain, in which we agree in the sequence of the rocks, excepting those at Bald mountain, which represents the limestones as dipping beneath the black slates, etc., instead of which, it is perfectly plain, that the limestones rest upon the slates. Beginning

* Transactions of the N. Y. Agricultural Society, page 820.

above the limestones, his section reads: 1, black slates with trilobites, answering to my section at Reynolds; 2, slates in the ascending order, argillaceous and silicious slates; 3, thin bedded limestone, equivalent to the first limestone in my section; 4, sandstones, answering to the first sandstones in my section; 5, silicious slate; 6, sandstone; 7, limestone; 8. green flinty slate; 9, slates with fucoids; 10, red slate; 11, sandstone, 100 feet thick, answering to my calcareous sandstone; 12, limestone; 13, slaty beds of sandstone, containing fucoids; 14, sandstones; 15, sandstones, these are the thick bedded sandstones of my section at Granville; 16, fucoids; 17, limestone; 18, chocolate colored slates; 19, glazed, and 20, black slates; 21, soft, green and variously colored crushed slates, which no doubt are the equivalents of the Columbia county roofing slates; 23, sparry limestones; 23, soft talcose chloritic slates The section terminates in the quartz rock dipping west, and conformably to the gneiss of the Green mountains, a representation, which if true, is certainly uncommon. Slate in the upper division, it will be seen, occupies more space than the shale or sandstones, and their condition is rarely changed sufficiently to obliterate the organisms if they were ever the receptacles of any. It is in slate that the many thin beds of limestone, calcareous sandstones, gray sandstones and brecciated conglomerates occur It is a system really protean from the numerous kinds of rocks which it contains. When the rocks are opened in quarries, we find them more protean than the unbroken surface indicates; thus in a slate quarry at Salem, four or five thin beds of limestone occur which were not seen before it was opened. The soil covers large areas and conceals much that would be important in assisting us to make out those details which are required for a full knowledge of its inorganic composition. Sections corresponding to those of Washington county might be described, also, in Rensselaer county. A brief notice of one passing from Bath to Alps, fifteen miles east, is all that I deem necessary to give at present. At and from Bath, three miles east, the rock is mostly concealed by Albany clay. A ridge of rock near Blomingdale crops out with a steep west dip and continues more than one-fourth of a mile eastward. These steep dipping rocks belong to the Hudson river group. Just east of Blooming Grove the Taconic slates crop out with a southeast dip, at an angle of about 35°. The slates contain a few hard silicious layers. The masses are so well exposed either in the road or adjacent fields that it is possible to obtain an unbroken succession of beds to East Sandlake, a distance of seven miles. The rocks are mainly slates, flags and thin bedded limestone without fossils. At Sandlake, the rock has become a thin bedded sandstone alternating with beds of greenish slates. These continue with a few variations to Alps, where we begin to obtain overlying conglomerates and breccias, more or less chloritic, which continue up the next rise of hill east from Alps. From these brecciated beds we pass to the lower Taconic series in Stephentown, consisting of talcose slates with limestones and marbles.

At West Sandlake there is a dislocation, but west from Blooming Grove to this place, a distance of four miles the sequence is unbroken. That a very minute examination might detect dislocations upon a small scale is highly probable. The average dip is 40°. We have therefore on this section the same facts which sustain the view already expressed, that this series is enormously thick. At one point on this section, there is a change of dip for thirty or forty feet, but the rocks resume their regular dip so speedily that it seems to be only a local variation which has no influence on general results.

A section in Prof. Mather's report of the geology of the first district of the New York survey and extending from Poughkeepsie east to Sharon, Conn., gives the same or nearly the same details. This section crosses the Taconic system, as we find in the series of rocks indicated upon the section, green, red and black slates, silicious slates, thin bedded limestones interlaminated with thin and chloritic slates, etc. The section is not designed to give a detailed illustration of the rocks over which it passes; but it contains enough to show that the protean group of slates, shales, sandstones and limestones, belong to the same series as those I have described in Washington and Rensselaer counties.

This series crosses the Hudson river between Poughkeepsie and New York and passes through Orange into New Jersey and onward to Pennsylvania and Virginia.

The next section which I shall describe, runs from Abingdon, Va., to Taylorsville in Tennessee. The rocks at Abingdon, Va., are limestones, alternating with beds of green, red and chocolate colored slates and gray and red sandstones. On the road leading to Taylorsville, the principal rocks are slates and limestones which are frequently exposed, but which generally are quite irregular in dip, being much bent or arched, so that frequent repetitions occur. I notice this route because I found a portion of the upper series much better exposed than in other sections, where the series are even much less disturbed.

At the ford on the middle fork of the Holstein, the limestone on the west side dips southeast with an angle of 80°. This is succeeded by heavy beds of green and black slates; but the beds are frequently concealed.

Thirteen miles from Abingdon, having passed over repeated repetitions of slate and slaty sandstone and limestone, there is a change in the rocks. The first intimation of a change is the occurrence of a low flat arch of reddish thick bedded sandstone alternating with slate; it overlies the slates and limestones which have been referred to. The sandstones are both reddish and gray, and continue exposed on the road, and constantly increasing in thickness for three or four miles, where the thick bedded sandstone is succeeded by conglomerates, breccia and interlaminated slates. The breccia is often feldspathic and resembles porphyry, but it is found to be on a strict examination a sediment, made up of or derived from a flesh colored granite. Both the sandstones, and brecciated conglomerates are chloritic, and hence this deep green trappean color. On comparing this formation with that of Grafton mountain, Rensselaer county, N. Y., I find there is a perfect similarity. It is impossible to distinguish specimens apart. On the Laurel a branch of the middle fork of the Holstein, 16 miles from Abingdon, this series attains probably its greatest thickness. Here the conglomerates and sandstone are at least 500 feet thick in a single cliff.

This section is important as it confirms the opinion already expressed relative to the succession of the rock of the upper division. Here the thick bedded sandstones and breccias are seen in position overlying beds of the thinner bedded slates and limestones.

In approaching the Iron mountain, as it is locally called, from the occurrence of brown hematite as in Berkshire, the thick bedded sandstones and conglomerates thin out, and we pass over a succession of blue and red slates with gray slaty limestone belonging to the lower division. These continue to Taylorsville. The lowest of the rocks of this division are the talcose slates and quartz, which are passed over on the route to Jeffersonville in Ash county, N. C. Stone mountain is an enormous development of these rocks and has already been referred to.

§ 43. *Great Limestone Valley of Virginia and Tennessee.*— The limestone valley of Virginia is celebrated the world over for its fertility and its wealth. The flowing outline of its hills, with their green slopes, forms a beautiful picture for the painter. It is no less attractive to the geologist, for here are some of the finest developments of the palæozoic formations which the country affords. The limestones belong in part to the lower Silurian and in part to the Taconic system. While the former maintain their usual developments, and resemble in almost all respects the same rocks in New York and Pennsylvania, the limestone belonging to the Taconic system seems to be much greater in extent and thickness than in the northern states. To illustrate in part this view I made a section in Wythe county, near the residence of Colonel Rapier, a gentleman well known in Virginia. The limestones and interlaminated slates and sandstones are so well exposed by the road and adjoining fields, that the true sequence is easily determined. Fig. 7 illustrates the secession or sequence of the rocks at this place. Their dip is nearly southeast, and usually it amounts to 40°. The section extends one and a half miles, nearly at right angles to the dip. The enumeration is in the descending order:

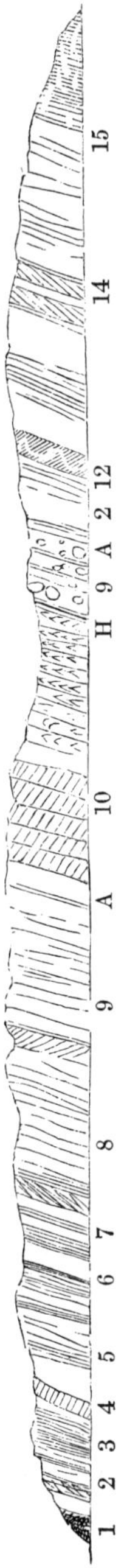

1, hard sandstone thirteen feet; 2, fine grained sand stone sixty-five feet; 3, slates 300 feet containing beds of limestone six feet thick; 4, blue limestone thirty-six feet; 5, purple sandstone 464 feet, and containing purple slates twelve feet; 6, decomposing drab colored slate 250 feet; 8, cherty limestone; 9, green and drab colored slates; A A includes the blue limestone; 10, concretionary limestone associated with brown hematite; the thickness of these limestones is 1400 feet; 12, red and purple slates 200 feet; 13, red and green slate 400 feet, containing a few beds of limestone: 14, chocolate colored slates 600 feet, containing a few beds of slaty sandstone; 15, coarse slaty sandstone.

It will be observed that this mass of limestone is bounded on each side by beds of slate and sandstone, with but few beds of limestone. This section I believe corresponds, at least in part, to the upper Taconic rocks of Washington county, New York. If so, the limestones which are known in the slates at the north must have become much thicker in their extension south. I am, however, unable to account satisfactorily for the development of limestone in Virginia and Tennessee and throughout this great valley. A question will probably arise, whether these limestones may not belong to the Silurian period. On this question I remark, first, that as a group it is not Silurian; second, considered palæontologically, it is not Silurian, inasmuch as it is destitute of fossils. After devoting some time to a careful examination of the weathered surfaces of the limestone and also to fractured surfaces, I was unable to discover a trace of organization. In this examination, I believe every bed received attention. I inquired of stone masons and of those who were constructing roads, if they had seen a fossil of any kind in these limestones, and

Fig. 10.

the answer was always in the negative; still, it seems probable that they will yet be found.

In respect to organisms, then, they stand out in strong contrast with the lower Silurian limestones of the neighborhood, for the moment these are examined fossils are found abundantly. The geologist may pass from the nonfossiliferous limestones and interlaminated slates in half a mile to the Trenton and Chazy limestones, and he will find there all the characteristic fossils which belong to them in the valley of the Mohawk. Again, it is not metamorphism, which renders the rock so perfectly barren of organized bodies, nor to the presence of chert or deleterious matter.

It seems, therefore, that the absence of fossils is significant of the period in which they were found, and this opinion should be received until it can be shown by the presence of characteristic organizations that they belong to the Silurian period.

I now propose to connect the last section with another (fig. 8), which includes the carboniferous and lower Silurian in the relation which they hold to each other, in the whole of the south-western part of Virginia.

1. Limestones of the Taconic system, which constitute a part of the series in fig. 7, and which are connected with slates, sandstones, &c. At this point they form the north-western part of a synclinal axis, which terminates against the carboniferous system, 2. On the west side of the fracture the systems are separated from each other only a few yards; 3, black calcareous slates of the Devonian period. The carboniferous and Devonian are represented by a series less than a thousand feet thick, and hence all the upper Silurian and most of the Devonian and carboniferous are absent; the Devonian is represented by a few beds only, which rest immediately upon the Niagara group, A. This is a cherty, concretionary and lumpy mass, a perfect confusion of beds. 4. Clinton group, with its beds of sandstone, slate and oolitic iron; 5. Medina sandstone; it always forms the crests of the mountain ranges of this region; 6. Lorraine shales, consisting also of green sandy marls,

even bedded, and weathering to a drab color; below them the black colored calcareous shales with chætetes, Leptæna sericea, alternata, &c.; 7, A mottled sandstone, terminating in beds of conglomerate, and which may be an equivalent of the Caradoc sandstone; 8, Trenton limestone, with Isotelus gigas, Ceraurus pleurexanthemus, &c.; 9, Birdseye and Chazy limestone, with Maclurea magna; 10, Calciferous sandstone; 11, Potsdam sandstone; F, fracture; 12, Carboniferous sandstones and shales. The distance between F F is about one mile. As members of the Silurian system, the series will require but few remarks at the present time, and in this place. They are brought in contrast with the Taconic series, and being rich in their characteristic fossils, it seems to me that this fact, taken in connection with the relations they exhibit to each other, places the independence of the Taconic system in a very strong light. It is plain enough that from 6 to 11, the series represent the lower Silurian most perfectly as it exists in New York, excepting the intercalated sandstone which contains fossiliferous bands, in which I found the Bellerophon bilobatus. As a physical group, the lower Silurian is in perfect accordance with New York rocks, and as an organic group it is very interesting to know that at a distance of one thousand miles from northern New York this series gives us such a perfect accordance also. The whole series from F to F does not exceed 4,000 feet in thickness, and that part of it included between 4 and 11 is quite as thick as in New York. A few miles eastward (about ten miles) the quartz rock, with its slate of the Taconic system, is thicker than the series included between 6 and 11. But I need not dwell on facts and arguments of this kind, especially when they may be regarded as inconclusive or not

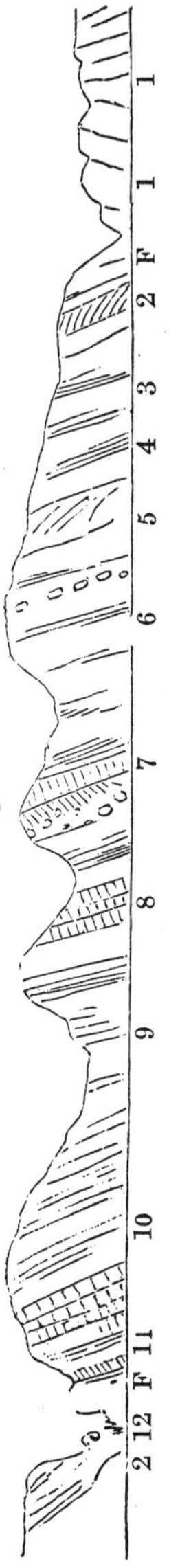

Fig. 11.

1 1 F 2 3 4 5 6 7 8 9 10 11 F 12 2

perfectly satisfying the evidence required. Still, I may add in this place to what I have already stated, that I believe that the Taconic series crops out from beneath the Potsdam sandstone at Shannon's on the north fork of the Holstein, sixteen miles from Wytheville; that the Taconic series plunges beneath the carboniferous at F, fig. 8, on the right, there is no doubt, and that it outflanks the lower Silurian is equally certain, inasmuch as it abuts against the saliferous and plaster series in Smith county. It, therefore, outflanks the Silurian or extends beyond it, at the western base of Walker's mountain.

There is probably no better section in Virginia than the one extending from Wytheville across the Iron, Little Brushy, Walker and Garden mountains, which furnishes that evidence in the form of contrast between the Silurian and Taconic system, which is very striking and conclusive. This section crosses the Alleghanies near the head waters of the Holstein, Clinch and one or two branches of New river. The easterly range of the Iron mountain is composed of quartz and the lower slates, from which we descend into the great lime stone valley of Virginia. Before reaching it we pass over many members of the upper division which crop out near its western base some two miles east of Wytheville. The latter place is upon the limestone commonly called the blue limestone, and which is finely exposed at numerous places in the valley. It evidently forms a synclinal axis, inasmuch as the rock east of Wytheville dips to the northwest, while near the base the dip of a similar series of beds is easterly at a steep angle. The dip, however, inclines towards the base of the Queen's knob, in the vicinity of which the most prominent rocks are carboniferous and lower Silurian, all of which dip southeasterly. The section, fig. 11, crosses the Little Brushy a little north of the Queen's knob. It passes only the first range. The ranges succeeding this as far west as Tazewell are repetition of the rocks of Little Brushy mountain.

THE FRENCH BROAD, HIWASSE AND OCOEE SECTIONS.

§ 44. The south furnishes many excellent sections of the upper Taconic rocks. The French Broad, after leaving the Warm springs in Buncombe county, cuts through a mountainous belt where the banks and cliffs upon the river show to good advantage the green, black and red slates; brown and gray sandstones interlaminated with slates and conglomerates. Beds of quartz nearly solid are not unfrequent, of which a remarkable one crosses the river about eight or ten miles west of the Warm springs; it is massive and thick bedded, and in consequence of its composition resists atmospheric agencies, and hence it stands out in bold relief on the mountain slopes. Hence it is a prominent cliff on both sides of the mountain which here encroaches upon the banks of the river. It may be traced up the sides or down one side of the mountain to the river's bank, forming a ridge where it crosses the road, and extending in an unbroken bed across it, it mounts up again on the other side to the top of the mountain to the height of 600 or 700 feet. The bed is fifteen feet thick, dips west or down the stream and makes with the horizon an angle of about 35°. The most remarkable fact is, that this massive quartz bed has been worn down so as to admit of the passage of the river. The part worn out is of the form of a horse shoe with the concavity directed up the stream. I had supposed that in almost all cases where rivers and streams flow through rocky gorges, that the rocks were first broken in the direction of the flowing stream but here there is no evidence of fracture. The river, it seems now, must have encountered this formidable barrier and have worn down the rock inch by inch, from 400 feet at least, down to its present bed.

The French Broad flows through a gorge for forty or fifty miles, or at least certain parts of its passage can be regarded in no other light than a gorge. The quartz rock which I have just described resembles a bed which crops out east of Lansingburgh, and which is known as the Diamond rock. Paint rock, six miles from the Warm springs, is a thin bedded gray sandstone, which resembles the rock between Blooming Grove and West Sandlake. Here thin bedded sandstone and shale pass into thick bedded ones with pebbly beds, succeeded by green and red slates and shales, in which there are beds of limestone. At Porrettsville, Tenn., black slates, and limestones alternating with them, are frequently repeated on the road to the Nolichucky, where the series seem to terminate in shales, shaly limestone, with intercalated silicious beds, similar to those in western Vermont. In the slates at Porrettsville, I found two or three species of graptolites. But I have been unsuccessful in finding organic bodies of any description in the rocks intervening between the Warm springs and Porrettsville. Black and green slates, sandstones. and conglomerates and brecciated beds, roofing slates, etc., wall up both the Ocoee and Hiwassee, where they cut through the mountains between North Carolina and Tennessee. In all the southern exposures of the upper Taconic rocks, conglomerates are much more common than at the north. With this exceptian the strata are physically the same.

In Virginia the rail road from Waynesboro to within two miles of Staunton, passes over a similar series of rocks, in which we find a perfect representation in the sandstones, slates and slaty beds representing those of the Hiwassee, Ocoee and French Broad and the Laurel, as well as those of New York in all of the foregoing routes which I have cited, succeeding the lower Taconic rocks, and consisting of the same rocks with the same sequence, from Canada to Tennessee.

I have already stated that the Taconic system on the east side of the Blue ridge in North Carolina, Rhode Island and Maine, is represented only by the lower division. In North Carolina it spreads out widely, but is more restricted in Virginia, while in Rhode Island it is very limited, though distinctly developed. In Maine the upper division extends seaward to an unknown distance, being easily recognized at the Fox Islands, twelve miles from Camden.

My object in the foregoing pages has been to describe in the briefest manner the rocks and their order of sequence, which belong to the system under consideration. The plan of the work will not admit of further details, excepting when it is necessary to compare a member of it with one which it is supposed it represents in the Silurian period. The end I have had in view, has been to show that as a series of sediments the Taconic system is unlike the Silurian; or when we compare the two physical groups with the lower Silurian, we shall not fail to perceive they have only a few characters in common, and those have little importance; hence they can not be brought into correlation without doing violence to geological principles.

REVIEW OF THE OPINIONS WHICH HAVE BEEN EXPRESSED RELATIVE TO THE TACONIC SYSTEM.

§ 45. I have shown in the preceding pages that the Taconic system, considered as a physical group, can not be brought into correlation with any part of or with any group of the Silurian system; that no individual member of the former can be placed in coordination with an individual member of the latter; that the lower Taconic rocks, as a whole, differ from the lower Silurian as a whole, and that the quartz of the former is not the Potsdam sandstone, nor the slates the Utica slate, nor the limestone of that group the Trenton or blue limestones, as has been maintained by many geologists of this country. According to the views of several eminent geologists, the rocks of Saddle mountain, in Massachusetts, belong to the Hudson river group; the limestone below, in the position which I have given it, is the Trenton, and the quartz rocks the Potsdam sandstone. The whole group of the lower Silurian is developed in this mountain according to this view. If so, how does it happen that high up, towards Graylock, the summit, there are no fossiliferous bands. If these immense masses which are piled up 3600 feet are sediments, and of course this is admitted, how can it be explained on palæontological principles, that there are no organisms at all in the mass of the so called Hudson river group. It is said of course, that they are metamorphic. But then, what evidence have we that they are metamorphic; of course it will be maintained, indeed it is affirmed that they are destitute of fossils, because they are metamorphic, and they are metamorphic because destitute of fossils. For there is no trap, no porphyry and no granite in Saddle mountain. We must therefore assume that after thousands of feet of sediment had been deposited in the then existing seas, this portion of the earth's crust was again heated sufficiently to destroy all vestiges of organization which belonged to the upper part of the lower Silurian series. Which view is the most probable, that which proposes to refer this formation to a period prior to the creation of animals, or to a period which abounded in life, of which the vestiges were destroyed by heat and which was communicated through an immense thickness of rock? Heat communicated to the degree required should leave its mark, but we find nothing in the whole range of the Saddle mountain which looks like an igneous product or an igneous change. In all other cases where it is probable that fossils have been obliterated by heat, some monument of its agency remains to attest the fact, but in the case under consideration it is all assumption without facts or phenomena to favor it.

Now it is not necessary to tell American geologists, that the lower Silurian is eminently fossiliferous. Mr. Hall's work contains eighty-eight plates and 381 species. It is therefore quite strange that all vestiges of this large number should have been obliterated. But again, how does this doctrine of metamorphism stand by the side of certain alleged discoveries. For example, we are informed in the beginning of the first volume on Palæontology of New York, that fossils are found in the crystalline quartz rock of Adams, at the base of the Green mountains. It seems then, after all, that these eastern rocks are not so much altered, but that fossils are preserved at the very base of the system. Geologists should reconcile their discoveries with their assumptions.

But again, if Saddle mountain is lower Silurian, or Hudson river, how does it happen that as a physical group it is totally unlike the rocks of Granville, Salem, Pittstown and Greenbush, or the group which lies immediately on the slope of the valley of the Hudson and lake Champlain, both of which are also claimed as Hudson river. Where is the consistency in maintaining that both divisions belong to the Hudson river group, since they are unlike each other in matter and arrangement; and at the same time, that both are unlike the Hudson river group in all the essential characteristics, considered as physical groups. To say that there are slates, limestones and sandstones, is futile, for if the mere presence of the lithological masses makes identity, then they may as well be called the upper Silurian or Devonian as the lower Silurian. But similar arrangements or sequence taken in conjunction with lithological characters is what makes similarity in physical groups, and if there is not even a similar sequence then it is unsafe to assume, that two such dissimilar groups belong to the same period. It is only the strongest palæontological evidence, that such dissimilar groups can be ranked in the same period. But, I shall show that this kind of evidence is also against the assumption, so that viewed in the light of physical and palæontological evidence the assumptions respecting the Taconic system can have no support.

Again; it has been asserted by my friend Mr. Hall that the so called Taconic system is situated in the midst of Silurian rocks. The inference to be drawn from this assertion is, that the existence of the Taconic system under these circumstances is an incompatible association, which of itself overturns my position. But this association is the very fact which really establishes my doctrine, and puts the question forever at rest. I shall soon go on and show my mode of reasoning and my application of the foregoing assertion. It would be a waste of time for a geologist to set himself about proving that the Devonian rocks belong to a later period than the Silurian. And why? Because we know that the Devonian rests upon the Silurian, it is the sequence of one period to that of another. It is precisely this kind of evidence which the presence of the Silurian rocks in the midst of the Taconic rocks gives us, and if this kind of evidence had been represented fairly there probably would have been much less controversy upon the question.

The first illustration of the use to which I shall put the Silurian rocks will be seen in figure 12.

Fig. 12.

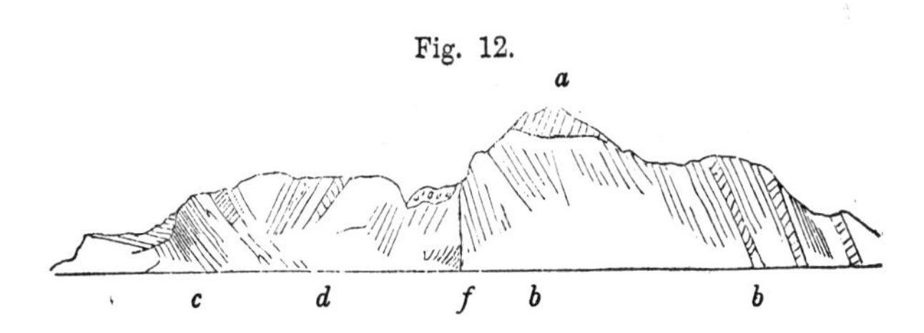

It is a section in a part of the Hudson river group, at Greenbush, passing through Cantonement hill. Just above the river but below the old red mill, the slates *c* are crushed, but it is evident they dip to the east. At the mill *d*, the blackish sandstones of the Hudson river dip also east; half a mile further, sandstone again crops out, dipping steeply to the west. Just beyond, the green Taconic slates dipping E. 10° S. support a heavy mass of calciferous sandstone *a*, and slaty Trenton limestone. Viewing the position of all the rocks, we find that there is an anticlinal axis running at the base of the ridge at *f*, supporting the limestone. This anticlinal is on the line of the great Hudson river fault. The limestone, which is the most important mass, rests unconformably upon the green Taconic slates *b b*. The beds dip in part west as represented in the cut at *a;* but they are not uniform in this respect. But this diversity in dip is proof that it is not interlaminated with green slates below. One of my friend's opinions which he at one time expressed, was, that the mass was thrust up through what he regarded as the Hudson river rocks. Since that opinion was expressed, he has constructed a section of the hill, and has represented this limestone as the Trenton, interlaminated with the Hudson river slates, as he calls those at *b b*.

Now some palæontologists are willing to admit, that a few fossils may go up from the Trenton into the Hudson river series, but I believe that this is the first time that a palæontologist is willing to transfer the whole of the Trenton limestone with all its contents into a higher group. But it unfortunately happens that the calciferous has to go with the Trenton in this case. That there is an error in Mr. Hall's view is evident from the exposure of the slate on the north and south sides of the hill, which proves that this bed of limestone is not an interlaminated mass in the so called Hudson river group. Both palæontological evidence, and that of superposition, forbid such an interpretation of the phenomena.

The true explanation is this, the limestone is an outlier of the Silurian system, deposited upon the slates which outcrop from beneath it. It was elevated subsequently, and undoubtedly was covered by the Hudson river rocks also, which have been carried away by diluvial currents leaving the present mass as it is now exposed. If this is the true explanation, and I see nothing to conflict with it, then as it regards these two systems the inference is too obvious to require to be stated.

It is worthy of notice, that this instance of superposition of the group of lower Silurian rocks, is upon the upper members of the Taconic system. It seems fitting in view of this fact, that I should now show that the rocks of the same period overlie and rest upon the lower division of this system.

I shall therefore select from the extreme eastern part of the county of Rensselaer an instance of an overlying mass of calciferous sandstone upon the Hoosick roofing slate. It has been very confidently set forth that the occurrence of Silurian rocks in this region proved the error of many of my views respecting the age of the rocks under consideration, while on my part I maintained that these rocks in consequence of the relation they sustain to the Taconic system, proved the soundness of my position. Before I refer to the figure which illustrates the true and real relations which subsist between the two systems, I shall give, in a few words, an explanation of the structure of the narrow belt of country of which Hoosick falls may be regarded as the centre. On the west, about three-fourths of a mile distant, there is a ridge some 400 or 500 feet high which stretches along the Hoosick river about four miles. The eastern face of this ridge is partly covered with a broken mass of calciferous sandstone, carrying as evidence of its age the Maclurea and several other fossils. It extends about four miles also, but its beds are broken and dip in a disorderly manner sometimes east, sometimes west, steeply sometimes, and only slightly in others. Its continuity is broken, yet it continues the distance I have stated. Its irregularity in dip and its fractured condition prove, that it rests on an uneven bottom, and that it has been unequally moved; and, furthermore, can not form a part of the Taconic rocks exposed below, upon the hills and in the valleys.

The limestone is discontinued before it reaches the river; on the east side of Hoosick falls is a ridge of about the same height running parallel with that on the west side. It is the ridge which furnishes the Hoosick roofing slates, quarries of which were opened forty years ago, and in which the graptolites occur. On the eastern slope of this ridge of roofing slate limestone occurs of the same period as that upon the west side. The limestone of this valley contains the fossils of the lower Silurian period, and hence we are not left in doubt respecting the most important palæontological facts. It remains now to determine how this well known limestone is situated; whether it does or does not form a part of the slaty group, or whether it is a part or parcel of it, in which case it would undoubtedly follow that the slates themselves are Silurian also, inasmuch as the limestone is Silurian, carrying the proof of its period in its fossils.

Now, it might well happen, that its true relations could not be determined by actual inspection; and in consequence of a nearly conformable dip with the slates, it might appear to form a part of the group with them. But at the outset, an honest and well meaning palæontologist might inquire what right has the calciferous sandstone to force itself into the company of the Hudson river rocks—what right has it to be interlaminated bodily between their beds? But leaving such impertinent questions to be answered by my friend the state palæontologist, who maintains that all these rocks, limestones and slates, are Hudson river rocks, we may proceed to determine the fact. It fortunately happens that this limestone lies upon the eastern face of the ridge upon which the state quarries are opened. Fig. 12 represents the situations of the limestone to the slate beneath.

Fig. 12.

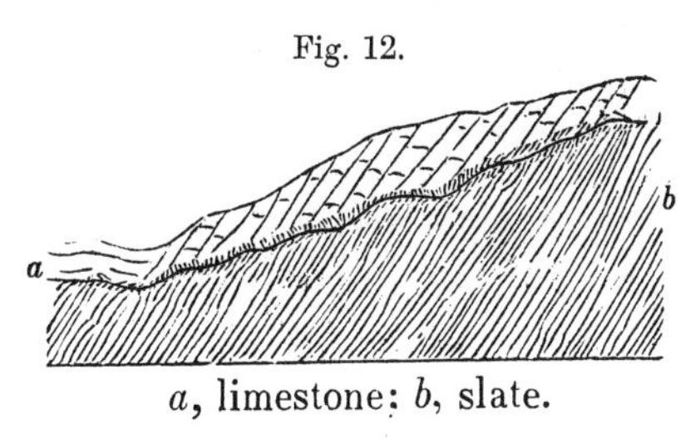

a, limestone: *b*, slate.

This mass of limestone lies upon the east slope of a steep ridge, and the first impression which would be received by any geologist would be, that it belongs to the group of slates. The dip to the eye is so nearly conformable to the slate that it would probably be regarded as an interlaminated mass. When, however, its position is ascertained partly by a natural and partly by an artificial exposure, by the removal of earth to obtain the slate for a firestone, its superposition is proved beyond a doubt, and the idea of an interlaminated mass at once disposed of. The rock had been quarried for lime many years ago, and the excavation for slate led me to examine it in this place, when it was not only found to be above the slate, but reveals other important facts along the junction of the rocks. The slate, for example, shows that it had been tilted up before the limestone was deposited upon it. The out cropping edges still preserve the stair-like arrangement, which we often see in slate beds which are inclined at an angle of from 5–7°. It was upon this unequal step-like surface, that the slate was deposited, and the consequence is, that the bottom beds of limestone are roughly crushed and partly perhaps concretionary. The junction of the rocks is so well exposed from top to bottom, that only one opinion can be adopted respecting the position of the limestone, of which the annexed cut shows most perfectly the relation of the two rocks. Now, in the absence of direct proof to the contrary, this limestone would have continued to be cited as a part of the group of the graptolite slate, notwithstanding the calciferous has no right geologically to occupy a position, or position above the Utica slate, for all the ranges of limestone west of Hoosick falls, as well as west of this place, belong to the calciferous sandstone; and it is quite fortunate that these beds are fossiliferous, containing stems of Encrinites, a small Maclurea, in great abundance, one which is found also at Bald mountain.*

If the Hoosick and associated slates and shales and thin bedded sandstones were what my friend Mr. Hall and certain other geologists say they are, then, to be consistent geologists, they should not group together rocks belonging to distant periods. It is true the limestone is the base of the lower Silurian, and the Utica slate near the top, but so orderly have the lower rocks been laid down that no instance occurs in this country or Europe where the two extremes are placed in contact. In these statements I am contending for a principle, and not to sustain the Taconic system. I have stated that the position of the limestone is fortunate, inasmuch as we can begin at the base of the ridge near Hoosick falls, and trace the slate step by step to a quarry containing graptolites, and then from the quarry we may trace the slate to the place on the eastern slope of the ridge, where the calciferous limestone rests upon this slate. It follows, then, that the slate is not the slate of Baker's falls, or the Hudson river series, which contains such an abundance of Graptolithus pristis as Mr. Hall is inclined to assert; but I shall show in its proper place, that the graptolite of the Hoosick slates is not the Graptolithus pristis. The extreme looseness and carelessness in observation is shown by the affirmation that the Hoosick roofing slate rests on Trenton limestone.† In the first place, there is no Trenton limestone in the neighborhood for it to rest upon, and, in the second place, the beds of calciferous

---

*The fossil alluded to is the Maclurea sorditda of Hall's Palæontology, the *Straparollus sordidus*, of D'Orbiny. Mr. Wilder, of Hoosick falls, who has accumulated a large store of information in his favorite branches of science, geology and mineralogy, has large slabs of limestone covered with this fossil. He accompanied me to the locality, and I am permitted to say that his views correspond with my own.

† Palæontology of New York, vol. I, p. 267; idem, p. 268.

sandstone west of Hoosick falls, which must, if any, be the underlying beds, *actually rest upon this same group of slates* which compose the slate quarries. The presence of the Trenton limestone, however, would not change or alter the conclusion which I have stated.

But the occurrence of the limestone, geographically and geologically, is important, as showing how observers may err by mistaking one limestone for another. In this neighborhood the marbles of the lower Taconic rocks occur not in immediate proximity, but sufficiently near to lead to error in their determination. These marbles and limestones it has been shown belong to the group of slates in which they are found, and it would not be strange if a bed of calciferous limestone with its fossils was found superimposed upon a bed of marble— not more strange than that the lower Silurian is actually superimposed upon an overlying slate of the same group. Hence we see that the doctrine of metamorphism would aid materially in confirming an error, if the position of the lower Silurian limestone was not determinable by actual inspection. But having determined it not only for once, but also in many instances, and shown that the limestone rests upon the group unconformably and therefore can not be a member of it, we are authorized to carry our generalizations farther, and say, that in all cases the lower Silurian holds the same relation to the Taconic groups (even in Canada) which I have shown they hold at Greenbush and Hoosick. That there may be no obscurity in these relations left to be cleared up hereafter, I have prepared several other sections which show how geologists have been deceived in respect to the superposition of these outliers of the lower Silurian system.

The section to which I shall now direct the attention of the reader is that of the Mettowee river, at North Granville. I select this, because I perceive that Dr. Fitch* was led into an

* Transactions of the New York State Agricultural Society, 1849, pp. 816, 907.

error respecting the true relations of the Chazy limestone, lying in a gorge of this river about one mile from the village of North Granville, and that I may not misrepresent Dr. Fitch's views, I shall first introduce his section, figure 13, with his explanation of the phenomena.

Fig. 13.

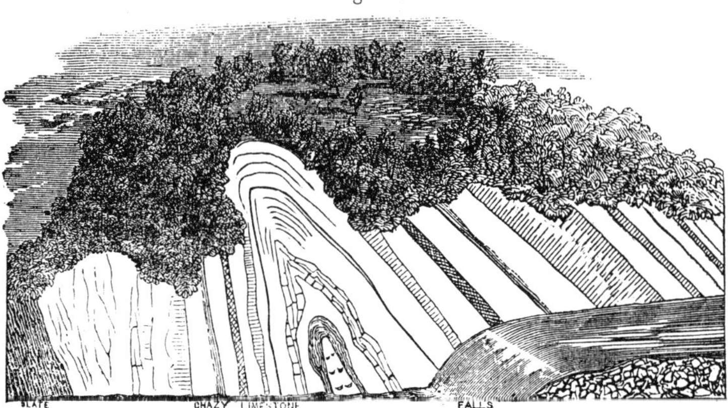

There is no disagreement between us respecting the name of the limestone represented in the cut. Dr. Fitch infers from the folded condition of the Chazy limestone that it has involved in it the slates which are in proximity, and hence, that all the cases of superposition I had relied upon, for proof that the Silurian belongs to a later period than the Taconic, was left unsupported. Now the fold in the limestone is correctly represented; there is a plication as the Doctor has shown. On the left, to use the Doctor's language, we have the black shale, or slate outcropping in the same situation, and in essentially the same manner that it does at Galesville. To this succeeds a thick molten mass of limestone, in which no distinct lamination can be traced. The synclinal, as the Doctor calls it, is finally reached, and the beds on the left side of the axis are less than half the thickness of those of the corresponding bed on the right side. The distance from the axis to the slate is forty feet; whilst to the right, thrice that thickness of the bed is passed over, without reaching the slate on the east, and the Doctor might have added more than ten times that thickness. The Doctor goes on to generalize, and remarks, that the Bald mountain range of hills is a fold, and not a fracture or uplift.

Fig. 14.

Without dwelling farther upon the illustration of Dr. Fitch, it appears from the fact exhibited on the south side of the stream, that the fold is confined to the Chazy limestone, and that it does not affect the underlying slates at all, and hence the fact of the plication of the limestone does not bear unfavorably upon the Taconic system, but as the phenomena really are, sustain the views I have adopted. To show how the limestone is placed with respect to the slates, I have introduced figure 14, which was constructed from the relations of the rocks as they are exposed on the south side of the stream, and a little below the great fold of Dr. Fitch's section. Thus 1, 1, Taconic slates; 2, Chazy limestone; the slate in this part of the section being concealed, but on the left three smaller folds of Chazy limestone are left standing obliquely upon their ends, and between which the slate is exposed. All of the standing parts of the folds, it is plain from inspection, may be removed from the slate, and besides the larger masses are seen also to rest upon them, though they incline downwards out of sight, but for forty or fifty feet it is evident the fold or plicated mass rests on the slate and can not form a part of the group to which the slates belong, but are simply crumpled up, or folded upon it. This view is sustained by the position of another insulated mass of Chazy limestone upon the hill north of this place,

Fig. 15.

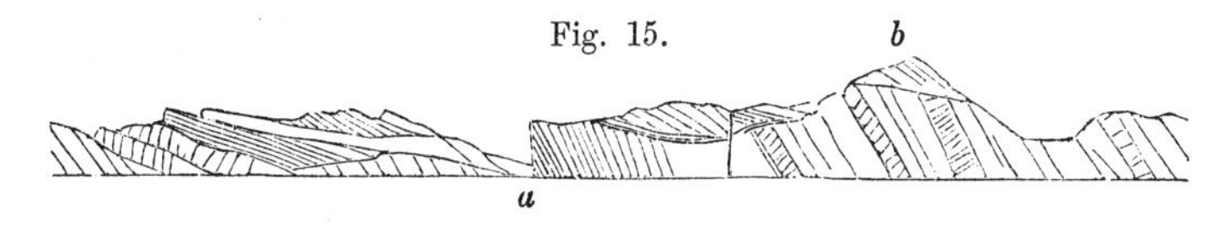

figure 15, *b*. My views are sustained also by other facts.

The rocks, on leaving Comstock's landing, fig. 15, are gneiss; 2, Potsdam; 3, calciferous sandstone; 4, Chazy limestone resting upon No. 3, but these rocks dip only at an angle of 5° until the Chazy, which extends east beyond the calciferous sandstone, dips so as to conform nearly with the Taconic rocks, on which it rests; the fold then only concerns the Chazy limestone, which is proof in itself that there is no general fold involving the plication of the slates, as has been maintained by distinguished geologists. Then again further down the stream, we find the calciferous sandstone abutting against the slates, figure 15, *a*, on a line of fracture; but this does not prove that the slates overlie the calciferous, for only a small part of the series of limestone appears at this place,* and besides, too, the highest member here, the Chazy limestone, *b*, is on the adjacent hill resting on the slates unconformably.

If for the sake of argument it should be admitted, or even it was proved, that the Chazy limestone is inclosed in a fold of the slates, which we know are beneath, what does it prove? Merely this, that it is an accidental occurrence, for the Chazy is only the subordinate part of the lower Silurian; it can not become, therefore, an associate of the slate group by itself. On the Mettowee it is separated from the calciferous by at least half a mile of outcrop of slates; and there is no calciferous below it at the fold. In conclusion, I say that the phenomena in the aggregate disprove the doctrine of such a plication as to affect the validity of my position respecting the Taconic system, and moreover the facts disclosed on the south side of the Mettowee prove that the folded Chazy limestone rests upon the edges of the slate, and therefore does not form a part of the

* This is an anticlinal perhaps which has passed into a fault; for on the one side is an older rock, Taconic, forced above a Silurian, the calciferous sandstone, the west side of the fracture being the most depressed, while on the east side the Taconic slates are the most elevated. We can not maintain consistently that the Taconic slates over-ride the calciferous; it would be physically impossible, seeing that the Chazy must at one time have been an overlying mass at this place.

slate group, and is in no way connected with the slate except by its superposition.

§ 46. The gorge of the Mettowee discloses also the manner in which rocks thin out. We began this section with gneiss. The Potsdam sandstone is the next older rock. It appears that this rock speedily thins out, but whether it laps on to the Taconic system I am unable yet to prove, but the next oldest, the calciferous sandstone laps upon this system and extends east farther than the Potsdam sandstone. The Chazy limestone succeeds, and this extends still farther east than the calciferous and rests on the Taconic slates in its most eastward prolongation, where it dips almost as steeply as the inferior slates upon which it rests. But at the west, these rocks had a dip of only 5°. When, however, these rocks lie near the great north and south dislocation or fault their dips become steep.

§ 47. Before I proceed farther in answer to objections which have been made to the Taconic system, I propose to explain an obscurity which I have often met with, which, unless it is cleared up, is calculated to mislead or to raise doubts respecting the relations which really exist between the Silurian and Taconic systems.

At Highgate, in Vermont, the Missisquoi passes through a gorge just below the bridge. The Calciferous sandstone lines its banks on both sides, but some distance below, the junction of the calciferous sandstone is entirely concealed, even at rather low stages of water, and from the inequalities of the slate which jut up, it appears that the limestone might form a part of the group, by plunging down between the beds of slate. Thus in fig.

Fig. 17.

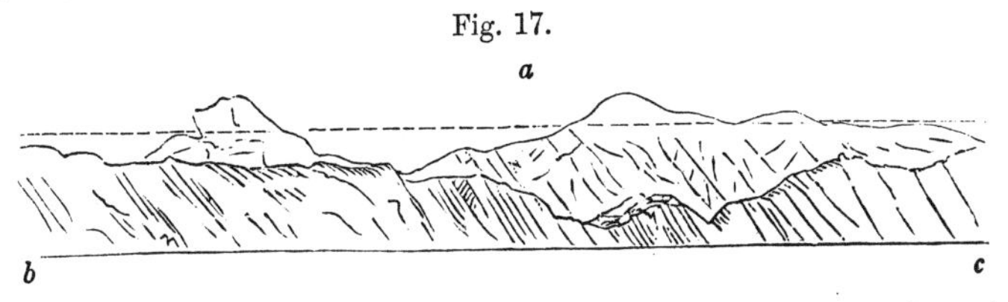

17 the dotted line represents the line of high water, where it would appear that the outcropping mass of calciferous sandstone, *a*, plunged between the beds of slate, *b*, *c*, which come up on the

right and left. On visiting this locality at a very low stage of water, I found that the calciferous rested in an irregular trough in the slate, and also upon its upturned edges. This fact explained all the obscurities which had formerly raised doubts in my mind respecting the relations of this rock to the slate; and, besides, the folia of slate are bent at the junction of the two rocks, fig. 18, which in this case I attribute to the force which has been communicated to the rocks, although it often occurs that folia of slate are puckered at the junction of a seam of calc spar with the slate. The phenomena at Highgate throw light on many other localities, where the lower Silurian rocks are in contact with the slates. Thus, one mile east of Troy, the junction of the calciferous resembles that at Highgate, but it is evident that the whole of the calciferous may be removed from the depressions, and from the unequal surfaces upon which it was deposited.

Another fact, however, at Highgate is worthy of attention, inasmuch as it proves the correctness of my view already expressed respecting the position of the calciferous sandstone. Thus, fig. 18, the spot selected for illustration is just below the

Fig. 18.

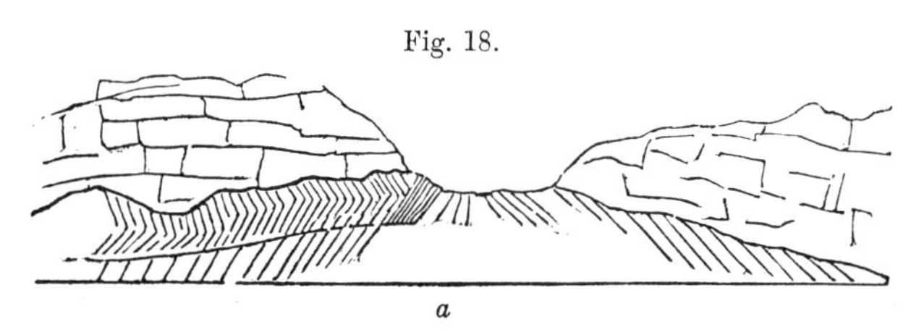

bridge. At very low water the slates appear in the bed of the river, and form an anticlinal at *a;* upon the upturned edges of this anticlinal the junction of the calciferous sandstone may be traced, with great distinctness, on both sides of it; on the right, the calciferous extends to and forms a mural cliff, on which a mill is erected. The possibility of the calciferous forming a part of the slate group is set at rest by this exposure.

I am now prepared to show how another error may be committed respecting the period to which a limestone belongs; an error which I suppose has been already committed by several geologists. At Orwell, Vt., see fig. 19, the calciferous sandstone, *c*,

Fig. 19.

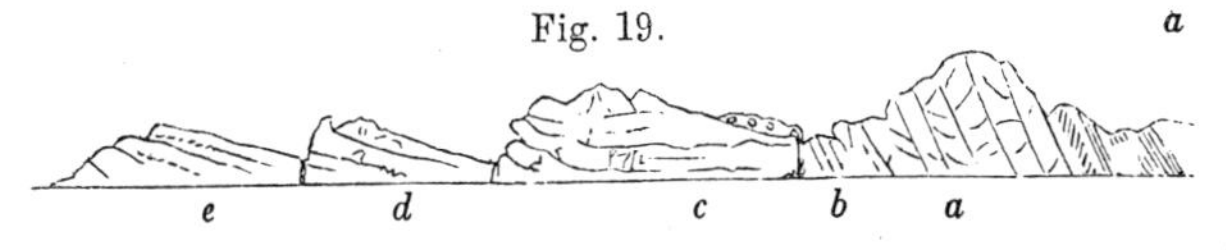

occupies the plain upon which the village is built. Eastward a few rods, a limestone, *a*, crops out in proximity with the former, but how many yards apart they may be I can not now state.

With only a superficial examination, they would be regarded as one rock, belonging to the same period. On a careful examination, however, it is clear that the eastern mass rests on a greenish slate, and by going east a short distance the evidence is plain enough that a slate, *e*, also overlies it, so that it is inclosed between beds of slate, and in this respect it is similar to the limestone which lies in the tunnel of the Western rail road near the state line, and which belongs to the slate group, as is clearly proved by observation. Having determined that the eastern limestone at Orwell lies between beds of slate, our convictions it seems should be, that it is not lower Silurian, nor Trenton, nor either of the masses into which the lower Silurian limestones have been divided. But what is the limestone on which the village is built? To determine this question, I traced it northward, keeping it in view for a few miles, and then turned directly west towards the lake. On passing over two or three rocky terraces composed of this rock, I came directly upon the Potsdam sandstone which cropped out from beneath this silicious limestone, and over which I had passed, and which I had traced from Orwell. The result of this examination proved that the rock at Orwell, which seemed almost to join on to the sparry limestone between the slates, is the calciferous sandstone. The two limestones therefore might be confounded, but without the aid of a fossil, the geological formation to which the calciferous sandstone belongs, is proved by simply determining its relations, and at the same time it is also proved that another limestone of a different period cropped out very near it, and might have been confounded with it. It proves, too, the fallacy of the doctrine, that the lower Taconic rocks, or the lower limestones, are but altered Silurian; we have two limestones of different ages together—one rests on the Potsdam sandstone, and is of the Silurian age, the other on slate, and belongs to the Taconic system. The theory of plications and folds it will be seen, will not explain the phenomena or the facts. It is not even a plausible hypothesis, when offered in explanation of the phenomena I have related. I have been unable yet to detect fossils in the sparry limestone of Orwell, but its condition is as favorable for their existence as in the calciferous sandstone. This locality proving the existence of a limestone which can not be placed in coordination with any of the limestones of the lower Silurian, throws the burden of proving the period to which it belongs on other shoulders, provided it does not belong to the Taconic system.

The sparry limestone at Orwell, when its relations are investigated, shows that what my friend Prof. Rogers claims as having proved is not yet established, viz: that the limestones of Berkshire are only the altered blue limestones of the lower Silurian, for this exhibits the same relations as the Berkshire limestones; and I go farther and remark, that the section of rocks from Orwell to the lake exhibit a series of dislocations and faults only; and though my section was not designed for this purpose, yet it shows the relations of the rocks as they succeed each other, beginning at Orwell and going west to the lake, in which distance there are no less than four dislocations; and the order of arrangement, beginning with the limestones is such, that we pass, by successive steps, from the oldest to the newest, terminating with those on the lake, the Utica slate and Trenton limestone, which we reach by a series of descending steps, though, geologically, we are ascending. Thus, figure 19, *a*, sparry limestone; *b*, *b*, slates; *c*, calciferous sandstone; *d*, Potsdam sandstone; *e*, Utica slate and Trenton limestone.

Bald mountain and the neighboring hills and ridges in Washington county, N. Y., furnish also many important facts bearing directly upon the relations the Taconic system sustains to the lower Silurian. This mountain is capped with the lower Silu-

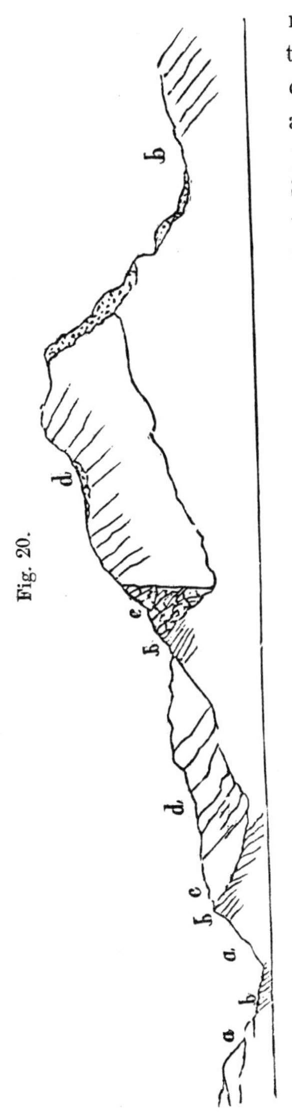

Fig. 20.

rian rocks; their position and situation are exhibited in figure 20; *c*, *d*, calciferous sandstone; the mass, *c*, is a compact black limestone, and appears unlike and distinct from the rock. Silicious and black masses perfectly well defined occur frequently in this neighborhood in the calciferous sandstone, but they are varieties of that rock notwithstanding their color. The Taconic slate crops out from beneath the calciferous at *b*, *b*, on the west or left hand side is slate rock; *b* appears on the east or right with a steep dip; at *c*, *c*, on the left, are fractures. The dark colored limestone, *c*, has been quarried for more than forty years, and the front has been worked back thirty or forty feet, down to the slate, *b* The question to be solved at this place is, whether the slate, *b*, on the left, is the Utica slate, or any other slate of the Silurian system; there is no difference of opinion respecting the mass, *c*, *d*, all agree that it is the calciferous sandstone. If the slate, *b*, is Silurian, it follows that at this place there is a plication, and the slate is folded under the limestone. The fact which militates against this view is, that on the south side, and at a point nearly under *d*, near the summit, an excavation was made in search of coal; the shaft was begun in a lateral shelf of the limestone, near the base of the mountain. This shaft was carried through the limestone into the slate, *b*, beneath it. Now, this fact taken in connection with another, viz: that the limestone does not penetrate the slate on its southern flank any where between *b*, *b*, on the right, proves incontestibly that the limestone merely caps the hill. If it was a part of the slate group, we should find it in the direction of strike in the slate.

Again, if a plication exists and the slate, *b*, is the Utica slate, then there are facts brought together which involve a physical impossibility, inasmuch as the limestone upon the slate should in that case be the Trenton limestone, which always underlies the Utica slate; the calciferous sandstone could not, therefore, in the case of an inversion be brought into immediate contact as it is with the slate, but it should be the Trenton limestone. The fossils of the rock resting on the slate, however, are those of the calciferous sandstone, viz: the Maclurea sordida, Hall's Palæontology, pl. 3, fig. 21. They occur upon the thin bedded limestone towards the top of the hill, as well as in the more compact beds just above the slate. If, on the contrary, there is no inversion and the phenomena do not indicate it, then here is a slate which is not Silurian; it crops out from beneath rocks which are known in this country as the lowest beds of the lower Silurian and besides may be traced beneath the mountain and beyond it eastward, forming a part and parcel of the series, *b*, cropping out in the figure on the right.

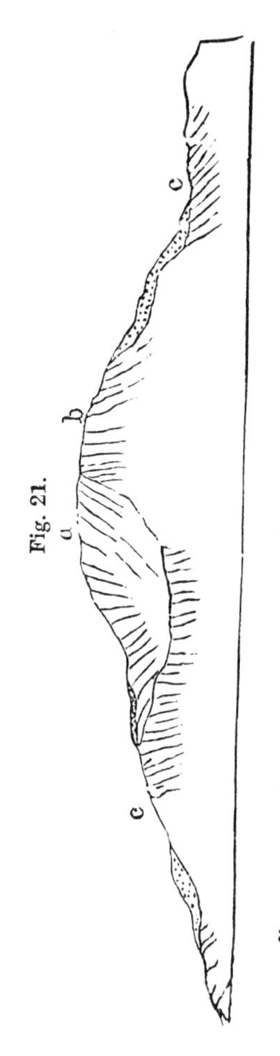

Fig. 21.

Now, all the knobs resemble each in this neighborhood in being capped in this way. Mount Tobey, fig. 21, not far distant, is an example of the same as figure 20, being capped with sandstone The lower part of the limestone dips at an angle of about 10° to the east; on the contrary, on the top, at *a*, it dips steeply to the west. The slate beneath may be traced around the mountain, without exhibiting the limestone upon its beds. It is, therefore, a mass resting on the top of the mountain unconformably to *c*, *c*, the slate on both sides of the mountain. The slate at *b* dips steeply east. The dips, were there no other facts, decide the question of superposition.

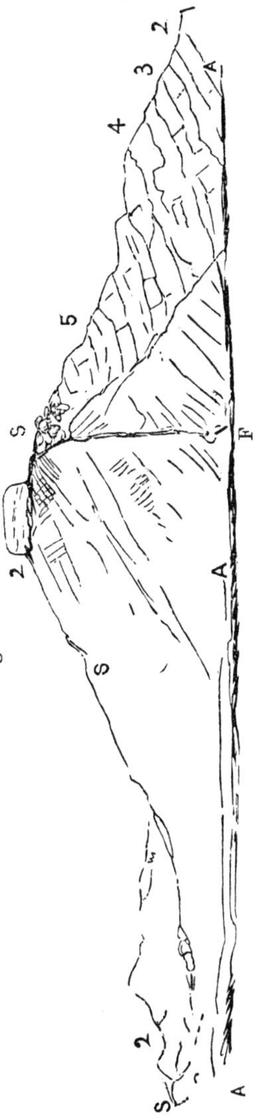

Fig. 22.

It has been stated in Silliman's Journal,* that Dr. Fitch found the Trinucleus concentricus in this region, adducing it in proof that the rocks belong to the Silurian system. This statement will not probably deceive any one. The fact, however, is important, as by it we prove the greater antiquity of the Taconic rocks than the rocks in which it was found.

§ 48. *On the structure of Snake mountain and the evidence its structure furnishes in support of the Taconic system.*—This little mountain in Addison county, Vt., is seven miles east of Lake Champlain. The rocks between the lake and base of the mountain are lower Silurian. The Trenton limestone with slaty beds lie upon the western flank of the mountain, fig. 22.

In the ascending order, 2, 3, 4 and 5, the rocks are calciferous sandstone, Chazy and Trenton limestones. And that I might exhibit the relations of the calciferous sandstone, 2, at the top of the mountain, it is made across its northwest angle, where in a few hundred yards we pass over to the slate S, S, which crops out both on the west and north sides beneath No. 2, and on the north side extends from the top to the bottom of the mountain. The Silurian rocks 2, 3, 4, 5, rest against the

* Silliman's Journal, vol. XIX, for May, 1855, p. 434.

mountain. At S and F there is a fracture and dislocation which may be traced in the direction of the mountain axis, four or five miles. The dip of the Silurian rocks is from ten to fifteen degrees; of the Taconic slates, S, S, twenty-five to thirty degrees. At the northeast base of the mountain, perhaps three-fourths of a mile from its summit, the slates crop out again below 2, on the left, beneath the calciferous sandstone.

The summit of the mountain is calciferous sandstone and is exposed in a perpendicular mural precipice for four or five miles. The debris and fallen masses from this bold front generally conceals the underlying slate, but it crops out beneath it at one or two places, while on the north side, 2, S, the whole slope is exposed and consists of one mass of slate from the calciferous sandstone to the bottom of the mountain.

The calciferous sandstone in this region is often red or chocolate color, especially the inferior part of it. The gray variety at Burlington graduates into the red, and the Potsdam which is used as a flagging stone, at Burlington and other places, is usually brown or chocolate colored also. But the blue and gray with a sparkling lustre are found in the masses composing the mural wall at Snake mountain. The junction between the Taconic slate and Trenton limestone and its upper slates, or the Chazy limestone on the west flank of the mountain, has not been observed.

The foregoing statement respecting the relative position of the rocks of this mountain seems to be all that is required to establish the inference I have drawn from them. I need not dwell on the error which has been committed by regarding the chocolate colored rock the Medina sandstone, or attempt to show that the plication theory will not adjust the rocks so as to make the black and greenish slates, S, the Utica slate or Hudson river group. It is one of simple dislocation, where the older rock on the east side is elevated vertically higher and above a newer series on the west. The rocks, in this case, are not engulphed upon the west side; all the phenomena seem to prove that the whole mass composing the mountain was raised vertically, but the east side was separated from the west by fracture, and elevated above it. The series between the mountain and the lake occupy a much lower position than those upon the flank of the mountain, proving that the latter have been broken from the former and elevated above them. We find in Snake mountain a fact of common occurrence, a fracture at the base of the ridge or mountain, and another running through it. Bald mountain, which has been described, is another instance of this kind. In fine, with respect to Snake mountain, the position of the mass 2, on the top of the mountain, and which covers the eastern slope, proves that it is an overlying mass and an inspection of the junction of the inferior beds which often jut over and beyond the slate immediately below, proves that it was deposited upon the slate, and as the two are unconformable both in the amount of dip and direction it is also evident, that they do not belong to one group or series. At this place the former is the base of the Silurian system. It is very silicious generally, and might be called a sandstone, yet it is rather a mass intermediate between the calciferous sandstone and the Potsdam sandstone. It is the same rock as that at Burlington, and at Sharp Shins, two miles northwest of Burlington, where the same black slate crops out as at Snake mountain. Those who wish to satisfy themselves of difference between the Hudson river group and the slate beneath the calciferous of this mountain, should explore the north end of it, where they will find a mass of slate from top to bottom laid bare by a small stream which takes its origin immediately beneath the jutting calciferous sandstone, and which by this little stream has been undermined for centuries, and from which huge blocks have been, and are still broken and carried down the mountain's side, and are found distributed far and wide upon its northern and western sides. This slate is uncovered in a continuous mass between 700 or 800 feet thick. I was unsuccessful in a search of a few hours for fossils, and yet it is similar to other exposures where I have found graptolites.

§ 49. *Position of the calciferous sandstone at Hoosick and elsewhere compared with Prof. Sedgwick's ideal sections of the Skiddaw Forest, and the calcareous hills of Westmoreland.**— The sections of the older slates, which have been given by Prof. Sedgwick, of the Cambrian rocks, contain important facts which have a bearing upon the question respecting the existence of a sedimentary series or system below the Silurian. The section through the Skiddaw forest is in point as it enables me to contrast it with sections through the strata belonging to the Taconic system. For example, the relations of the Coniston limestone, 4 *a*, and the calcareous flagstones, 4 *b*, to lower Silurian of this country. It appears to have been determined as long ago as 1847–8, that the Coniston limestone, which is a part of the series in the section of the Skiddaw forest, contains lower Silurian fossils, but still lower a vast series of slates marked No. 3 in the section are found to be fossiliferous, and Prof. Sedgwick regards these lower series of slates (Skiddaw slates) in England, as the oldest fossil group of the British isles. This group, however, is conformable to the Coniston limestone which contains lower Silurian fossils. It is the position of the Coniston limestone compared with the position of the calciferous sandstone to which I wish to direct the attention of the reader; the former is a part of a physical group conformable with the lowest fossiliferous beds; while the latter is not a part of the group of slates, etc., of the series at all, being neither in coordination with a member or conformable to any part of the Taconic system. But as American geologists have watched the discussions going on in England respecting the claims of the Cambrian system, and as all the representations of its strata carry the idea of conformability, not only among themselves but also with the Silurian system and as Silurian fossils characterize these lower beds, the opinion has gained ground that the Cambrian is only a modified condition of the lower Silurian, and hence the same view has also been taken respecting the Taconic system. The representations which

* Proceedings of the Geological Society, vol. iv., p. 216.

have been made have conveyed precisely the Cambrian type. Whereas the facts proving the relative age of the Taconic system contain another order of proof entirely. The Silurian limestones in the Taconic districts do not form, as I have shown, a subordinate part of the Taconic series, but they are always overlying unconformable strata, as represented in all my sections, the most instructive of which is the Hoosick section, fig. 11. It is in consequence of that misrepresentation of facts which has prevented the adoption of the Taconic system here and abroad.

Before I close my remarks on the objections which have been brought forward in opposition to my former views of the Taconic system, I feel bound to notice those which have been urged by Prof. Rodgers on several occasions when the question came up for discussion in the meetings of the American Association.

I have anticipated in the main Prof. Rodger's objections. But by stating the points at issue in a formal manner, I shall be able to clear up the objections to his and my own satisfaction. The principal fact is admitted, viz: that there are rocks cropping out from beneath the lower members of the Silurian system. If these rocks which crop out from beneath the Silurian beds, are not in their original relation or position, then the question of age is debatable. If these inferior rocks are in their original relation and position, then it follows that in this country there are fossiliferous rocks older than the Silurian system, for it is settled beyond a doubt that the Potsdam sandstone is the base of the system in this country. Prof. Rodgers maintains that the inferior slates, those at Bald mountain and upon that range, those at Snake mountain and at Burlington, Vt., Greenbush, Troy,* and many other places, are not in their

---

* I should state that Prof. Rodgers, in his remarks, has never given localities upon which he has based his opinions. He has referred to Snake mountain, but it was in terms which satisfied me that he had seen but a small part of the series, and hence was unprepared to express an opinion.

original position. It is sufficient to say that at all the places I have named, this slate crops out from beneath rocks of the lower Silurian age. This is admitted by all geologists.

Now the explanation which Prof. Rodgers gives is this, viz: that wherever these relations exist which I have stated, they are due to an inversion of the lower Silurian rocks; and the superior masses are folded beneath the older; and as the strata are plicated in mass, and as the Hudson river group succeeds the limestones, the former consisting of black slates and sandstones in the plication, the latter are folded beneath the former; and hence this black slate which I have had occasion to speak of, is no more than the Utica slate, or a slate near its horizon.

Of this opinion, I do not propose to express doubts as to its being theoretically possible, neither to deny that it may exist as a fact. But I do say that from Georgia to Maine, there are no plications which create the least intricacy. I speak of phenomena and the interpretations which they themselves suggest, as rational interpretations which we should put upon them.

We have only to examine the localities in detail where the supposed plications exist, to be satisfied that the plication or fold has not reached the slate, and if the slate beneath has suffered a movement, it has extended its influence to the mass as an overlying one, prior to that movement; it has crumpled up the limestone between two ridges, as at the Mettowee, where the fold is still visible in the limestone, but not in the slate, or it is not such a fold as to place one rock beneath the other.

The objection to Prof. Rodgers' explanation must be considered also in a mechanical light, and attended with effects powerful in proportion to the masses displaced and laid in an inverted position. I maintain that the fitting mechanical effects of an inversion should be among the most prominent phenomena; whereas none exist. In the next place, admitting an inversion of all the Silurian rocks in a fold, so as to place the slates at the bottom, then upon the slate in the ascending order, we shall find the Trenton limestone, the next the birdseye and Black river, and lastly the calciferous and Potsdam sandstone, if the latter is present. We shall look then, if the theory is true, for the Trenton limestone, as the rock which rests immediately upon the slate; but what do we invariably find? It is calciferous sandstone, there is no exception, or Potsdam sandstone when it is present. This fact can not be reconciled with the theory of an inversion, inasmuch as it is a physical impossibility to arrange the rocks in the order we find them. Admitting again the theory of plication, I say it can not deceive us, or lead us into error. The theory supposes a succession of close plications or folds. In this case, every alternation in the fold corrects the error of the inversion. If the true order is determined, then, the series is a check upon its own errors. Besides, on the southeastern slope, before the inversion takes place, if there is one, the rocks must occur in their true order of superposition, and by the exposures on these slopes will prove a check upon their inverted position on the northwestern side. In proof that plications can not involve us in the error which has been supposed, and mislead as to order of superposition, I have found the Taconic slates on both sides of a ridge, and the lower Silurian, confined to the crest of the ridge, showing that the mass of the ridge is slate, with only a crest or capping of lower Silurian, which it seems to me proves there never was a fold or inversion, but a simple dislocation.

Facts are always useful aids when debating important questions. In the midst of the most disturbed districts of Virginia, at the head waters of the Clinch and Holstein, the lower Silurian never furnishes an instance of an inversion or plication by which the Lorraine sandstones and shales, &c., are folded beneath the Potsdam sandstone or calciferous sandstone. See fig. 8, where their order is represented. The insuperable difficulties, then, which attend this theory require its dismissal; besides the slate beneath the calciferous can not be regarded as the Utica slate, or any of the beds next above it, in the Silurian series. If it is carefully examined across the line of strike or dip, we find it in conformity with beds whose fossils are unknown in the Silurian system.

A theory which may be true in the abstract, and perhaps may be sustained by facts disclosed at certain localities, still when it fails to explain the phenomena at other places and is even entirely at variance with the principal facts, should be abandoned. Its conditions require that the rocks when plicated should lie in a certain order, they must lie in a certain relation which the theory supposes, but it is almost needless to say, that the order in which they actually lie upon the lines of fault in Vermont, New York and Virginia, is entirely at variance with the position the theory requires. Indeed, when the theory is applied to the rocks under consideration, as explanatory of their present position it involves a physical impossibility.

SUMMARY OF PROOF CONTAINED IN THE FOREGOING REVIEW OF OPINIONS.

§ 50. 1 The proofs are found in the constant relation which the lower Silurian rocks hold to both divisions of the Taconic system, the relations being those of an older and newer system, inasmuch as the evidence of superposition in consequence of succession is an incontrovertible fact.

2. It has been shown that where the Silurian rocks were folded or plicated, the fold itself did not extend to the slate upon which the Silurian rocks repose. The fold is confined to the overlying and unconformable rock, so that in the cases of plication the evidence of different periods to which the rocks belong is strengthened.

3. Again, I have proved that when a member of the lower Silurian seems to occupy a conformable position to the slate beneath, and hence might belong to it as a member of a group, yet it is still proved that the mass is really unconformable and rests on an inclined surface of the slate, as at Hoosick falls.

4. It has also been proved that troughs in which a member of the lower Silurian group was deposited may deceive by conveying the impression that the included limestone belonged to slate as a member of the group, as at Highgate, Vt.

5. I have also made it evident beyond a doubt that the fact of the existence of Silurian rocks in the midst of the Taconic system furnish by their presence and their relations, the highest possible evidence of the existence of the Taconic system.

6. The strength of the evidence is increased (if possible) when it is considered, that these masses of Silurian rocks though really isolated, overlie both divisions of the Taconic system, that they are found at the western and eastern borders of the system and always occupy a position superior to and unconformable with the Taconic rocks upon which they rest. These overlying outliers dip in all directions, sometimes west and sometimes east and northwest, etc., according to local circumstances; another proof of the general unconformability with the system.

7. I have shown that the theory of plications has certain physical impossibilities to perform when it is attempted to apply it to any part of the Taconic system where the overlying rocks are of the lower Silurian epoch, the order in which the plicated mass must lie, being totally different from that which exists along the lines of fracture as at Bald and Snake mountain, etc.

THE TACONIC SYSTEM CONSIDERED AS THE REPOSITORY OF THE METALS AND AS A PERIOD OF ERUPTION.

§ 51. It is a rare combination or phenomena which circumscribes an epoch of geological events so clearly and so closely that it can be referred to a subordinate part of a given period.

Indeed most of the references to periods are rather approximations than determinations, and probably the references which I may make of the epochs of veins and eruptive rocks in the Taconic system may partake more of the former than the latter. Still, when all the facts are brought together and weighed, I am disposed to regard the veins and eruptive masses which traverse this series of rocks as really belonging in part to the

Taconic period. This result which I have stated is rather confirmed by the fact that the presence of eruptive rocks is confined to the earliest part of the succeeding period, and hence it may appear that the eruptive period to which the veins and dykes of greenstone and porphyry belong may be referred with greater exactitude to both periods, the first part of the Taconic and the first part also of the Silurian period.

§ 52. *Veins which may be regarded as belonging to the Taconic period.*—The lower slate rocks of the Taconic system are remarkable in some parts of the country for the frequent occurrence of veins of milky quartz. In New England and New York they are white opaque masses traversing the rocks with very little regularity. They usually appear to thin out rapidly, running out from a large mass in thin strings, and to terminate very soon in the neighboring rock. Besides, they not only end speedily in a lateral direction as I have stated, but appear also to thin out beneath and to terminate in thin strings. So constant are these veins or irregular masses, that the talcose slates may be said to be characterized by them.

The minerals which are associated with this kind of quartz, are chlorite, sulphuret of iron in crystals, carbonate of iron and stains of manganese. They can not be regarded as metalliferous, notwithstanding we frequently find in them a few of the metallic combinations. But veins possessing the character of true veins also belong to the talcose and chloritic slates, which is their veinstone also, and I am unable to perceive that the quartz in its mineralogical characters differs from that already referred to. The veins, however, of the latter extend laterally and in depth to an indeterminable distance; and when they divide into strings, the fact may be regarded as an exception to the general law or rule which they obey.

The latter are metalliferous; and to this class we may refer both the auriferous and cupriferous veins of Virginia, North Carolina and Georgia. The rock in which the gold occurs in the southern states is regarded by many as the primary talcose slate, which is associated with hornblende, gneiss and mica slate. In North Carolina and Virginia these auriferous rocks are associated with breccia and conglomerates, and such are their relations that it can be no longer doubted that the formation of the auriferous veins comes within the epoch of the oldest sediments of the Taconic system; for although the porphyries and metallic copper veins occur in the lower Silurian, still auriferous veins appear to belong to a much earlier period. But again it is true, that many of the auriferous and cupriferous veins occur in the talcose and hornblende rocks of the Blue ridge, still as the veins run in a direction parallel with those of the Taconic system, they should therefore be regarded of the same age or epoch, and the auriferous veins of Somerset, Vt., should also be referred to the Taconic period.

Native copper occurs rather sparingly in certain porphyroid rocks in Chatham county and may be also cotemporaneous with the auriferous quartz vein.

§ 53. In addition to the foregoing metalliferous veins of this epoch, it seems to be established that it is also the period to which the veins of magnetic and specular iron belongs. In North Carolina both species of ore are found in veins traversing the slates which are associated or connected with beds of conglomerates as in Chatham county, four miles northwest of the gulf. Magnetic iron also occurs in Randolph county, N. C., in the same kind of rock as in Chatham.

In the first part of this work I have shown that the epoch of the production of veins of specular iron belonged in part to the lower Silurian period, inasmuch as the Potsdam sandstone is disturbed where the beds crop out. In northern New York, however, serpentine is the accompanying eruptive rock. The period to which a portion of the trap and porphyry belongs in North Carolina, is equally well determined as that respecting the auriferous quartz veins.

In this state, for example, there is probably one of the most singular belts of trap in this country. It traverses Guilford county, and the western part of Alamance, and pursues a northeasterly and southwesterly course across the state. This belt consists of numerous veins of trap, granite, quartz and feldspar which cross each other in various directions, and in their frequent crossings they form so complete a net work that the rocks traversed can scarcely be distinguished. That this singular net work of eruptive rocks, should be referred to an epoch as late as the Taconic period, is evident from the fact that it penetrates these rocks along its western border in Alamance, Guilford and Davidson counties, and it occupies in this system a belt of it, six miles wide, at least, and extends also into the adjacent primary district, so that these veins occupy a space between the two systems. But it may be questioned perhaps, whether these eruptive rocks may not belong to the Triassic period, or to one near the close of the palæozoic period. When we compare this net work or mass of dykes and veins with the trap of the last part of the palæozoic period, or a period extending from the last part of this period to the Triassic, we can not fail to discover a remarkable difference in the material composing them. In the latter the traps are not accompanied with veins of quartz, granite and feldspar; at least, I have not observed them. In the kind of matter and in the circumstances, there seem to be those differences which indicate that the two periods of eruption are distinct; and hence I am induced to regard the net work of eruptive rocks which occupy a belt between the Taconic system and the pyrocrystalline rocks as belonging to the Taconic period. We know that it comes within the period of sediments. But there are no positive data to fix it with certainty.

THE TACONIC PERIOD ONE OF ANIMAL AND VEGETABLE LIFE.

§ 54. This system is not less thoroughly peculiar in its organisms, than in its physical characteristics. It is true that the number of its fossils is small when compared with the Silurian period; but as far as they go they stamp upon it a distinctiveness which is as marked as that of the Silurian and carboniferous. But this is not all: we have a right to consider the absence of certain

Silurian fossils, a fact which looks favorably towards the view I have attempted to sustain. I admit that even the Silurian system is not equally rich in fossils in all its groups. In some instances their absence is accounted for on principles upon which all geologists are agreed, and which are considered as good and sufficient reasons for their absence. There are instances, too, of their absence, for which we can not assign a satisfactory reason. As a general rule, however, the fossiliferous bands occupy nearly the same horizons, and they are so rarely absent that the palæontologist always expects to find them. It is not so, however, in the Taconic system; there is a general barrenness of life and vitality, which is not accounted for, unless it is regarded as due to the period in which the rocks were deposited. Their rarity is not local; it is coextensive with a certain series of rocks. While the Silurian carries its characteristic fossils for more than a thousand miles, the Taconic system is equally comparatively barren for the same distance. Again, the scarcity of fossils can not be explained on the ground that the rocks have not been examined. This series of rocks have been under the eyes of geologists since 1817; they have been examined minutely in Rensselaer and Washington counties, N. Y., and Berkshire, Mass., and with more or less care over the whole area of western Vermont. A few fossils only have been discovered over this large area Of the fossils which these rocks have furnished, marine vegetables are the most common, but they are limited to a few obscure species; the thickness of the bed in which they they occur is at least 2,000 feet. Graptolites rank next in numbers; they even exceed the marine plants in the number of species which have been found. In addition to the foregoing. there are three species of trilobites and some four or five of molusca.

The plants, it will be observed, occupy a wide and vertical space; it is the reverse of this with respect to the animals. The graptolites of the Hoosick roofing slate are confined to beds whose thickness scarcely exceeds two feet. The trilobites are quite limited also, and the little Staurograpsus is confined to the thickness of half an inch.*

I propose now to describe the fossils of this system in the following order: 1. The Marine vegetables; 2. Graptolites and the supposed foot-prints of Molusca, or, as some regard them, as animals allied to graptolites; 3. Molusca; and 4. Trilobites.

The Marine vegetables may be separated into three divisions: 1. The flat leafy expansions, using the term leafy in its ordinary meaning; 2. The elongated and rounded flattened chord like bodies lying in convoluted folds; 3. Stem like bodies usually short and rounded.

From the imperfect condition of all the vegetables in this system, it is impossible to classify and arrange them in a satisfactory manner. Hence, the most which can be done, is to give them some name by which they may be known. In my report of the agriculture of New York, I applied the common appellation fucoids, then in use. In Mr. Hall's report, two of those vegetables were named generically, Buthotrephis. I have no objection to the name; I shall therefore adopt it.

1. MARINE PLANTS.

Buthotrephis rigida, *pl.* 2, *fig.* 1.

Fucoides rigida.

Frond rather narrow, branching and only slightly curved or flexuous. It occurs at numerous places in Rensselaer and Washington counties on the black flags and slates. It is much less common than the following.

* If we assume the Hoosick roofing slates to be the repositories of the most ancient graptolites, and then trace the beds upwards and into the places where the graptolites and fucoids occur, we can not fail to be satisfied that these low forms of life and vitality are distributed through a much greater thickness of rock than I have stated in the text.

B. FLEXUOSA (*Hall*).

Fucoides Flexuosa (*Emmons*).

Frond wide, flexuous and branching, consisting of a very thin expansion of vegetable substance. The two species differ much in size; the first has much more substance, and appears stiff and rigid, and the frond is scarcely more than one half the breadth of the flexuosa, and is quite uniform in this respect. The flexuosa is eight or nine inches high, as the stem appears in the best preserved specimens, though the frond is often broken or interrupted by layers of slate. There is not much doubt that the two are distinct species.

I have observed two other kinds also in the rocks of Rensselaer—one with a very narrow frond, less than a line in width, with only a few branches, and another whose width is the same as the B. rigida, but had no branches on the part exposed, which was about six inches.

BUTHOTREPHIS ASTEROIDES (*Fitch*).

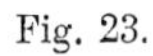

Fig. 23.

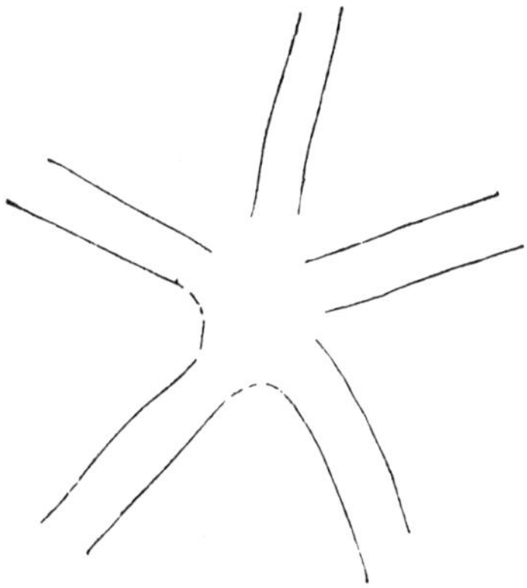

Frond stellate, having five branches radiating from a centre; Salem or black thin bedded slates.

Dr. Fitch refers this singular, though imperfect, organism to Mr. Hall's genus, Buthotrephis; perhaps correctly. It is, however, a remarkable form for a vegetable, and it may turn out to be one of the graptolites, inasmuch as discoveries looking that way have already been made. The edges are without cells or crenulations, but the extremities of the arms were much longer, it would seem, from the manner in which

* Transactions of Agricultural Society, 1849.

they terminate, than they appear at present on the specimen. Four or five of these singular plants, or graptolites, were observed on a single slab of slate two or three feet square.

Mr. Hall describes a vegetable fossil which occurs in the neighborhood of Union village and Salem. It is supposed to consist of fragments of succulent stems of plants, more or less compressed, but of a rounded form, and may be tubular or solid; they are about three inches long; they are referred to the new genus Palæophicus; see Palæontology N. Y., p. 263. It is the Palæophycus virgatus. Another variety or kind of stem was found several years ago in the green shales of Rensselaer county. The stem is twisted, see fig. 24, and may belong to the same genus or family as the foregoing.

Fig. 24.

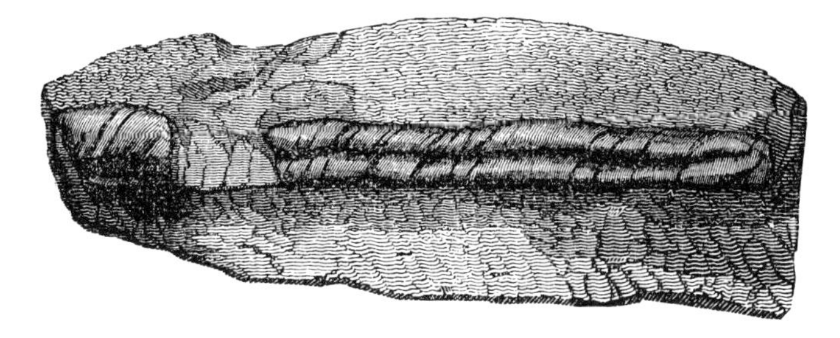

These fragments are evidently casts of parts of vegetables, but their characters are too imperfect to enable the palæontologist to do more than refer them to the class of marine vegetables.

The rounded and slightly flattened chord-like fronds (if a frond), which are rarely if ever branched Professor McCoy has expressed the opinion that they are also marine plants, and has given them the generic name Palæochorda.

Generic char.—Frond, very long, cylindrical, chord-like. very slowly tapering at each end, surface smooth rarely dichotomous (*McCoy*).

PALÆOCHORDA MARINA, *pl.* 2, *fig.* 8.

Gordia marina (*Emmons*).

Helminthoidichnites marina (*Fitch*).

Frond very long, sinuous, slightly flattened by pressure, and one line in diameter. It occurs in flags at McArthur's quarry, in Easton, Washington county. A small part of the fossil appears in the figure.

PALÆOCHORDA TENUIS, *fig.* 25.*

Helminthoidichnites tenuis (*Fitch*).

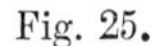

Fig. 25.

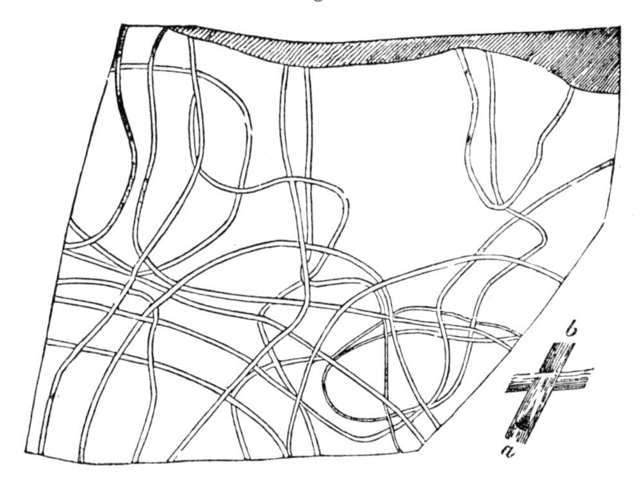

The frond in its convolutions is similar to the marina, though rather more complicated; its diameter is about half a line.

Dr. Fitch has given the above name to this animal or plant, footprint or whatever it may be, on the ground that it is a footprint or mark of some marine worm. Dr. Fitch's specimens of the tenuis were found by him at Middle Granville, Washington county, N. Y.

---

* Transactions of the Agricultural Society, 1849.

2. GRAPTOLITES.

The discoveries which have recently been made respecting the relations which exist between the serrated arms or bodies usually known as Graptolites and a flat membranous central disk, and also the discoveries of other forms belonging to this family, will no doubt render it necessary to reclassify and rearrange the species so as to form and harmonize this curious family of animals. Without attempting, however, any thing of the kind at the present time, I shall satisfy myself with following Dr. H. B. Geinitz's arrangement as far as it has come to my knowledge. In regard to the affinities of the family, it seems that it partakes more of the asteroid than the hydroid type of Zoophites. But it may turn out that it contains species whose affinities may be in one case hydroid and in another asteroid; for it is certain that the central membranous body which has an asteroid type in one or two cases, is not always formed upon this plan.

Family—GRAPTOLITHINA (*Bronn*).

Genus 1. *Diplograpsus* (*McCoy*).

Serrations on both sides of the stem; the stem provided with an axis.

DIPLOGRAPSUS SECALINUS, *pl.* 1, *fig.* 11.

Graptolithus secalinus (*Hall*).

Fucoides secalinus (*Eaton*).

F. Simplex (*Emmons*).

Straight; serrations sharp or pointed, cells rather distant oblique to the axis; the serration equal in length to one-sixth or one-seventh of the width of the stem. The upper or young part of the stem is three-eighths of an inch wide, and the number of serrations is twenty-four to an inch. It narrows towards the base where the serrations are rather obtuse and more distant than those above, and is ten inches long as exposed upon the slate. It is confined to the Hoosick roofing slate.

Mr. Hall maintains that it is the G. pristis expanded, or widened by pressure. The intermediate varieties referred to, especially those upon the slate from this locality, belong to the inferior part of the stipe. The width of the species is variable, that is, from one-half to three-eighths of an inch, along the wider parts of the stem. Besides, the number of serrations differ from those of the pristis, the latter having thirty-six in the length of an inch and the former twenty-four; besides, the cells it will be seen from the figure which is of the natural size, differ entirely from those of the pristis. The differences, therefore, can not arise from any amount of pressure they have received; and, besides, this principle is inapplicable to the case.

D. RUGOSUS (n. s.), *pl.* 1, *fig.* 26.

Stem straight, thick, central column deeply and transversely corrugated and occupying all the space between the cells; the large cells appear to meet near the middle of the stem. In this species twenty cells are developed in an inch, and when they are removed by disintegration a broad depression is left, which gives a rugose appearance to the stem. It occurs at Parrottsville, Tennessee, in a black slate, which weathers to a soft light drab.

D. DISSIMILARIS (n. s.), *pl.* 1, *fig.* 5.

Straight; cells dissimilar on the different sides of the stem; on one side they open at right angles to the axis, on the other obliquely; axis nearer to the margin of the oblique serrations. The stem, excluding the tips of the serrations is 3-16 of an inch wide. The number of serrations or cells in an inch is twenty-six. The figure is taken from the base of the stem, and is the only part which has been found. It occurs in reddish slate, in Augusta county, Va.

D. CILIATUS (n. s.), *pl.* 1, *fig.* 19.

Straight, thin and ciliated; ciliæ, bulbous and jointed or transversely marked proceeding from the point of each serration;

serrations unequal, the intervening smaller serration rounded, the larger prolonged and run into the base of the ciliæ, axis distinct.

The specimen is imperfect, but probably, from the character of the column, it was free. The entire width of the column embracing the extended lateral ciliæ, is one-fourth of an inch, the membrane is rather less than one-eighth of an inch wide, the margins appear to be dissimilar. In another specimen the end is rounded and complete, and furnished like the sides with ciliæ. Found in Augusta county, Virginia.

D. OBLIQUIS (n. s.), *pl.* 1. *fig.* 22.

Straight, serrations turned obliquely outward, exposing the mouth of the cell.

The substance of the graptolite is olive green, thin and membranous. There are twenty-four cells in an inch. The sides are similar; axis, if any, concealed. Found in Augusta county, Virginia.

D. FOLIOSUS (n. s.), *pl.* 1, *fig.* 13.

Serrations prolonged, pointed and leaf like; expansions directed backwards or towards the base of the stipe at this part of the column; the upper end, the lengthened points, are directed upwards. The length of the free parts of the serrations is greater than the width of the column. This graptolite has forty cells in an inch, the substance green, or similar in color to the preceding. The cells appear as if they were arranged in a circular or spiral manner around an axis. It is found in soft, reddish slates in Augusta county, in Virginia.

GEN. MONOGRAPSUS.

Serrations confined to one edge of the stem; axis none.

M. ELEGANS, *pl.* 1, *fig.* 27.

Outer edge of the serrations straight and nearly parallel with the opposite edge; depth of the serration equals one-half of the width of the stipe. Fig. 14 enlarged. The width of the stipe is about one-sixteenth of an inch and there are twenty-four cells in an inch. The substance of the graptolite is green and coriaceous. This beautiful species occurs in Augusta county, Va., soft whitish slates.

M. RECTUS (n. s.), *pl.* 1, *fig.* 28.

Straight, serrations pointed, upper edge of a serration oblique to the axis of the stem. Width of a serration equals one-half of the width of the undivided part of the stem; width of the stem one-eighth of an inch, and twenty-two crenulations in an inch whose edges are perfectly straight and not curved so as to leave a curved space between the notches.

From the point of each crenulation, there runs an oblique ridge which meets a longitudinal one, the latter runs nearer the straight than the crenulated margin. This species occurs in Columbia county, in the Taconic shales and is closely allied to G. latus of McCoy.

GEN. CLADOGRAPSUS (n. s.)

Serrations or cells arranged on the outer edges of a branching stipe; axis none.

C. DISSIMILARIS (n. s.), *pl.* 1, *fig.* 15.

Outer edges of the stem dissimilar, serrations rounded, mouths of the cells appear to open at the base of the serrations on one side; on the other, at the apex.

The thin upper membrane when removed, discloses curved canals leading from the cells to the middle of stem. The substance of the graptolite is green and coriaceous.

C. INEQUALIS (n. s.)

Stipe very narrow and arcuate; serrations obsolete, being indicated by a waving edge; the cells open on the margin of the stipe just at the anterior edge of the faintly developed serration; the cell is an elongated oval, lying slightly oblique to the margin. The branches in this specimen are somewhat dissimilar, one being rather wider than the other. It is found at Parrottsville, in Tennessee, in a soft decomposing slate.

Glossograpsus (new genus).

Column free; thin membranaceous, ligulate, extremeties rounded, axis distinct.

G. ciliatus (n. s.), *pl.* 1, *fig.* 25.

Straight linear crenulations faintly developed and prolonged into ciliæ, equal in length to the width of the ligulate body, ciliæ surrounding the whole body or membrane. The axis is prolonged beyond the membrane forming the column or stem. Length one inch. This graptolite seems to be one of the perfect forms under which this family is sometimes developed. It is well known that these serrated bodies, the Diplograpsus, etc., are but the mutilated parts of the animal, which radiate from a central membranous scolloped disk in some instances.

The Glossograpsus I regard as a perfect animal with its axis extended beyond the body; and this organ may serve to attach it to other bodies, or it may have floated freely. This form of axis often appears among other fragments of graptolites and hence like them may be regarded as a distinct species. The Diplograpsus, ciliatus and crinitus, may also belong to this sub family Glossograpsus, inasmuch as we know they are rounded at one extremity, but the character of the other is not determined and hence I have placed those in the genus Diplograpsus. The genus occurs in the dark colored shales of Columbia county, N. Y.

Staurograpsus (new genus), *pl.* 1 *fig.* 21.

Disk free, cruciform, arms four, dichotomous cells terminal, substance membranaceous, free and furnished with an axis.

S. DICHOTOMOUS.

Disk, provided with four arms each, of which is divided near the extremity and forms a cell. Surface of the arms uneven and it is possible cells existed on the sides of the arms as well as at their extremities. This small and remarkable graptolite belongs to the asteroid type, but it preserves its relation to the graptolites by the presence of an axis which is visible where the branches are separated from the arms. The arms might have been reckoned as five instead of four, inasmuch as one divides just beyond the point where they cross. The largest of these asteroid graptolites is represented in the small figure by the side of the enlarged one. It was found in the black Taconic shales of Rensselaer county, N. Y.

NEMAGRAPSUS (new genus).

Axis elongated and thread like, simple or compound branches round at the base and flattened at the extremities; cells appear to be arranged on the flattened part of the axis instead of the margin.

N. ELEGANS (n. s.), *pl.* 1, *fig.* 6.

Axis arcuate and sending off numerous branches from the convex side, branches round where they leave the arcuate axis but expand upwards; margin even.

This species is extremely attenuate where the axis becomes flattened, and hence, through the particles of shale are fine, they break and obscure the continuity of the membrane. It is evident, however, there are no serrations upon either margin, and under the microscope the surface is slightly dotted so as to give the appearance of the mouth of cells. It is found in the Taconic slates of Columbia county, N. Y.

N. CAPILARIS (n. s.), *pl.* 1. *fig.* 7.

Axis long, convoluted and furnished with a few short branches; under the microscope it appears annulate, but whether the rings are due to structure or to fractures, it is difficult to determine. The substance resembles the axis of a graptolite. It is, however, a veritable fossil, though I am unable to discover the cells proper to graptolites or other appendages for nutrition and growth. Taconic slates of Columbia county.

NEREOGRAPSUS (*Geinitz*).

"Biserial, the stem having no central axis or a very soft one."

N. JACKSONI, *pl.* 2, *fig.* 2.

Convolute, crenulations large, rounded and rather oval, the depression on each edge separating them, meeting in the middle on the line of axis. The body is half an inch wide and less than four crenulations in an inch. This large Nereograpsus occurs in the green talcose slates of Waterville, Me.

N. LOOMISI.

Convoluted, narrow crenulations, lanceolate terminating in an axis, or rounded ridge. There are twenty-eight crenulations in an inch. Waterville, Me.

N. DEWEYI, *pl.* 2, *fig.* 3.

Convoluted crenulations, small, rounded, and terminating in the middle of the stem, which on the side exposed produce a groove. Crenulations in an inch, twenty; width of the body one-eighth of an inch. This is one of the finest of the Nereites* in the series belonging to the Waterville slates.

---

* I have many doubts respecting the class to which the agents belonged which produced the singular impressions which have been regarded as foot prints at one time, and at another as belonging to a class of annelids, and which have received the name of Nereites, Myriantes, etc. But as Dr. Geinitz has seen the open mouths of cells, it is probable they should be placed in the family of graptolites.

N. GRACILIS, *pl.* 2, *fig.* 6.

Convoluted, crenulations numerous, narrow, scarcely pointed, extending to the middle of the body. The number of crenulations in an inch is twenty-two. Waterville, Me.

N. LANCEOLATA, *pl.* 2, *fig.* 4.

Convolute, centre of the body has a narrow ridge to which the lanceolate crenulations extend. The width of the body is five-eighths of an inch, and there are ten crenulations in an inch. Waterville, Me.

N. PUGNUS.

Convoluted, crenulations large, long, oval, extending to the middle. A part of the specimen is cylindrical and without crenulations.

In addition to the foregoing, there are two species of Myriantes, at the Waterville locality. These singular bodies are confined to a thickness of slate not exceeding four or five feet.

N. ROBUSTUS (n. s.), *pl.* 2, *fig.* 7.

Convoluted, crenulations nearly round, terminating in a strong, narrow ridge in the middle. There are about eight crenulations in an inch; body one-fourth of an inch wide. This Nereograpsus was found in the Taconic slates of Columbia county, N. Y., thus proving a similarity or parallelism with slates of Waterville, Me.

3. MOLUSCA.

Most of the molusca of this system belong to the family of Brachipods, all of which are so minute that it is difficult to discover their most important characters. Their shells are so extremely delicate and thin, that it is impossible to succeed in exposing those parts of the shell upon which their specific characters are founded if they are concealed in the rock. It is highly probable, therefore, that the references I have made may be incorrect. The figures of the forms and markings embrace all the characters which can be relied upon for their determination.

GENUS LINGULA.

L, striata, pl. 1, *fig.* 17 (n. s.)

Minute, oval, beak acute, concentric lines few, radiating lines distinct and numerous, comparatively wide at the extremity opposite the beak; it tapers rapidly to an acute beak, from a point about two-thirds the distance from the opposite extremity; it is extremely thin and attenuate; occurs in the white fragile shales of Augusta county, Va.

L. ELLIPTICA (n. s.)

Small, elliptic, extremities rather rounded, lines of growth faintly preserved, margins gently and regularly curved and alike; extremities subequal. The shell is extremely thin and delicate, and nearly one-fourth of an inch long. It occurs in the white fragile shales of Augusta county, Va.

LINGULA, *pl.* 1, *fig.* 9.

In this fossil there is a departure from the common characters of this genus. The obliquity, however, of the figure, on a careful examination of the specimen, is rather exaggerated; the apex is subcentral, or rather the shell is inequilateral.

GEN. ORBICULA.

O. excentrica (n. s.), *pl* 1, *fig.* 4.

Small, extremely thin, ovate, apex excentric and acute, rather elevated, rising from a nearly flat expanded border, the base of which is marked by a sharp ridge; concentric lines or lines of growth obsolete in front, distinct behind. Found in the white fragile shales of Augusta county, Va., associated with lingulas, graptolites, &c.

Gen. OBOLUS, *pl.* 1, *fig.* 10.

I refer to this genus, pl. 1, fig. 10. It has no teeth which can be discovered, but it is slightly inequilatral, and the groove for muscular attachment does not appear in any of the casts of this shell. The external form of the shell and its striæ are well preserved. It occurs in the whitish shales of Augusta county, Virginia.

CYPRICARDIA, *pl.* 1, *fig.* 1.

This fossil is referred to Cypricardia from its form. It is impossible to discover the essential characters in the teeth, if it has any. It had a very thin shell, and the lines of growth are rather prominent. A single muscular impression is preserved.

4. TRILOBITES.

The trilobites which have been discovered in the Taconic system, belong to a family which has been known in the lower Silurian rocks for many years. The species, however, are unknown in the Silurian period. They were first discovered in fragments, and hence there were reasons for difference of opinion respecting them. The Eliptocephalus and Atops were discovered in Washington county, N. Y., in dark-colored shales. I published figures in my reports of the foregoing genera, which I regarded as new at the time, and subsequent discoveries of specimens both species have confirmed the opinions which I then expressed. I have since discovered a minute trilobite in the shales, in Augusta county, Va., which I regard as older than those of Washington county. They are not, however, far removed from the same geological horizon. The condition of those in Washington county renders their characters somewhat obscure. They are not, however, distorted, but simply flattened, but not so much as to obliterate the stronger lines upon their surfaces.

1. ELIPTOCEPHALUS ASAPHOIDES,* *pl.* 1, *fig.* 18.

Olenus asaphoides (*Hall*).

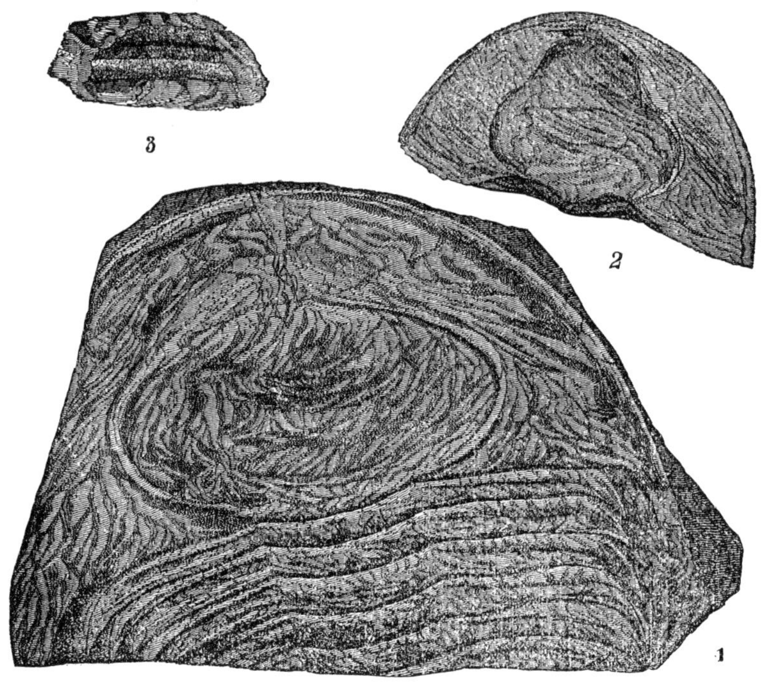

E. asaphoides.

Triangular, buckler, semi-eliptical, terminating posteriorly in elongated sharp shines. Suture indicated in part by an elliptical outline of the glabella, which in young individuals is lobed; eyes none, or entirely undistinguishable in the specimens. The number of ribs is unknown, but the number which can be distinguished is fourteen, which is evidently less than belongs to the species; pygidium unknown.

The Eliptocephalus has been referred to the genus Olenus by Mr. Hall. The ground upon which this reference is made is unquestionably insufficient, as the latter has been constituted.

* The genus Elipsocephalus was unknown to me at the time of the publication of the above; the name, from its similarity, is no doubt objectionable, but I am disposed to retain it for the present.

For comparison I have annexed a figure of the Olenus gibbosus Dalman, the one which is best known of the species. The buckler and its spines are much shorter, and from which the glabella is separated by a furrow. The length of the head of the Eliptocephalus is proportionally much greater than in the Olenus, to say nothing of the eyes; and the suture which, according to Mr. Hall, is similar to the Olenus, there is really no mark or line by which it can be determined, and probably if its direction in the specimen could have been seen, it would have been described, instead of saying in general terms that "it is as in the Olenus," for this is not an available description. The number of ribs is also greater. The figures, however, are placed side by side and may be compared by the student. I have also added a small figure of a Paradoxides, in which the comparative length of the head is much greater than in the Olenus.

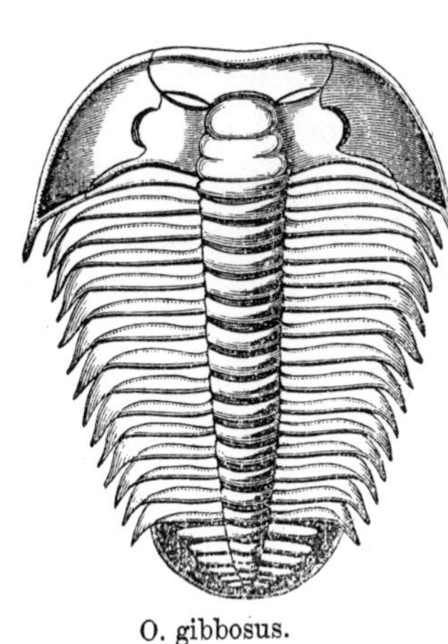

O. gibbosus.

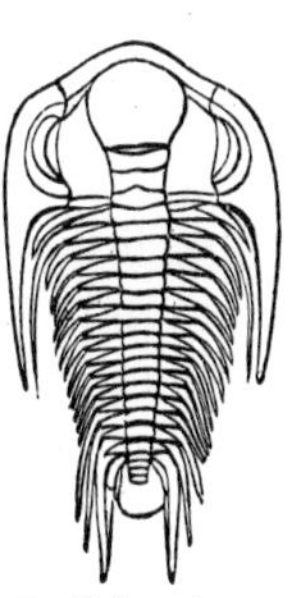

P. Bohemicus.

2. ATOPS TRILINEATUS, *pl.* 1, *fig.* 16.

Calymene Beckii (*Hall*).

Crust granulated, cephalic shield semicircular, with its anterior and lateral edges turned upwards; posterior angles rounded, facial suture, beginning at the outer angle of the cephalic shield, and runs nearly parallel with the anterior margin to the middle lobe, when it turns at a right angle and runs parallel with that lobe; eyes undistinguishable, body composed of seventeen or eighteen rings, narrowing very gradually to the caudal extremity; pygidum a flat expansion of the crust and is provided with a single ring; axis narrower than the lateral lobes, rings seventeen, each of which is separated by a groove about as wide as the rings. Axis armed by a row of short spines; lateral lobes provided with a row of tubercles or prominences along the median line; margins of the rib groove run parallel as far as the tubercle, when they diverge; tubercles become obsolete towards the tail; caudal shield very small and provided with one, or at most two rings.

This fossil has been confoünded by Mr. Hall with the Triarthrus Beckii, Greene, or Calymene Beckii. He has been misled by the row of short spines along the middle lobe or axis, which it is well known exist in that species. It differs, however, from the Triarthrus in every other character; in the number of abdominal and caudal rings, the proportions of the subdivisions of the head, the granulations of the integuments and the row of tubercles along the lateral lobes. So palpably different were the heads of the Atops and Triarthrus, that a committee of the American association decided they were different from the heads alone: and since this decision was made the specimen figured has been found, which sets the question at rest.

Both of the foregoing are found at Reynolds, about seven miles north of Union Village, Washington county, N. Y., in the shales of the Taconic system.

3. Microdiscus (new genus), *pl.* 1, *fig.* 8.

Microdiscus quadricostatus.

Minute, oval, middle lobe of the cephalic shield strongly developed; ribs of the body or abdomen, four; of the tail, four or five. The form of the cephalic shield is only obscurely indicated; the size of this trilobite is shown in the small figure. It is found in the white fragile shales of Augusta county, Va., associated with minute moluscas and graptolites.

GENERAL DISTRIBUTION OF THE TACONIC SYSTEM IN THE UNITED STATES.

§ 55. The Taconic system forms a belt on each side of the Blue ridge. On the west side the belt is continuous from Canada East to Georgia. On the east side, the belt is wider at certain places than on the west side, but its continuity is broken and I am inclined to believe it never embraces rocks which belong to the upper division. It is more distinct in North Carolina and Virginia than in Maryland or in any of the states farther north. In Rhode Island and Maine it occurs in areas of small extent. The quartz and its associated talcose slate is the base of the system, and constitutes the largest part of the lower series. On the east side of the Blue ridge the Silurian system is unknown, and hence we have no data by which we can directly determine the epoch of this sedimentary belt, and hence we are obliged to rely upon lithological characters for its determination, and as these agree in the main with those upon the west side, which are overlaid by the rocks of the Silurian system, we refer them accordingly to the same period. The only sediments which overlie the Taconic system, are those belonging to Professor Rodgers' middle secondary period, and hence from the time the lower Taconic were formed up to the middle secondary period, the eastern side of the Blue ridge was dry land, or rather that section which is now known as the Blue ridge. At Haverstraw, the new red or middle secondary is formed of materials derived in part from the lower limestone of the Taconic system. On the west side of the Blue ridge, the lower rocks, quartz, talcose slates and limestones have a strike of N. 60° E. if reckoned on the data furnished by the southern extremities of the series. Thus from the southwest corner of North Carolina, in Cherokee county, the line of outcrop of the quartz rock as far north as Harpers ferry, on the Potomac, the strike is N. 60° E. In Massachusetts and Vermont the strike of the same rocks is nearly north and south. The width of the belt is greatest between Troy, N. Y., and Adams, Mass., where on an air line it is about thirty-five miles wide, though it is about as wide when measured on a line from Abingdon, Va., to Taylorsville, Tenn. The Taconic rocks are divided on the Hudson river by the highlands. The western or upper division crosses the river near Poughkeepsie. They appear on the west bank about five miles above Newburgh, and from this point traverse the county of Orange. It is near Poughkeepsie that the series begins to be deflected to the west, and in crossing the northern part of New Jersey into Pennsylvania, the series has passed into the same strike that it has between the south-west corner of North Carolina and Harpers ferry. I have not traced it farther south than north bounds of Georgia. I have examined and proved the system in Canada and Maine, and am able therefore to state from personal observation that it extends at least 1200 miles in lines parallel with the Blue ridge.

In Michigan, in the Lake Superior region, on the north shore of Lake Huron, the Taconic system is largely developed. It has been briefly described by Mr. Logan, and has also received the name Azoic system by Messrs. Forster & Whitney, United States geologists It was, however, first noticed by the lamented Houghton. I alluded briefly in my geological report to a verbal communication which was made me, expecting, however, a fuller account of the northwestern series at a future time. I have received a series of rocks from the northwestern region, which furnish me all the lithological characteristics of the series as it is developed in Washington and Rensselaer counties, N. Y.

Having never seen the Taconic system in the northwest, I shall content myself with what I have said in regard to it. Of its existence then, in the region specified, I have never entertained a doubt since I received the information I have already stated.

As facts have accumulated from time to time, I have become satisfied that the Taconic system will be recognized in other parts of the United States. It would, however be premature to attempt to identify it in sections which have been imperfectly explored.

Former extent of the lower Silurian eastward of the Hudson. The existence of insulated outliers of the lower Silurian system, some fifteen or twenty miles east of the continuous masses at the Cohoes, near Waterford, indicate that there was once a continuous sheet of them. Those isolated masses usually occur upon the knobs and ridges, or upon their slopes. In the valley of the Hoosick, at Hoosick falls, the ridge of the calciferous sandstone is some four or five miles long, and is well characterized by its fossils. The ridge west of Hoosick has escaped denudation in consequence probably of its being upon the eastern slope, for the northwest sides were exposed more directly to the action of denuding forces; it was the *struck side.* The effect of denudation furnishes important facts respecting the relative age and position of the two systems. Thus, in many places the Taconic rocks are swept perfectly clean, and are laid bare, while a knob or ridge in the line of strike is capped with the remains of lower Silurian rocks. These patches or remains of Silurian rocks differ greatly in thickness, a difference which arises partly from activity of the denuding force. But it is also probable that the sheet of the overspreading lower Silurian was never as thick upon the eastern outskirts of the system, as in the Mohawk valley, and besides the limestones which are well defined there, are in this region of outliers feebly represented and poorly defined. I have found at Greenbush the fossils of the calciferous sandstone, the Chazy, birdseye and Trenton limestones. Yet the two predominant masses are the calciferous rous and Trenton; and yet there are small masses of limestone which are scarcely determinable upon the spot which contains the Maclurea and Fucoides demissus of Conrad, the characteristic fossil of the birdseye limestone. The dip of the outliers of the Silurian limestones is exceedingly variable. At Hoosick, the dip of the calciferous sandstone coincides very nearly with the slate upon which it rests. It is an instance where, if the junction of the two systems could not be seen, it would be maintained that the two are conformable, and hence would lead the observer, as similar instances have already proved, to adopt the conclusion that being conformable masses, they belong to the same period.

§ 56. *Is it necessary to assume that the gneiss, mica slate and hornblende or the laminated pyrocrystalline rocks are altered sediments?*

The discussion of this question involves that of another, whether we have evidence that the Taconic system contains the sedimentary base I have claimed for it. Without debating this point at this time, I remark, that inasmuch as we are unable to recognize in the oldest palæozoic strata gradations into the laminated pyroctystalline rocks, and as it has been shown that there is a series of sediments far below any trace of an organism, it may well be inquired if they are not in that condition which may be rationally expected, for representing the oldest sediments upon the globe. The laminated do not, in any part of the globe, pass into the oldest sediments. Geologists do not find pebbles, obscurely defined, in laminated rocks, or other phenomena, which indicate that they were originally sediments. The possibility of changing sediments into laminated masses, in which all traces of the aqueous origin shall disappear and become obliterated, will not be denied. But on the ground, that of possibility alone, is it necessary to assume that the laminated rocks were originally sediments? seeing, too, there are no intermediate masses which prove a passage from the sedimentary to the laminated rocks, can it not be proved that sediments have a beginning in a clear and well defined base which forms also a distinct boundary between what we know to be sediments and those rocks whose aqueous origin is only assumed, because there are forces in existence which can change sediments into the likeness of the laminated masses. It seems to me unnecessary, therefore, to regard the laminated rocks as altered sediments.

Is the Taconic system established on facts and principles which are received and acknowledged by geologists of authority?

For the benefit of the student, I shall copy the views of Mr. Murchison as he has expressed them in a communication to the Geological Society of London, and which will be found in its proceedings, vol. 3, p. 38. Mr. Murchison says:

"That it is not by finding, after several years of elaborate research, a few undescribed and rare British palæozoic forms, that the age of rocks can be determined. The true tests are the order of superposition, and the common prevalent fossil types; for, if amid forms peculiar to one or two localities, the prevailing typical shells of a previously named group should occur in lower or thicker strata, or, if the band in question can be followed into other tracts, where the usual types abound, the point is determined."

The doctrine contained in the foregoing quotation is evidently this, that where there is no superposition of rocks, and where the fossils throughout a series of beds belong to the prevalent fossil types, we have the proof that the system is one system, though it may contain a few new and unknown species which are discovered by diligent search.

But, then, the doctrine is equally clear that if we find overlying rocks of one system and the fossil types do not belong to the common typical forms, but differ as a whole, then we are warranted in regarding two series thus distinguished, as belonging to different systems or different epochs. The foregoing principles form the basis upon which my convictions have rested. It is, first, upon superposition, as exhibited at Hoosick, Greenbush, Bald mountain, Galesville, North Granville, Snake mountain, Highgate, at Sharpshins, and at many other places where the relations are the same, that prove what I have contended for the Taconic system in all its length and breadth. But superadded to this, I have shown that the species of fossils differ also from the Silurian types; that is, the species are different, and they are not intermingled with the well known Silurian types, for of the latter none are yet known, and this last is a significant one. If the beds which contain fossils contain none which are Silurian, what are we to infer, especially if this is a general fact; certainly, that if fossils are to have weight in determinations of this kind, it goes strongly against the doctrine that the series is Silurian. If it is an established doctrine, that rocks which are separated in vertical space and also in time, will be characterized by different fossils, then that doctrine should govern our opinions. If we can account for the absence of fossils in a fossiliferous series, on established principles, their absence becomes of no account in questions of age; but when the fact is general, and it prevails for a thousand miles, it becomes significant. A fossil period will furnish fossils somewhere on lines so extended, and if on lines thus extended, they are not found, then we are justified in the belief that the period was not one of life and vitality. The lower part of the Taconic rocks belong to this period; they are, every where, so far as discoveries have yet been made, devoid of fossils.

The Taconic system rests, then, on the following points:

1. Its series, divided into groups, are physically unlike the lower Silurian series.

2. It supports unconformably at numerous places the lower Silurian rocks.

3. It is a vital system, having been deposited during the period when organisms existed.

4. As a natural history system, it is unlike the lower Silurian, first, in containing fossils yet unknown in the lower Silurian, and, second, in the absence of the typical forms which are prevalent in the lower Silurian.

5. In the Taconic system we have the palæozoic and sedimentary bases; the former comes in far above the latter, or at a period long subsequent to the time when deposits began to be formed.

6. The Taconic system carries us back many stages farther in time when life gave vitality to its waters than the Silurian. It represents a period vastly longer though it may occupy a less superficial area.

THE SILURIAN SYSTEM.

IT IS WIDELY EXTENDED IN NORTH AMERICA. ITS PHYSICAL FEATURES ONLY SLIGHTLY DISTURBED BY IGNEOUS FORCES. THE CHAIN OF ORGANIC BEINGS IS THEREFORE THE MORE PERFECT.

§ 1. The Silurian, which fills so large a volume in geologic history in Europe, seems to be still more full and complete in America. It extends from Canada on the north to Alabama in the south. The northern highlands lying between the St. Lawrence and Lake Champlain bulge up in an irregular dome, from which it is thrown off in every direction, but in a more important sense, in two directions; the first, towards the northeast; and the second, towards the southwest.

The dip of the rocks on the northeast side, indicates the existence of a great basin widely spread out in this direction which might be called the Laurentine basin of this system. Following the dip to the southwest, the indications are equally clear, that in this direction also there is another basin extending over an immense area, having its southeastern base in the Appallachian mountains, and hence may be called the Appallachian basin of the Silurian system.

The rough measurement over this anticlinal which separates these great basins of Silurian rocks which extend on the one side far to the northeast, and on the other to the southwest, gives at least twenty degrees of latitude. Following the base of this system from the northern extremity of Lake Champlain to the St. Lawrence and then tracing its course upon the irregular borders of the great lakes to the waters of the Mississippi, above the Falls of St. Anthony, it has a basal line nearly 1500 miles long.

This vast extent of base, taken in connection with the developments of the system northeast or southwest, or on both sides of the anticlinal shows the magnificent scale upon which this system was laid down in North America.

§ 2. When we take into account the great extent of the system as stated in the foregoing paragraphs, it will no doubt be regarded as a remarkable fact, that over these wide areas it is comparatively unbroken by igneous injections and the regular succession rarely broken up or displaced by outbursts of pyrocrystalline rocks; the sediments therefore, when examined on a large scale, seem to have been quietly deposited, or their orderly accumulation scarcely interrupted. This freedom from breaks and interpolated igneous masses, has an important bearing upon the existing regularity in the arrangement or succession of the organic stages which belong to this system; and from these favorable conditions for the prolongation or preservation of life, we may attribute the more perfect representation of these stages during the palæozoic period. The two facts seem to harmonize so well, that they may be connected together or related to each other as cause and effect. From the foregoing, it will probably follow also, that the time when important species were created may be more exactly determined, as well as their derivation and time of their disappearance or extinction. If so, the palæozoic period will be more interesting, as it will furnish a fuller and more complete history of life and organizations, especially when taken in connection with the vast area over which they are spread, furnishing thereby a more varied and more favorable condition for its manifold developments. Its natural history will be far more complete and full, and its connections preserved better than it possibly could have been, had this period been remarkable for extensive dislocations and disturbances.

SUBDIVISIONS OF THE SILURIAN SYSTEM.

§ 3. The present twofold division of the Silurian system gives the student all the advantages and facilities for investigation which are necessary and essential to his purpose, or which could be secured by a farther subdivision. The established divisions are *upper** and *lower*. The latter embraces the old geographical division of New York, the Champlain group. With this modification, the Oneida conglomerate should be removed to the base of the upper division, or upper Silurian.

The lower Silurian, to which I have had so many occasions to refer, includes the following rocks: Potsdam sandstone, calciferous sandstone, Chazy limestone, Black river limestone, Trenton limestone, Utica slate, Lorraine shales and sandstones. The last in this subdivision, has been frequently called Hudson river group. The only reason assigned for the name was, that this subdivision presented certain peculiarities arising from a disturbance it had suffered along the Hudson river. The Hudson river region, however, presents no facilities for the examination of the upper part of the lower Silurian; it is only in Lorraine or Pulaski, and the neighborhood of Rome, in New York, that this part of the series can be examined satisfactorily.

---

* The writer, while engaged in the New York survey, proposed the following geographical divisions, or groups, viz : Champlain, Ontario, Helderberg, Erie and Cattskill, or old red sandstone group. These groups were not considered at the time as good Natural History divisions, but as a large number of rocks were to be described, it became necessary to propose subdivisions so as to introduce some method into reports. The Champlain and Erie groups could be recommended, on the ground that they were natural, at least in part ; the Ontario group is also a natural group.

It has frequently been intimated that the subdivision of the New York rocks was too minute, and also founded on unessential characters, as well as those of an unequal value. The reason for this minute division was, that all the rocks named contained respectively many fossils which seemed to be confined to these, and hence it facilitated very materially a reference in each case ; it saved an indirect reference and circumlocution.

LOWER SILURIAN SYSTEM.

Its members and its distribution; compositoin in different parts of the Union; rocks which seem to be unimportant from the limited space they occupy, contain important memorials of the past; points of view under which rocks should be examined.

§ 4. The lower Silurian palæozoics seem to be fully represented in New York and along the eastern rim of the great Appalachian basin as far southward as Alabama. In New York and Canada the following rocks may be regarded as their representatives, viz. Potsdam sandstone, calciferous sandrock, Chazy limstone, birdseye limestone, Black river limestone, Trenton limestone, Utica slate, Lorraine shales, terminating in a thick bedded gray or greenish gray sandstone. The foregoing rocks are not of equal value or equally important, considered simply as members of a physical group. This arises from two or more causes; the first of which is the limited space they occupy when present; the second, from their frequent absence, and lastly, when regarded as the repository of fossils, they contain less which is important than their associates. Thus, the Black river limestone rarely exceeds fifteen feet in thickness, but as it furnishes at one or two places, and perhaps more, a fine black marble, it becomes necessary to give it a name and place in the series. So the birdseye rarely exceeds thirty feet, yet its fossils and the excellent lime which it makes, render it necessary to speak of it as a distinct rock. The Utica slate is scarcely recognizable in the western states, yet in New York it exists in a distinct mass, at least seventy-five feet thick.

To the foregoing list of rocks it may be proper to state in this place, that in Wythe county, Va., and other parts of the southwest, slates and sandstone are important masses, as they are intercalated between the Trenton limestone and the Lorraine shales. This rock is a soft, reddish sandstone, mottled with green spots, but contains narrow calcareous bands from ten to eighteen inches thick, which are heavily loaded with fossils, which, as a whole, resemble the Lorraine shales. It is from 150 to 200 feet thick, and near its junction with the Trenton limestone has beds of conglomerate. I deem it useful to notice even those rocks which appear unimportant, as masses, when placed by the side of those which are several hundred feet thick. Those items of information which may be gathered from all points, and upon which the geologist makes up a full statement of geologic history, become important in filling up gaps and intervals both in time and space.

In consequence of the interest which invests the study of the palæozoa and the rocks which contain them, their meteorological characters should not be neglected, as they usually are, excepting in those points which are general, or which are recognized at a glance, the minor characteristics being regarded as useless in geological reasoning. It is, however, certainly better to examine, with a good degree of minuteness, the physical constitution of all the rocks of a group. It puts us in possession of data which may be employed in interpreting an important class of phenomena which have their influence upon the organisms which they contain, and moreover they throw light upon those meteorological conditions which prevailed during the time of their deposition.

Rocks may be examined and studied, first: as to their composition and the variation which at distant points change with the circumstances under which they were deposited; and secondly, the physical and chemical changes which they have undergone since they were consolidated; and thirdly, as the repositories of the metals which belong to the period of their formation as fossiliferous masses, which contain the most valuable memorials of change and progress which the crust of the earth has been undergoing in past ages.

INDIVIDUAL AND SPECIAL CHARACTERISTICS OF THE LOWER SILURIAN ROCKS.

§ 5. *Potsdam sandstone.*—Formation No. 1, of Prof. Rodgers.—Considered as the base of this system, it is suggestive of many inquiries with respect to the period which it represents at its beginning. Its composition is uniformly silicious, but in texture and color it is variable. At Potsdam it is a firm and even bedded, even grained, yellowish brown sandstone. In other places in New York, as at Mooers and Moira on the Provincial line, it is equally even grained and even bedded, but is white and friable, especially at Mooers. At Chazy it is a deep red rock, and contains many particles of undecomposed magnetic iron, and at Whitehall, Keeseville, Corinth and Hammond it is white or brownish white, but is more or less vitreous. At Burlington and Charlotte, Vt., it is a jointed chocolate colored flagging stone; in other places in the valley of Champlain and Hudson river, it is a brown and rough bedded mass and quite thin. In Canada east, at the Falls of Montmorenci, it is a coarse sandstone loose in texture, and stained with carbonate of copper. It here reposes on gneiss and is not more than ten feet thick.

In Iowa and Wisconsin it is a light colored, soft sandstone, intercalated with argillaceous, and argillo-calcareous deposits. It is therefore variable in its coherence.

In Virginia, and the southwest part of the union it is also a light colored sandstone with an open texture, rarely vitreous, but its upper beds alternate with calcareous layers of considerable thickness, as at the head waters of the Holstein and Clinch rivers.

At many places its inferior beds are pebbly and very coarse, but they are by no means always present, and the gneiss on which it reposes seems to be changed gradually and imperceptibly into a fine sandstone with thin beds of mica between the strata, as at Corinth and Whitehall. The Potsdam sandstone, although in northern New York it is 400 feet thick, is sometimes absent, as at Little falls, where the calciferous sandstone reposes upon the primary rocks. It often thins out rapidly as at Whitehall and Comstock's landing, where in two or three miles, and even less, a mass two or three hundred feet thick gives place to the next rock above.

In the interior of the region lying between Lake Michigan and the Mississippi river, the Potsdam sandstone crops out from beneath the calciferous sandstone. It would be traversed obliquely in crossing from Green bay to Prairie du Chien. On both sides of the Mississippi and the lake, it is overlaid by the succeeding rocks, though upon the Mississippi and St. Croix, the latter is cut through by numerous streams down to the sandstone; the Potsdam sandstone of Wisconsin and St. Croix is about 500 feet thick; the upper mass is a white friable sandstone. It covers a large area, but its extent has not been determined, in consequence of a deep covering of drift towards the southern shore of Lake Superior. This great mass of sandstone is fossiliferous. It is so even near its base; Dr. Owen having discovered five or six beds or strata containing trilobites, lingulas, obolus, and crinoids, and bivalves belonging to the family of the Orthidæ. This mass of sandstone is probably nearly continuous with that upon the north shores of Lake Huron, where it rests unconformably upon the Taconic system; and where, too, the latter system is the repository of copper, and is traversed by traps and porphyries, as in North Carolina.

I have already alluded to the position of the Potsdam sandstone, in the southwestern part of Virginia, and at certain intermediate points; and from what has been said it will be inferred that it does not immediately flank the Blue ridge on the west, but occupies a position, even in the latter state, many miles from the base of this range; thus in the great valley west of the Blue ridge, it is some twelve or fifteen miles from its base. In Augusta county, about one, or one and a half miles west of Staunton, the Potsdam sandstone crops out from beneath the overlying calciferous sandstone, and in going west to the Buffalo gap, the series is lower Silurian, and is repeated twice or three times, the Potsdam being brought to the surface in each repetition. The Trenton limestone, fully identified by its fossils, crops out for the last time nine miles west of Staunton and one mile east of the Buffalo gap. This rock is reddish white and reddish brown, and usually alternates once or twice with the calciferous, or rather, beds of a silicious limestone occur, interstratified with beds of sandstone. One of the places where the Potsdam will be found west of Staunton, is at the western slope of a ridge, just beyond West View, about six miles west of Staunton.

This rock throughout this region, both to the northeast and southwest is not a vitrified quartz, but a sandstone more or less granular and sometimes pebbly. So also in Wythe county, Va., the Potsdam sandstone crops out near the base of Little Brushy mountain, some six or eight miles west of Iron mountain whose rocks belong to the Taconic system. Here, then, it is still farther removed to the west of the Blue ridge. But it should be distinctly stated that this rock appears at the base of the Walker and Garden mountains in its true geological relations. I am, however, unacquainted with it farther west and southwest than Jeffersonville, Tazewell county, on the head waters of the Clinch river.

From the foregoing statement of facts, it will be perceived that we are not to look for the base of the Silurian system upon the flanks of the Blue ridge, or at its base; it no where, I believe, touches this range; and even the South mountain, where it comes up into Pennsylvania, has no connection with the Silurian base. I have already shown that, at Harper's ferry, the quartz rock is not Silurian, but belongs to the Taconic system.

I have alluded to the Trenton limestone, one mile east of the Buffalo gap of the North mountain. The Potsdam is not brought up to view between the Trenton and this broken range near the road, but it may probably be found not far from this place northeast or southwest. The lower Silurian, consisting of the Trenton and the inferior members seem to overlie the Clinton group, a group which constitute the North mountain at the gap, and I mention this for the purpose of saying that the lower Silurian holds here the same relation to the superior rocks, that the Taconic rocks do at the Queen's Knob, in Wythe county; and also along the range of Little Brushy mountain.

The position of the Potsdam sandstone in its southwestern prolongation may be very accurately fixed by conceiving it to lie near the eastern flank of Little Brushy, in Wythe county, and North mountain, in Augusta county, Va. It will occupy the western side of the great limestone valley of Virginia, following the westward deflection of the Appallachians as they approach the state of Tennessee. But the range of the Potsdam is its most easterly one. It appears still farther west, as I have stated, at Jeffersonville, in Tazewell county, and hence, it will probably be found in all the great valleys at the base of the mountains which traverse Virginia. It holds also the same relations in Tennessee, and will be found by tracing from Virginia its lines of outcrop in the direction of the strike of the lower Silurian. It is everywhere a sandstone, that is, it is not vitrified, but more or less porous and is also more or less interstratified with calcareous beds in its upper part.

§ 6. *Palæozoa of the Potsdam sandstone.*— The Potsdam sandstone had been examined with considerable interest, at several places which appeared favorable for the preservation of fossils, without success. The writer, however, succeeded in finding lingulas in the first instance, at Birmingham, in Essex county. They occur in the rock which forms a mural precipice on the Ausable, at a place known as the High Bridge. The rock is scarcely disturbed, the sandstone itself is somewhat vitreous and thin bedded, and between the beds a thin slaty matter intervenes, upon which the lingulas are found. The dark slaty matter is just sufficient to give a dark stripe to the layer. These fossils are distributed through a thickness of seventy feet. Since this discovery many other fossils have been found; indeed they have become so numerous that it may be regarded as rather rich in organic remains. Near its junction with the next rock marine plants are quite abundant. Trilobites, crinoids, Orthidæ, etc., are among the fossils of this rock; thus, the base of the Silurian system has proved to be fossiliferous at the lowest beds. It is particularly so where carbonate of lime enters even in a small quantity into its composition. The fossils will be described in another part of the work.

From what has been said, it will be perceived that it is widely distributed; that it is quite uniform in its composition. In northern New York it is harder and more vitreous than at the west and southwest.

§7. *Economical uses of the Potsdam sandstone.*—The typical form of the rock occurs at Potsdam, where it is very even bedded, breaking with facility into pieces of almost any size, even of the size of a brick. At Moira and Malone, N. Y., it is also even bedded and suitable for flags and for building. Houses which are constructed of this rock look remarkably neat and finished. It has an advantage over the compact limestones, that moisture is not condensed upon the inner surface of the walls. The rock is also suitable for glass making at many localities. A white, friable variety suitable for this purpose is found at Moira, in Clinton connty, N. Y.

The fine grained varieties usually stand the fire remarkably well, and hence have been rather largely employed for the hearths of furnaces.

§8. *Calciferous sandstones.*—Formation No. 2, of Messrs. Rogers' Geo. Rep.—This rock has a variable composition. The name was first applied by Mr. Eaton to a rock consisting of carbonate of lime and fine grains of sand so intimately blended that it appeared homogeneous. This is a gray rock, and the grains of limestone give a sparkling surface. It always contains small masses of calcite, or calc spar, intimately blended in the mass. Subsequent examination has proved that this rock passed into a pure carbonate of lime, sometimes gray, and at other times black. In many places it contains beds of magnesian limestone and a small quantity of iron.

There are, therefore, when composition is regarded, three varieties; the silicious, the magnesian and the pure carbonate of lime. Of course there are intermediate grades of composition.

Leaving out of view its composition and turning to its structure, it is proved to pass from compact to granular; the granular becomes porous, and frequently this variety has cavities which are lined with crystals of quartz and calcareous spar.

Instead, however, of a cavity lined with numerous points of crystals, a single perfect crystal of limpid quartz will be found nearly filling the space. These cavities also contain a solid coal-like substance, in the form of globules and drops. The composition of this substance has never been determined by analysis. It exfoliates in fine particles in the flame of a candle, leaving a smooth, conchoidal surface; but it neither burns nor exhales a bituminous odor. The substance is sometimes contained in the quartz crystals and in the calc spar also; the quartz is also sometimes in globules. The structure of the rock is often oolitic; the rounded grains are usually the size of mustard seed, but sometimes again they are six or eight inches in diameter, when the structure is regarded rather as concretionary than oolitic. Strata, sometimes six or eight thick, are concretionary, forming alternating beds with those of the ordinary kind. The quartz, however, is not always in grains or in crystals, but is more commonly cherty and frequently very abundant in the lower beds. This rock, as it usually occurs, is dull and would not admit of a polish. The dullness is due to the presence of a yellowish earthy sediment between the strata. It is a tough rock, especially when thick bedded, or when it passes into the Potsdam sandstone.

At Chazy the following strata occur in a series of beds, beginning with the lowest:

1. Silico-calcareous beds more or less interspersed with sparry

masses, thirty to thirty-five feet; fossils rare and more or less cherty.

2. Crinoidal mass composed almost entirely of disjointed and broken columns of encrinites in which plates of the cistidea may be recognized, twenty feet.

3. A dull, gray, earthy mass, ten feet, without fossils, and passing into oolitic beds.

4. Another crinoidal mass similar to the first, though its color is of a brighter red, 15 feet.

5. A mass more earthy and silicious, and more massive; it contains the Isotelus aud Illænus, twenty feet.

6. Mineral character similar to the foregoing; the fossils are mostly Orthidæ, of which individuals two or three species are very numerous.

7. Red crinoidal mass, with less earthy matter, and is susceptible of a fine polish; fifteen feet thick.

8. Drab colored, thin bedded, magnesian beds, suitable for hydraulic lime. The drab colored layers rarely contain fossils; at Glenn's falls, however, they contain fucoids. Towards the top of the rock it is blue and frequently cherty and oolitic with concretionary beds. These upper masses are variable in thickness at different places, but usually they are from 20 to 30 feet thick.

In the state of Wisconsin, the calciferous sandstone is well developed, according to Dr. Owens's report. Upon the Mississippi and other western rivers, it is the rock which gives character to the landscapes. It rises in castellated terraces, which look like ruined structures of by gone ages. It rises up in steep mural precipices from the water's edge or from a fine slope of luxuriant sward, clothed with grass and showy flowers.

These steep escapements are composed of jointed columns, which stand out from the main mass in strong relief, a structure which is due in the main to the wear of atmospheric agencies, and not unfrequently to the direct undermining effects of water.

This rock, in Wisconsin, is surmounted by a white sandstone, which has received the local name of St. Peter's sandstone. It differs from the same rock at the southwest only in being thicker and exposed more prominently to view. It is superimposed upon the inferior beds; it stands prominently out in the mural escapements of the St. Peter's river. It is not proposed to separate it from the calciferous. It merely forms one of its prominent subdivisions, and is noted as No. C, of the calciferous sandstone, of which it is regarded as the terminal member.

In the southwest, or in Virginia, Tennessee and Kentucky, the calciferous sandstone at Shannon's scarcely differs from that of northern NewYork. At the western foot of Walker's mountain the calciferous includes a sandstone which comes in near the top of the series, and is analagous to the St. Peter's sandstones of the northwest.

Chemical Composition of the Rock.

§ 9. Dr. Owen gives the following result derived from Mr. Norwood's analysis of this rock. It is the magnesian variety, as the analysis shows:

	1. From Lake Pepin.	2, Oolitic, from Winnebago.
Carbonate of Lime,	52·0	50·93
Carb'te of Magnesia,	42·2	41·13
Insoluble Earthy Matter,	4·3	
Oxide of Iron and Alumina,	0·9	1·74
Water and loss,	0·6	0·86
	100·0	100·00

The average thickness of the calciferous sandstone is about 300 feet in New York.

At the mouths of Vermilion and Wisconsin rivers, Dr. Owen states its thickness at 225 feet, and the St. Peter's sandstone at from forty to ninety feet.

There is a peculiar variety of the foregoing rock, which, although it comes under the concretionary kinds, yet requires a few additional remarks in order to convey a correct idea of its condition. It is a lumpy, concretionary mass, similar to a breccia; indeed the lumps are frequently so much angulated that it may be mistaken for the latter rock. Still, it must be regarded as concretionary, as its most common condition is that of a concretionary mass. It occupies a place near the bottom of the series.

The only real breccia or brecciated conglomerate which has fallen under my notice is upon the Montmorenci, in Canada East, where these beds are made up in part of boulders.

The concretionary kind which I have described occur at Greenbush Cantonement hill, one mile east of Troy, the gorge at Highgate, Vt., and at other places on the east side of Lake Champlain.

§ 10. *Distribution of the calciferous sandstone.*—The calciferous sandstone is more persistent than the Potsdam sandstone, hence, it is the base of the Silurian system when the latter is absent. It surrounds the irregular dome of primary rocks, which form the northern highlands of New York, overlying the sandstone. It is better represented at Chazy than at any other place situated upon the flanks of this dome. At Middleville, in the Mohawk valley, it contains fine limpid quartz in its cavities together with calcareous spar and globular unbituminized coal, both in the cavities and in the interior of the quartz and spar. They also contain brown spar.

The concretionary variety is found at Little Falls near the top of the cliff of this rock, where it rests in gneiss. On both sides of the St. Lawrence it follows the quartz or Potsdam sandstone, and may be traced from Kingston to Lake Huron; it disappears under the waters of Georgian bay.

On the west side of Green bay, Lake Michigan, it crops out from beneath the Trenton limestone or its equivalent, and appears in a long narrow belt running down to Jamesville on the Rock river, where the belt runs westward to Prarie du Chien, and finally it passes up both sides of the Mississippi; near to Fort Snelling, it inclines again to the right and left, the right fork going up the St. Croix, the other follows the St. Peter's southwestward to the forks, near the Marrah Saukah lake. It has been traced upon the routes of these large rivers. On the Mississippi it is denuded on both sides, exposing to view the lower sandstone. This rock accompanies the Potsdam sandstone southwest; passing through Orange county, N. Y., into New Jersey, Pennsylvania, Virginia, Tennessee, keeping company with the sandstone beneath. I have stated that the Potsdam sandstone does not flank the Blue ridge, but follows the middle or the western side of the great southwestern valley, passing west of Winchester, one mile west of Staunton and about 2 miles west of Wytheville, at the base of Brushy mountain, and the western base of Walker's mountain. The calciferous pursues the same route and following the same valley it passes into Tennessee, but I am unable to say whether it may be traced entirely across the latter state into Alabama.

§ 11. *Mineral contents of the rock.*—This rock is not regarded as rich in metals. Dr. Owen however reports localities where lead ore (galena) has been found. Thus, on the west side of the Mississippi, fifteen miles above the mouth of Turkey river, large quantities of galena were taken from openings in this rock. Numerous localities have been observed where this ore has been found in the northwest. A few seams of copper have also been discovered by the government surveyors, but nothing which promises much up to the present time. Sulphuret of iron and blende in small lumps occur in the rock in New York, but it is unknown as a metalliferous rock; in the southwest it is equally barren.

Its Palæozoa.—This rock is rich in fossils, and contains many at certain localities. At Chazy, particularly, most of the strata are rich in organic remains. Marine plants, corals, brachiopods, gasteropods and crustaceans are abundant in this mass at Chazy. No less than three beds of lingulas, neither of which is less than ten feet thick, occur in this ancient formation. Silurian forms and types had already become numerous, and though it can not be said that the species were numerous, yet individuals of species were remarkably so. Strata are frequently made up of organic remains, among which we find only a few species; the simple plaited orthidæ are the most common in the fossiliferous beds.

§ 12. *Chazy limestone.*—In New York where this rock is clearly developed, it is a dark colored, irregular, thick-bedded limestone. At Chazy it is a rough, cherty mass; the fossils are imbedded in a flinty matrix. At Essex, on Lake Champlain, this rock is a thick bedded limestone, and contains less foreign earthy matter than the rock just described. This rock is not the one known as the Black river limestone, though it has been referred to as such; it is a distinct rock, one which the writer was the first to recognize by its peculiar fossils as well as by its position. In southwestern Virginia and in Tennessee it occupies the same position and contains the same fossils as at Chazy and Essex. Its thickness is 130 feet. The most prominent fossil is the Straparollus (Maclurea magna) of Le Sueur, *pl.* 4, *fig.* 15. It seems to be absent in Wisconsin and Iowa.

§ 13. *Birdseye limestone.*—This limestone is close grained and frequently compact like flint; it is brittle, and breaks with a conchoidal fracture. It is black, dove-colored, taking sometimes a yellowish tinge. The rock always has a compact structure, in which it is unlike the other limestones of the group.

§ 14. *Distribution.*—It accompanies the lower Silurian limestones in northern New York and Canada. It exists in the Champlain and Mohawk and Black river valleys. In the Mohawk valley, at a few places, it is only one or two feet thick. In Canada, it is a beautiful, light colored, compact rock, sufficiently light colored and fine grained to be employed for lithographic purposes. It is associated in the southwest with the preceding rocks. Its fossils are not numerous; the most common one is the Fucoides demissus of Conrad, *pl.* 4, *fig.* 12, 13.

Isle La Mott marble.—Is a thin, black, fine grained rock; it is a pure limestone, free from foreign matter. It will receive a fine polish, and in consequence of its fine grain and color, is highly esteemed as a marble. At Isle La Mott it is thicker than at Watertown, but its greatest thickness does not exceed twelve or fourteen feet. It merits a passing notice on account of its value as a marble. It contains a few obscure and rather broken fossils, which injure it for marble, as they make white spots. It is unknown in the north or southwest. Its fossils are similar to the Birdseye, or those below it, rather than those of the Trenton limestone which succeeds it.

§ 15. *Trenton limestone.*—This rock in northern New York and Canada is black and fine grained, or else it is a grey subcrystalline limestone. It is massive, or in thick beds, as at Trenton; in other places it is rather thin bedded and alternates with black slate as at Chazy. The upper part is shaly, and passes into the Utica slate. It is therefore not uniform in its composition and in the formation of its constituent strata. This limestone is often bituminous. In the quarries of Montreal its surfaces are adhesive from bitumen. This rock is impure, from the presence of shale or agillaceous matter, while the calciferous sandstone is impure from the presence of silex. This rock is white in southwestern Virginia.

§ 16. *Distribution.*—It is the most persistent of all the lower Silurian rocks except the calciferous sandstone. It is prevalent in northern New York and Canada, the valleys of the St. Lawrence, Champlain and Mohawk. In the northwest, in Wisconsin and Iowa, it is described by Dr. Owen as the shell limestone, but seems to be less important than the calciferous sandstone. It is also present in Pennsylvania, Virginia and Tennessee. In southwestern Virginia it is a white crystalline rock, though loaded with organic remains. Its average thickness is about 400 feet.

§ 17. *Its Palæozoa.*—Is its highest claim to notoriety. We have no rock so rich in fossils as the Trenton limestone. It contains gasteropods, brachiopods, lamellibranchiates, crinoids, and five or six species of crustacea as the Isotelus gigas, Calymene senaria, &c.

§18. *Utica slate.*—The Trenton limestone passes into this rock by becoming more slaty or shaly and losing its calcareous beds.

It is itself a calcareous shale, more or less bituminous. In New York the dividing lines between the limestones below and the Loraine shales above, are not distinct. In the western states it is merged in the blue limestone. In the southwest it is not recognizable. The calcareous shales in the gorges of Loraine, Jefferson county, in which there are neither limestone nor sandstone beds, are about seventy-five feet thick.

§ 19. *Loraine shales terminating in a thick bedded gray sandstone.*—The rock succeeding the Utica slate passes from the latter into the former by its beds of sandy shale and thin bedded sandstone. The argillaceous matter loses its lime also, and in passing upwards changes affecting its palæontology are very soon perceived. These shales consist, in the inferior parts at least, of thin alternating beds of gray sandstone and slate; the sandstone becoming thicker at the superior part, the shale diminishing, the rock finally passes into a gray even bedded sandstone. The typical rock is displayed in the gorges of Loraine and Rodman, and not upon the Hudson river. In the latter region the rock is crushed, and is by no means in a condition suitable to give character to a group, hence it should be referred to only as a modified condition of the Loraine shales.

§ 20. *Distribution of the Loraine shales and sandstones.*—In Jefferson and Oneida counties, in New York, St. John's and its neighborhood Canada East, are the most important points at the north where this group is developed.

On the Mohawk, at Cohoes, the shales and sandstones in a crushed condition, are tolerably well exposed. They may be traced to Schenectady and Saratoga, where they lie in a horizontal position. At the west near Cincinnati, this group is calcareous throughout, and is called the blue limestone. In Virginia, in Wythe county, the group consists, first, of a mottled sandstone, pebbly at the bottom and becoming marly at the top, and contains fossiliferous bands which identify it in part with the Loraine shales. Above this reddish mottled sandstone were found beds of calcareous shales and thin bedded limestone, and above, still, olive green sandstones and marls, with Pterinea carinata of Conrad. This group, therefore, differs physically here from the western group near Cincinnati, and the same group in New York and Canada. The fossils, however, are the same, with the exception that among those which occur west and north, there are a few species confined to the south. The palæontology is, however, almost identical. The rock, equivalent to the Oneida conglomerate is not universal at the south, neither are the upper beds so constantly thick bedded as in Jefferson and Oneida counties.

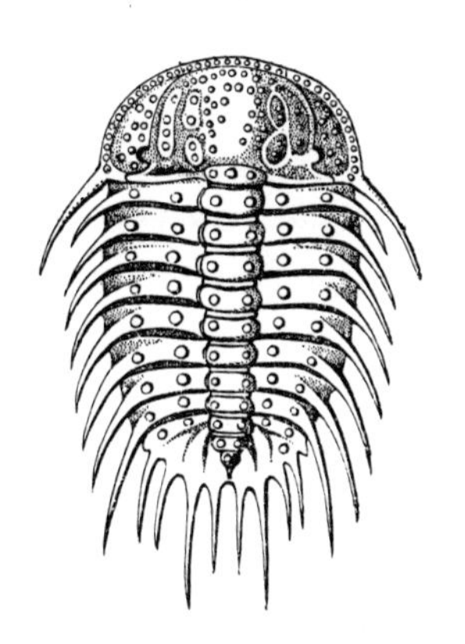

THE PALÆOZOA OF THE LOWER SILURIAN SYSTEM.

§ 21. The lower Silurian rocks of this country form one palæozoic group. The distribution of the palæozoa is such that no subdivision between any of the members would secure on either side a good natural history group. We have, it is true, been in the habit of speaking of the calciferous and Hudson river groups, but the use of the term has arisen from a restriction of a few species of fossils to those beds, which, when coupled with certain lithological peculiarities of the rocks themselves, led to this mode of grouping. If a general subdivision was necessary, the lower Silurian might be grouped under three heads, as has been virtually proposed by Prof. Rodgers; thus the Potsdam sandstone forms the first mass designated No. 1, the limestones by No. 2, and the upper rocks consisting of shales and sandstones equivalent to the Loraine shales and sandstones of New York, by No. 3. This division is applicable to New York, Pennsylvania and Virginia, but in the west the calcareous matter is continued into the upper group or No. 3. The second and third divisions, therefore, are not clearly separated, and as the Potsdam sandstone often passes into the calciferous sandstone, we find the distinctive lines between the three divisions nearly obliterated. The lower Silurian is one group only, because the palæozoa are restricted to its limits. This fact I pointed out many years ago, and was led to designate it the Champlain group, and subsequent researches have served to confirm this view. The discovery of a few dilapidated and imperfect specimens of fossils belonging to this division in the upper part of the Medina sandstone, does not affect the generalization. The occurrence of the Exogyra costata in the Miocene of North Carolina does not prove that it lived in the Miocene period, so neither does the discovery of a few weathered Champlain fossils in the Medina sandstone prove that they lived in that period.

The order which I propose to follow in describing the fossils of the group will accord with the modern systematic arrangements which have been generally adopted by zoologists. The student will find many advantages resulting from a classification which embraces both the living and extinct species, as he will thereby aid his conceptions of the latter by means of the former.

MOLUSCA.

§22. As the name indicates, are animals whose bodies are soft or a peculiar fleshy substance, as in the well known animals the clam, oyster and snail. They have no hard or bony skeleton, but are usually provided with external hard parts called the shell, which is a secretion from the skin, consisting mostly of carbonate of lime. Their nervous system is distinct, but contains but a small amount of medullary matter. They have also a circulation and all the organs of sense. The place assigned them in the rank of beings is between the articulata, as the spiders, crabs, &c., and the radiata, or the corals. This subkingdom is divided into five classes: CEPHALOPODA, GASTEROPODA, ACEPHALA, TUNICATA and BRYOZOA. These classes are subdivided into two sections; the first embraces the more perfect Molusca, all of which have a nervous gangliated cord surrounding the œsophagus; they have also the sexual organs of the higher animals, and are never aggregated or produced in family groups. The lower division, on the contrary, and which is called *moluscoides*, has only a rudimentary nervous system, no distinct ganglia, and are reproduced by buds as well as from eggs, and they are also small and live in families in a mode similar to the corals, to which they have a strong resemblance. They embrace the two last classes mentioned in the foregoing classification.

1. Cephalopoda.

Animals, some of which are destitute of an external or an internal shell, others are provided with an internal shell only, and others still with an external chambered one; head distinct and provided with eyes; instruments of locomotion and prehension numerous, and arranged about the head, and which are armed with suckers or hooks. The cuttle fish, nautilus and spirula are examples of Molusca belonging to this order. All the known living species, together with many extinct ones, were provided with a dark colored secretion stored up in a sack, the contents of which can be emitted at will, which discolors the water and thereby favors their escape when pursued by their foes. The external shell may have one or more chambers; the former are called *monothalamacea;* the latter, *polythalamacea.* The transverse partitions of the chambers are called *septa,* and the septa are perforated by a tube called a *siphon* or *siphuncle,* which may be continuous or interrupted.

Those of the class which are provided with external shells may be divided as follows: 1. Those whose shells are rolled up so as to form a circular disk; 2. Those whose shells are merely curved, or arcuate; 3. Those whose shells are straight.

Before I proceed to describe the fossils belonging to the foregoing divisions, it will be useful to the student to illustrate by figures the forms of the septa and position of the siphon, as both furnish important distinctive characters. Thus, septa are

Fig. 26.

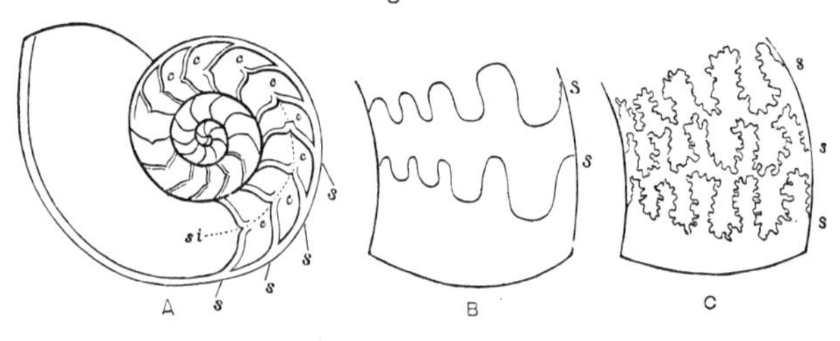

simple as in the Nautilus, fig. 26, A, undulated as in certain kinds of Ammonites, as in B, arboresecent as in C, and angulated as in the Goniatites, fig. 27, A, B.

The siphon is central as in fig. 26, A, or in other genera it is subcentral, ventral or dorsal.

Fig. 27.

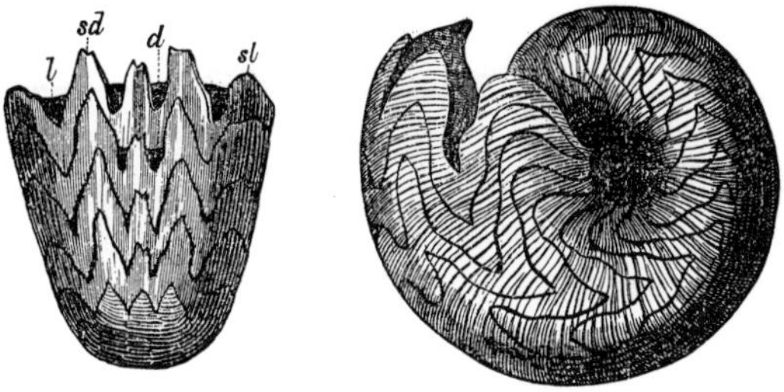

A B

The measurement of symmetrical conical univalves is shown in fig. 28. The width is measured across the aperture, the length in the direction of *l l*, and *d d*. The figure explains sufficiently the other measurements.

Fig. 28.

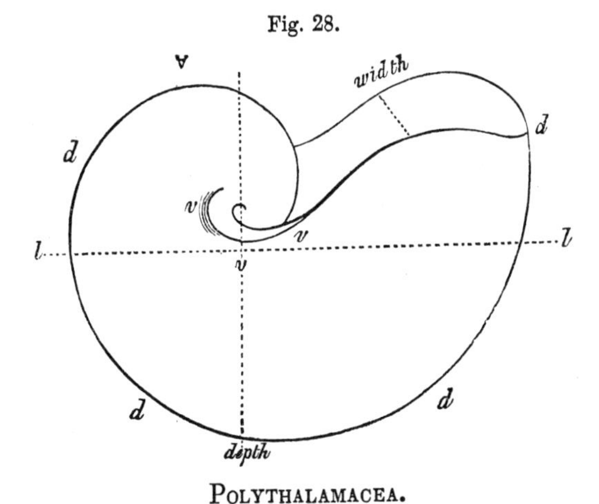

POLYTHALAMACEA.

Synopsis of the order.

I. Shell *discoidal.* Lituites. Trocholites Hortholus.

II. Shell arcuate; Cyrtoceras, Oncoceras.

III. Shell straight; Orthoceras, Gonioceras Actinoceras Gomphoceras Endoceras, Cameroceras, Melia.

D. *Discoidal, septa simple.*

Genus, LITUITES.

Shell discoidal; volutions contiguous; siphon central.

L. UNDATUS, *pl.* 5, *fig.* 14, 14, *a.*

Inachus undatus (*Conrad*).

Shell rolled up with its convolutions equally depressed on either side; the outside as well as the inside of the volution flattened; surface marked by strong oblique ridges between which it is finely striated; striæ upon the back form an imperfect arch. It occurs in the Black river limestone.

HORTHOLUS (*Montfort*).

Convolutions disjointed or somewhat distant in the young shell, and become more so after two or three volutions, becoming in the end nearly straight towards the aperture.

H. AMERICANUS.

Lituites convolvam (*Hall*).

Volutions, two or three, scarcely contiguous; surface smooth, section circular, septa plane or moderately convex; direction oblique from the inner side of the volution outward and upward (*Hall*); position of the siphon undetermined. Occurs in the Black river limestone at Watertown.

TROCHOLITES.

Volutions nearly in the same plane, ventral disk more concave than the dorsal; ventral surface flattened and even slightly concave; transverse section obtusely lunate; siphon small, ventral.

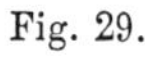

Fig. 29.

T. AMMONIUS (*Conrad*), *pl.* 12, *fig.* 14 *c*, 15.

Volutions about four; enlargement gradual; surface ornamented by oblique ridges, which form a distinct arch upon the back, and marked by fine waving interrupted striæ, and very obsucre longitudinal ridge; aperture suddenly enlarged as in fig. 29

Trenton limestone and Utica slate

T. PLANOBIFORMIS (*Conrad*).

The shell is rather larger than in the preceding section; more elliptical; mouth lunate; transverse ridges form a curve rather than a distinct arch; surface longitudinally striated. Loraine shales.

B. *Shell arcuate, septa simple.*

CYRTOCERAS (*Gold*).

Shell moderately curved; section circular; siphon dorsal.

C. MARGINALIS (*Conrad*).

C. Macrostomum, *fig* 30 (*Hall*).

Fig. 30.

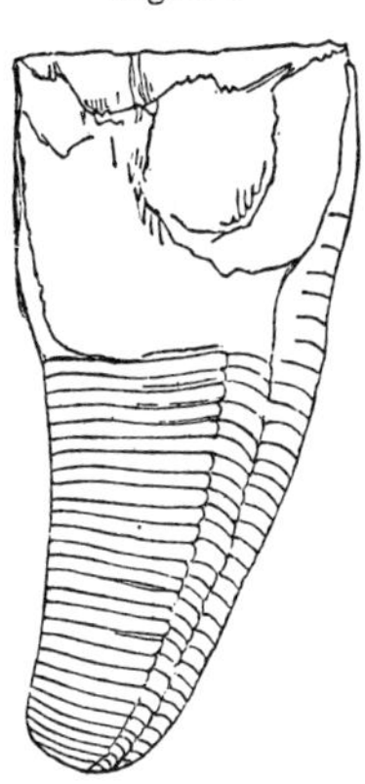

Shell moderately curved, enlarging rapidly towards its aperture: septa numerous; it has twelve to thirteen septa in an inch towards its small extremity. Belongs to the Trenton limestone.

C. ANNULATUS (*Hall*).

It has a central siphon, and hence should probably be placed in another genus.

C. LAMELLOSUM (*Hall*).

Shell gently curved; surface ornamented with sharp undulating lines, which are arched downwards towards the apcol extremity. Trenton limestone.

C. CONSTRICTOSTRIATUM (*Hall*).

Shell nearly straight; surface ornamented by fine transverse striæ.

C. MULTICAMERATUM (*Hall*).

Septa close and numerous, and curved at the apex; septa about thirty in an inch.

C. ARCUATUM (*Hall*).

Shell gently curved, and tapering slowly towards the apex; septa about twelve or thirteen in an inch.

C. CAMURUM (*Hall*).

Curve greater than in the preceding species; septa about nine to the inch. The three last are not marked by striæ. They all belong to the Trenton period.

ONCOCERAS (*Hall*).

Shell arcuate, swollen near the middle, siphon small dorsal; septa nearly flat.

O. CONSTRICTUM, *pl.* 12, *fig.* 2.

Shell short; constricted near the mouth, and rapidly contracting towards the smaller extremity. It is always imperfect at the smaller extremity, being apparently truncated. Trenton limestone.

C. *Shell straight.*

ORTHOCERATA.

Synopsis of the straight Cephalopoda.

I. Shell regularly tapering; siphon small, central or subcentral; section circular. Orthoceras.

II. Shell subconical, siphon ventral; section circular; siphon with interseptal swellings. Actinoceras.

III. Shell subfusiform; siphon eccentric. Gomphoceras.

IV. More than one subconical body enclosed in the shell; siphon large, ventral. Endoceras.

V. Siphon large, lateral and spiral; section elliptical. Cameroceras.

VI. Siphon large; septa waving; section an elongated ellipse. Gonioceras.

VII. Siphon large, ventral, and without interseptal swellings; section subcircular. Melia.

I. ORTHOCERAS.

Shell straight; tapering; having its siphon central or subcentral; section circular.

O. PRIMIGENIUS (*Hall.*)

Shell gently tapering; septa thin and numerous; proximate, calciferous sandstone.

O. LAQUEATUM (*Hall*).

Shell tapering slightly, and longitudinally fluted. Calciferous sandstone.

O. RECTIANNULATUS (*Hall*).

Shell tapering very gradually, and marked by transverse ridges. Chazy limestone.

O. SUBARCUATUS (*Hall*).

Septa rather distant; slightly curved. Chazy limestone.

O. BILINEATUS.

Tapering very gradually; low or rather flat; septa rather distant; surface marked by coarse distant longitudinal striæ. Calciferous sandstone, Greenbush. [See Palæontology, N. Y., by James Hall, p. 35. "Trenton limestone thrust up through the Hudson river slates." ! !

The fact in regard to the rock at Greenbush is, that the calciferous sandstone forms the greater portion of the calcareous mass. The upper part is the Trenton limestone consisting of thin beds of black limestone interlaminated with shale or slate, as at many localities in the Mohawk and Champlain valleys. In these beds of limestone we obtain the fossils of the Trenton limestone. Besides, it is quite difficult to conceive how one rock can be thrust through another without making a hole in it. If the specimen on pl. 7, fig. 4, is compared with those of pl. 43, fig. 14 and 15 it will not fail to strike any one that either the figures are very badly drawn, or else the fossils are quite different. To admit the existence of the calciferous sandstone below the Trenton at this place, would be equivalent to the admission of the Taconic system.

O. TERETIFORME (*Hall*).

Shell marked by subspiral rounded close annulations; surface longitudinally striated; septa quite convex; siphuncle small. Trenton limestone.

O. ARCUOLIRATUM (*Hall*), *pl.* 12, *fig.* 4.

Shell slender; ornamented with transversely spiral and slightly elevated undulating ridges; separated by spaces equal to their breadth; surface finely striated longitudinally.

O. TEXTILE (*Hall*).

Shell rather small slightly tapering evenly and finely striated longitudinally; undulations circular; proximate.

O. CLATHRATUM (*Hall*).

Small transverse ridges angular and finely striated and distant one half the diameter of the shell. Trenton limestone. Rare.

O. ANELLUS (*Conrad*).

Shell elongated; tapering very gradually; annulations angular; approximate and in width equaling one fourth the distance of the shell, which is marked by fine undulating longitudinal striæ. Trenton limestone. Greenbush.

O. UNDULOSTRIATUS (*Hall*).

Annulations subangular; arched upon the back and direct upon the ventral side; back transversely striated; septa quite convex.

O. AMPLICAMERATUS (*Hall*).

Shall rather large, long and tapering very gradually; siphuncle eccentric; septa distant and very convex.

O MULTICAMERATUS (*Hall*), *pl.* 4, *fig.* 8.

Elongated terete, slender, smooth; septa numerous, and distant from each other from one fourth to one half the diameter of the tube; siphon central. Birdseye limestone.

O. STRIGATUS (*Hall*).

Shell terete; surface marked by distinct and rather undulating longitudinal striæ; septa quite convex, and distant about one fourth the diameter of the shell.

II. ACTINOCERAS (*Bronn*).

Ormoceras (*Hall*).

Shell straight; subconical; siphon eccentric and swollen, or extended beyond the perforation by which it appears to be large.

A. TENUIFILUM (*Hall.*)

Shell gradually tapering, and the larger end marked by transverse striæ; siphon ventral; septa rather distant; moderately convex, double; the upper being deflected so as to enclose the swollen siphon. I have observed that when the outer part of the shell was weathered it shows annulations of growth or striæ. Distance between the septa about one-fourth the diameter of the shell. Common in Jefferson county, N. Y., in the Black river limestone, to which it seems to be confined. Williams's College Museum.

O. GRACILE.

It is less conical, or more cylindrical, than the foregoing.

O. CREBRISEPTUM.

This shell tapers gradually; the distance between the septa half as great as in the tenuifilum.

III. GOMPHOCERAS.

Shell very large; straight; subfusiform.

G. HALLII (*d'Orb.*).

Shell enlarged near the middle; contracting towards the aperture; septa approximate and rather flat; siphon large; eccentric. Black river limestone.

IV. ENDOCERAS (*Hall*).

Shell straight; siphon subcentral and large and contains two or more subconical bodies in the same tube.

E. PROTEIFORME, *pl.* 13 *and* 16, *fig.* 1 *a*, 3 *a*.

Shell circular and very gradually tapering; surfaces marked with transverse and longitudinal fine striæ; septa rather distant; inner shells subconical; finely striated; the striæ often preserved upon casts; siphuncle large and subcentral. This species is common and frequently large. It is found in the Trenton limestone at Middleville and numerous other places in northern New York and Canada.

E. SUBCENTRALE.

Shell elongated, cylindrical, and marked by distant spiral ridges; septa distant; siphuncle large; subcentral. Black river limestone. Rare.

E. LONGISSIMUM.

Cylindrical; extremely elongated and tapering very gradually. Black river limestone.

E. MULTITUBULATUM.

Contains many concentric tubes one within the other; tubes thin, smooth. Black river limestone, Watertown.

E. GEMELLIPARUM.

Large; tapering rather rapidly; septa moderately convex distant.

E. ANNULATUS (*Hall*).

Shell large tapering very gradually; annulations wide flattened; distant one-fifth the diameter of the tube; septa quite convex; siphon large.

E. *Arctiventrum*, *angusticameratum*, *magniventrum*, *approximatum*, *duplicatum* and *distans* (*Hall*).

V. CAMEROCERAS (*Conrad*).

C. TRENTONENSE.

Elongated; section elliptical; diameters as five to seven; septa distant, and with low and oblique annulations.

VI. GONIOCERAS (*Hall*).

G. HALLII (*dOrb.*).

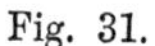

Fig. 31.

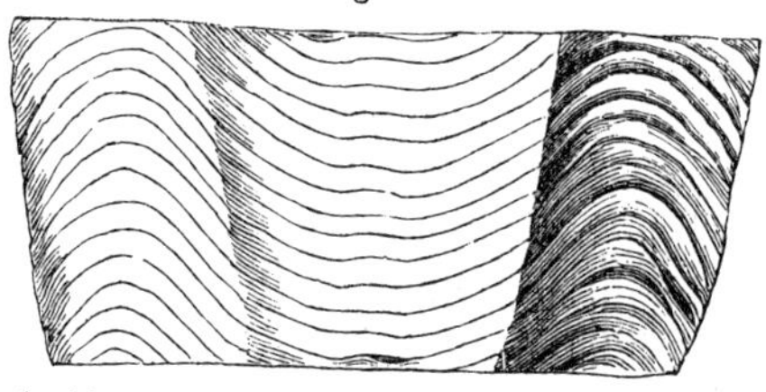

Compressed sides expanded; divergence of the edges 25°; sections are unequally flattened; ellipse; septa numerous, sinuous and double; siphon subcentral, with interseptal swellings Fig. 30. Birdseye.

VII. Melia.

Siphon large; ventral; shell large; section elliptical.

M. cancellatus (n. s).

Shells large; elongated; sides only slightly convergent; siphon large; surface marked by longitudinal and transverse striæ, giving it a pitted appearance. Loraine shales.

GASTEROPODA.

This class takes its name from the circumstance that the species it contains moves upon a large fleshy disk attached to the abdomen, which has been called a foot; the slug or snail represents this fact: In this class the head is more distinct. It is provided with tentacles which are regarded as organs of touch, and upon which the eyes are sometimes mounted. The mouth is below them, with its jaws armed with teeth. Their respiratory organs are in part fitted to perform their functions when in the atmosphere or water. The sexes are separate, or as is sometimes the case, are united in one individual. Certain species belong to fresh, others to salt water, and some are destitute of an external shell. Their shells are mostly unsymmetrical, being obliquely rolled into a spire, as fig. 32, of which *l l* is the length; *b b* the breadth; *a* the mouth, and and 1 2 3 4 the volutions forming the spire. We have an illustration in the shell of several whorls or volutions which together form a rather elevated spire.

Fig. 32.

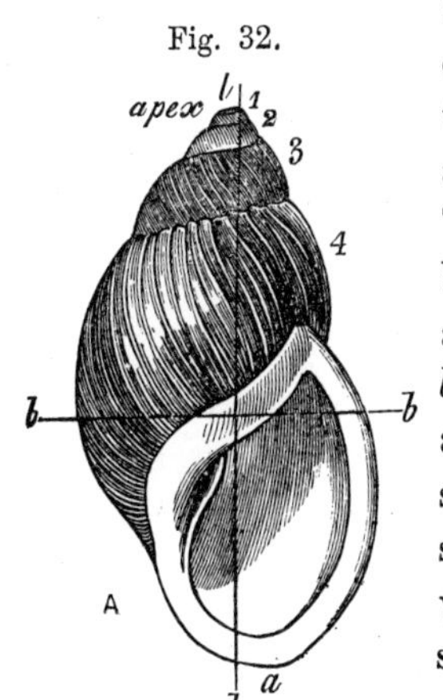

The spire is often depressed or flattened, or does not rise above the plane upon which the spire is rolled, thus fig. 33, the Euomphalus of Sowerby (Straparollus

Montfort), is a flattened spire. Shells of this form are said to be discoid. In this case, the volutions are angular and terminate in an angular aperture, A. The depression in the middle

Fig. 33.

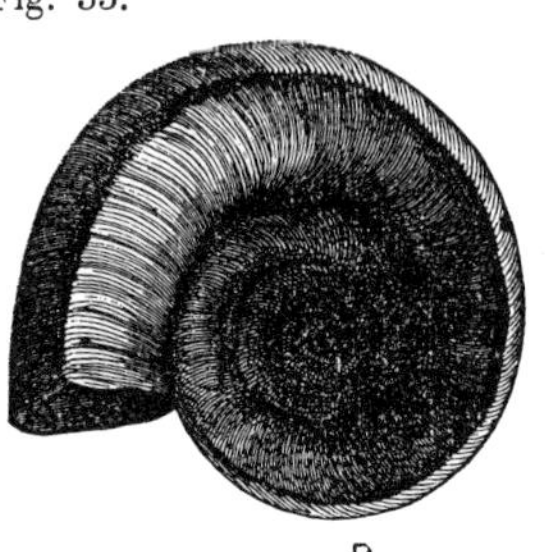

of the disk, is the umbilicus. These belong to the symmetrical kinds. There are still other forms of univalves which are symmetrical, where the spire is replaced by a nonspiral, subconical elevation, with a wide open aperture, upon which the animal rests. Thus in the fig.* 34, the line *a p* divides it into

Fig. 34.

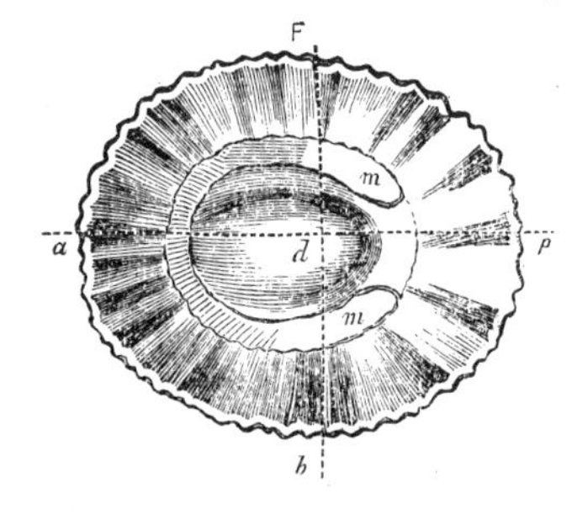

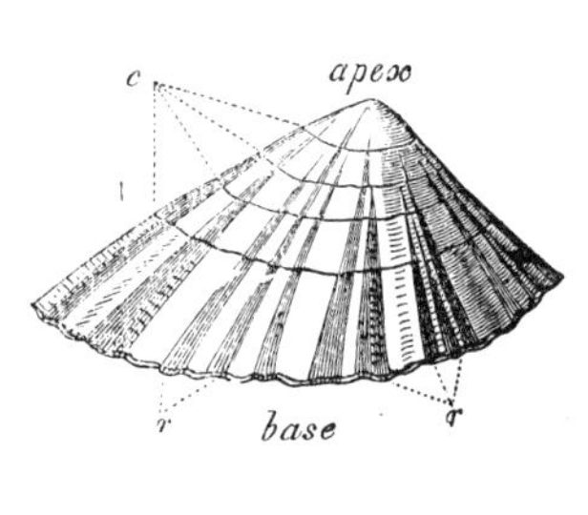

two equal parts; *a*, the anterior; *p*, the posterior extremity. The muscular impression terminates at *m m*, between which there is a space left for the head or anterior parts of the animal. Shells of this form are said to be patelliform.

The class Gasteropoda embraces seven orders:

First order, the *Pulmobranchiata* — they respire in the atmo-

* Sowerby, Conchological Manuel.

sphere; if provided with a shell, it is an oblique spiral. They inhabit the land and water, and are represented by the snail or Helices and Lymnacea.

Fig. 35.
3

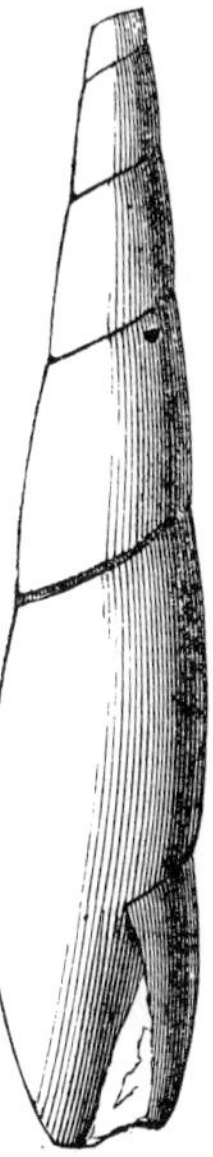

The second order contains the *Pectibranchiata*. A part respire in the atmosphere, and a part respire in the water by means of pectinated gills. Loxonema, fig. 35, Turbo, Macrocheilus and Scalites among the fossils belong to this order. Their shells are rather thick. Of the living fresh water shells the Paludinas are examples. Several genera appear in the lower Silurian stage.

The third order composes the *Scutibranchiata*. The shells are conical, spiral or sub-spiral, and are often symmetrical; Helcion, Cyrtolites and Bellerophon are among the fossils which belong to this order; fig. 36, is the Cirtolites biloba. They appear in the lower Silurian stage.

Fig. 36.

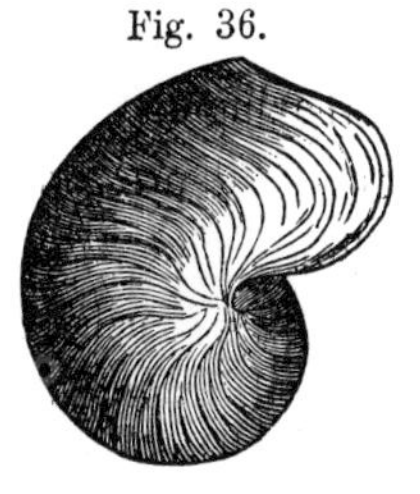

The fourth order comprises the *Tectibranchiata*. The shell is unsymmetrical. The Bulla and Umbrella are examples of shells belonging to the order.

The sixth order contains the *Pteropoda*. The shells are thin, fragile and remarkable for their uncommon forms; the Conularia is an example. The order contains only a small number of genera. The Conularia appears in the lower Silurian.

Synopsis of the fossils belonging to the *Pectibranchiata*, which have been found in the lower Silurian:

1. Shell *discoidal*. Straparollus.
2. Shell *auriform; spire depressed*. Stomatia.
3. Shell *turbinate; volutions ventricose*. Turbo.
4. Shell *subturbinate; volutions angular*. Scalites.
5. Shell *turrited; rather elevated; aperture subquadrate; volutions rounded; surface transversely striated*. Plueratomaria.

6. Shell *turrited; middle of the volution ornamented with a spiral band; the middle of the mouth is notched.* Murchinsonia.

1. *Discoidal.*

STRAPAROLLUS (*Montfort*).

Spire much depressed; umbilicus large and grooved, and not crenulated. Aperture subquadrate.

S. MAGNUS.

Pl. 4, *fig.* 15.

Maclurites magna (*Lesuer*).

Maclurea magna (*N. Y. Geol. Rep.*).

Shell obtusely carinated on the upper edge; whorls rapidly increasing in size; aperture on the left, irregularly and horizontally depressed above; lips not deflected; surface transversely striated.—*Le Sueur.*

Chazy limestone; it is confined to that horizon, yet widely dis tributed. New York, Canada, Virginia, Tennessee, Kentucky.

S. SORDIDUS, *pl.* 3, *fig.* 6.

Maclurea sordida (*Hall*).

Spire scarcely elevated; containing about three volutions; surface smooth. It is usually about one inch in diameter. It is upon the weathered surface that they become known to us, and herein exhibit considerable diversity of form; found at Chazy in the calciferous sandstone, Bald Mountain and Hoosic Falls, and in the same rock also in Wythe county, Virginia.

S. MATUTINA.

Maclurea matutina (*Hall*).

It is usually a cast, but sometimes a distinctly striated shell, and is larger than the foregoing.

S. UNIANGULATUS (*d'Orb.*).

Enomphalus uniangulus (*Hall*).

Upper and outer edge of the volution angular; rounded below; surface striated.

S. LEVATUS *pl.* 5, *fig.* 1.

Ophileta levata (*Vanuxem*).

Volutions numerous; thin, and without an expanded mouth; angular below; surface smooth. Calciferous sandstone.

S. COMPLANATUS, *pl.* 5, *fig.* 2.

Ophileta complanata (*Vanuxem*).

Volutions numerous; uniformly thin and slender, increasing in diameter. It belongs to the calciferous sandstone.

S. LABIATUS (*Emmons*), *pl.* 4, *fig.* 2.

Outer volution rapidly widening below to a wide, angular aperture; surface smooth. Birdseye limestone, Chazy.

S. ANGULATUS (n. s.)

Shell discoid, smooth; upper surface of the outer volution nearly flat, excepting that it has a wide, shallow groove traversing it near the outer edge, forming, with the oblique inferior and outer face, a marginal ridge. Section of the outer volution, somewhat triangular.

Diameter of the disc, one inch. Upper part of the calciferous sandstone at Chazy, N. Y.

2. *Shell auriform, spire depressed.*

STOMATIA.

Suborbicular, oblong, auriform; spire drepressed; aperture wide, entire and oblique.

S. AURIFORMIS (*d'Orb.*).

Capulus auriformis (*Hall*).

Spire scarcely elevated; whorls about three, the last wide and ventricose; aperture semilunar.—*Hall.* Trenton limestone.

3. *Turbinated.*

TURBO.

T. BILIX.

Pluentomaria bilix (*Conrad*).

Volutions from four to five, larger one traversed by fourteen or fifteen spiral ridges, intersected by numerous longitudinal but oblique striæ; between the three or four first ridges there are one or two low, interrupted, parallel ones; apex acute; aperture triangular, subtransverse.

Blue limestone, Cincinnati.

T. OBLIQUUS.

Pl. 5, *fig.* 18; 18 *a;* 18 *b;* 8.

Holopea obliquus (*Hall*).

Height and breadth nearly equal; spire oblique, acute at the apex; volutions three or four, the last ventricose; aperture transverse, entire; surface smooth or finely striated.—*Hall.*

T. SYMMETRICUS (*d'Orb.*).

Holopea Symmetricus (*Hall*).

Height rather greater than the breadth; apex acute; volutions four; aperture subrotund; surface longitudinally striated by unequal lines of growth. Trenton and blue limestone.

T. VENTRICOSUS.

Pl. 6, *fig.* 19, *a*, *b*.

The volutions swollen, especially the last; heighth and breadth subequal; volutions three.

T. AMERICANUS (*d'Orb.*).

Pl. 6, *fig.* 20.

Holopea paludiniformis (*Hall*).

Spire higher than wide; volutions four, rounded, full; aperture, subangular, preserving an ovate form; casts smooth, in which condition it has been found.

4. *Shell subturbinated; volutions angular.*

SCALITES (*Conrad*).

Raphistoma (*Hall*).

Shell subturbinate; spire depressed; volutions angulated; aperture oval, with the upper and transverse margin straight.

S. ANGULATUS, *pl.* 4, *fig.* 21.

Spire low; volutions four, and angulated; aperture rather narrow, triangular, and terminating before in a rather long, curved canal. Occurs in the calciferous sandstone at Chazy, Clinton county, N. Y.

S. STRIATUS, *pl.* 4, *fig.* 30.

Raphistoma striata (*Hall*).

Discoidal; spire only slightly elevated; volutions four or five; outer margin angulated; surface coarsely and obliquely striated. Found in the calciferous sandstone, Chazy.

S. PLANISTRIA, *pl.* 4, *fig.* 16, 17.

Raphistoma planistria (*Hall*).

Subturbinate; spire formed of from three to five angular volutions; surface marked by distinct, longitudinal flat striæ, and arched in the middle part of the whorls; aperture narrow, trigonal. Calciferous sandstone.

S. STAMINEUS (*d'Orb.*).

Raphistoma staminea (*Hall*).

Spire depressed; volution externally sharply angulated, and welted upon the margin; surface marked by curved striæ above and below, separated by the sharp, welted edge.

Calciferous sandstone.

5. *Shell turrited; rather elevated; aperture subquadrate; volutions rounded; surface striated.*

Pleurotomaria.

Turbinated, spiral; aperture subquadrate, with rounded angles; outer lip with a deep slit near its union with the spire. Sowerby.

P. LENTICULARIS (*Conrad*), *pl.* 6, *fig.* 13, *a. b.*.

Shell lenticular; spire depressed; outer volution flattened above, rounded beneath, and its two surfaces forming an obtuse edge; section somewhat triangular; volutions four, gradually enlarging; umbillicus rather large; surface marked by fine striæ. This fossil is usually obtained in the condition of a cast when it is smooth; the shell is thin and not deeply marked. Common in the Trenton limestone.

P. SUBCONICA (*Hall*), *pl.* 17, *fig.* 9, 96.

Spire elevated, and terminating in an acute apex; volutions 4, 5; only slightly convex above; last carinated; the others subcarinated near the outer and lower edge; beneath, the outer vo lution is convex or swollen, and marked near the middle by a low ridge; surface ornamented by striæ, which are reflexed near the carina and intersected by longitudinal ones, which together form a network of lines, or a cancellated surface; aperture subquadrate; height of the spire one and a quarter inches. Trenton limestone, extending up into the Loraine shales, where it is the most common.

P. QUADRICARINATA.

This is described as having four caririna upon the last whole, and its surface marked with zigzag striæ. Birdseye limestone.

P. PERCARINATA (*Hall*).

Spire conical, subacute, ornamented with numerous spiral ridges, which are crossed by light undulating striæ; volutions three to four; umbillicus none.

P. TURGIDA.

Height and breadth subequal; volutions subangular; the last whorl inflated. Calciferous sandstone.

P. UMBILICATA, *pl.* 5, *g.* 4, *a. b.*

The shell is discoidal or very much depressed; volutions four; angular; flattened above; convex below; suture caniculated; aperture quadrangular; umbillicus large; surface marked by waving striæ. Birdseye and Trenton limestone.

P. VARICOSA (*Hall*).

Volutions four; somewhat angular; ventricose; surface strongly striated. Birdseye limestone.

P. NODULOSA (*Hall*).

Is described by Mr. Hall as being marked by nodulose spiral ridges, or as having a short· ascending spire. Birdseye limestone.

P. ANTIQUATA (*Hall*).

Large; subconical; wider than high. A cast only has been observed in the Chazy limestone.

P. ROTULOIDES, *pl.* 6, *fig.* 10.

Volutions about four; spire depressed; wide again as high; outer volution concave above; convex below; margin angular, subacute; aperture subquadrate; umbilicus small; surface striated. Resembles the P. lenticularis.

P. INDENTA (*Hall*), *pl.* 5, *fig.* 5.

Volutions three; spire acute; enlarging rapidly towards the aperture; aperture angulated or subrotund; surface finely striated. Trenton and Watertown, N. Y.

P. AMBIGUA, *pl.* 5, *fig.* 5.

Volutions about four; spire subacute; outer volution ventricose angulated; height and width nearly equal.

P. UMBILICATA (*Hall*), *pl.* 5, *fig.* 4, *a*, *b*.

Shell scarcely discoidal; spire small and low; outer volution flattened above upon the back; below it is scarcely convex; umbilicus large. The angularity of the outer volution and the large unbillicus forms its most striking peculiarities. Abundant at Watertown in the gray Trenton limestone.

6. *Shell turreted; middle of the volution ornamented with a spiral band.*

MURCHINSONIA.

They embrace those Pleurotomarias whose height exceeds their breadth, and which are ornamented with a spiral band and bent striæ.

M. VENTRICOSA (*Hall*).

Turreted; volutions about five, swollen; angulated above; surface striated; striæ bent at the angle of the volution; volutions taper rather rapidly from the larger one. Birdseye and Trenton limestones.

M. BICINCTA (*Hall*), *pl.* 5, *fig.* 6, *a*, *b*, *c*, *and* 11.

Spire rather elongate; subacute; volutions four or five; angular; last tricarinate; surface finely striated and bent at the carinæ. Trenton limestone.

M. UNIANGULATUS (*Hall*).

Spire elongated; volutions about five angular; a single spiral ridge ornaments the shell, which is also traversed by fine longitudinal lines.

M. BELLICINCTA, *pl.* 5, *fig.* 1 *and* 16 *and* 12.

Elongated, acute, formed of about five or six volutions; section of the volutions rounded convex; shell ornamented by a flat spiral band placed in the centre, and traversed longitudinally by zigzag striæ, which are angulated at the spiral band. Usually in casts and common in Trenton limestone in Jefferson county, N. Y.

M. GRACILIS.

Narrow; very much elongated; volutions six or more; apex acute; the shell is ornamented by an obtuse carina, of which no marks are left in the cast.

M. ABBREVIATA (*Hall*), *pl.* 4, *fig.* 9 *and* 13 *and pl.* 4, *fig.* 11.

The shell is marked by sharp angles, between which the surface is convex and striated. Rare in the Birdseye limestone.

7. *Spire much elongated; acute; surface covered with longitudinal threads; aperture prolonged before.*

LOXONEMA.

Spiral; turriculated; whorls convex; upper edge adpressed against the next above; without a spiral band; mouth oblong; attenuated above, effused below, and with a sigmoid edge to the right lip; no umbilicus (?); surface covered by longitudinal threads or ridges generally arched.—*Phillips.*

L. SUBFUSIFORMIS (*d'Orb.*).
Murchinsonia subfusiformis (*Hall*).

Subfusiform; spire elongated; volutions about four or five; outer lip prolonged below; the casts are all compressed. It is five or six inches high; greatest breadth of the lower whorl about one and a half inches. The peculiar longitudinal threads of the Loxonema have not been observed, as the species has been found only in casts, but the form of the lip and aperture are those of the Loxonema. Trenton limestone.

L. SUBELONGATA (*d'Orb.*), *pl.* 6, *fig.* 20
Subulites elongata (*Conrad*).

Spire high or elongated; apex acute; volutions about five; surface flat, or only slightly convex; aperture narrow; its exact form remains undetermined; it is nearly three inches long, and scarcely half an inch wide. It is distinguished by its great height and narrow and flat volutions. Trenton limestone.

L. VITTATA.

Elongated; subfusiform; subacute; volutions four or five; rather more convex than the foregoing; aperture elongated. Trenton limestone.

Analysis of the *Scutibranchiata*, belonging to the lower Silurian:

1. Shell patelliform or depressed, apex turned forward. Helcion.

2. Shell convolute subglobose or discoidal, and ornamented with a dorsal band. Bellerophon.

3. Convolutions contiguous or disjointed; dorsum without the band. Cyrtolites.

I. HELCION (*Montfort*).

Carinaropsis. Metoptoma (*Hall*).

H. subrugosa (*d'Orb.*).

Metoptoma rugosa (*Hall*).

Patelliform, mouth elliptical, sides nearly parallel, apex acute and directed forward.

H. PATELLIFORMIS (*d'Orb.*).

Pl. 6, *fig.* 1.

Carinaropsis patelliformis (*Hall*).

Patelliform, mouth wide and oval, apex acute, incurved compressed towards the back, forming rather an obtuse carina which terminates in an incurved apex.

H. ORBICULATUS (*d'Orb.*).

Carinaropsis orbiculatus (*Hall*).

Form approaching that of an orbicula, with an acute subcentral apex, finely striated concentrically.

2. *Shell convolute, subglobiform or discoidal and ornamented with a dorsal band.*

BELLEROPHON.

Shell convolute, symmetrical, and ornamented with a single or double dorsal band.

B. EXPANSUS (*d'Orb.*).

Bucania expansa (*Hall*).

Pl. 6, *fig.* 7.

Volutions about four, the last expanding rapidly into a wide everted semicircular mouth, obtusely carinated upon the back; surface of contact with the inner volution concave; section somewhat pentangular. Trenton limestone, Watertown, N. Y.

B. BIDORSATUS (*d'Orb.*).

Bucania bidorsata (*Hall*).

Pl. 6, *fig.* 8-27.

Volutions about three, outer one expanding abruptly near the wide everted aperture; dorsal carina grooved on each side of a narrow sharp ridge; surface ornamented with striæ, which are arched upon the back; aperture twice as wide as high; umbilicus deep. Trenton limestone, Middleville, Watertown, N. Y.

B. SULCATINUS (*Emmons*).

Bucania sulcatinus (*Hall*).

Pl. 4, *fig.* 4.

Sides angulated; aperture round oval; surface ornamented by undulating, longitudinal and transverse lines, the latter meet upon the back, forming with each other an obtuse angle. Chazy limestone, Clinton county, N. Y.

B. PUNCTIFRONS (*Emmons*).

Pl. 12, *fig.* 11 *and* 12.

Bucania punctifrons (*Hall*).

Shell small; volutions about three, enlarging gradually; outer, rounded or obtuse, back broadly rounded, covered with superficial lozenge shaped punctures; carinal band narrow and sharp; umbilicus deep. The direction of the obscure oblique undulations, or lines of growth, indicate the existence of a dorsal sinus, of the form of that belonging to the Cyrtolites bilobus. Trenton limestone.

B. ROTUNDATUS (*d'Orb.*).

Bucania rotundata (*Hall*).

General form rounded; outer whorl expanding towards the mouth, angular at the sides; surface transversely striated. The surface markings distinguish it from the Sulcatinus.

B. INTEXTUS.

Dorsal band narrow and sharp; surface ornamented by intersecting striæ, of which the longitudinal ones are rather strong; volutions rounded. Trenton limestone, near Watertown, Jeff. county, N. Y.

B. CANCELLATUS.

The outer whorl expanded, forming two lobes, between which there is a dorsal sinus; surface striated or cancellated by fine striæ. Loraine shales.

B. RUGOSUS (n. s.).

Fig. 37.

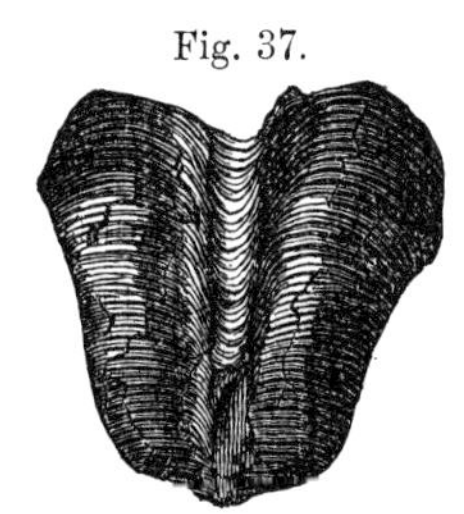

Volutions about three, larger angulated and the larger part covered with striæ, sharply arched upon the wide, dorsal, grooved band at the curve, both the striæ and band are discontinued and replaced by rather distant waving lines; aperture undetermined. (Rare.) Loraine shales and sandstone, Loraine, Jefferson county, N Y.

3. *Volutions contiguous or disjointed; dorsum often smooth, or only marked by ridges or a superficial sinus; no continuous dorsal band.*

II. CYRTOLITES (*Conrad*).

This genus is restricted by d'Orbiny to those shells which are destitute of a dorsal band, or have a shallow marginal sinus only.

C. BILOBATUS (*d'Orb.*).

Pl. 6, *fig.* 2.

Rotund and transversely flattened; height and breadth subequal; mouth large; bilobed, faintly striated. Trenton limestone, Loraine shales. Often transversely corrugated upon the back in old shells; but which in the Loraine shales is never marked in this way.

C. ACUTUS.

Pl. 6, *figs.* 4, 5.

The shell is compressed so as to form an acute angle upon the back; surface marked with fine striæ.

C. SUBCARINATUS (*d'Orb.*).

Pl. 6, *figs.* 25, 26.

Somewhat patelliform; compressed or subangular towards the base; apex incurved; mouth widely expanded. Trenton limestone.

C. COMPRESSUS, *pl.* 12, *fig.* 10, *b.*

Discoidal; volutions slightly compressed and also slightly disjointed; sides rounded; back sharply carinated; surface ornamented by sharp zigzag lamellæ, which only penetrate through the shell. Trenton limestone.

C. TRENTONENSIS (*Conrad*), *pl.* 5, *fig.* 22.

Arcuate; curvature somewhat variable, sometimes it forms a short curve, in others nearly a circle, as in the annexed fig. 38: Section triangular; aperture slightly compressed, and transversely striated, and with a shallow sinus upon the dorsal margin. The shell of this species is quite thick, and the constriction is just below the aperture.

Fig. 38.

PORCELIA ORNATA, *pl.* 17, *fig.* 2.

Cyrtolites ornata (*Conrad*).

Shell thick, consisting of two or three whorls wholly exposed and increasing rapidly in size; whorls sharply carinated; ornamented on each side with a row of twelve or fifteen ribs, which extend from the keel to the lateral angle; back crossed by elevated rows of tuberculated lines arranged nearly at right angles to the keel; aperture irregularly quadrangular.

C. FILOSUM (*Conrad*), *pl.* 12, *fig.* 4.

Shell tapering gently to a point; volution a semicircular arch; surface finely and thickly striated; striæ arched upon the back; section circular. Trenton limestone (rare).

ACEPHALA.

In this class the head is wanting, it is also destitute of the organs of vision and of hearing. The mouth is unfurnished with teeth, but is supplied with fleshy lips and tentaculae which occupy the middle of the large mantle. Their shell or habitation consists of two valves. The mouth and the anal orifice being both upon the same plane but at the opposite extremities, the valve is naturally unsymmetrical, or in other words, is inequilateral; clams and oysters belong to this division of the Molusca

As a reference to the parts of the shell is constantly occurring, it will be convenient to exhibit in this place the position of certain parts of it, which are more or less employed in the description and determination of species.

This figure, from Sowerby's Conchological Manual, exhibits

Fig. 39.

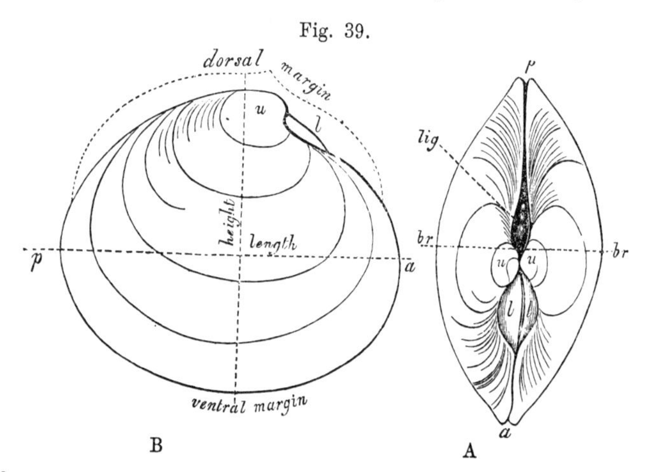

the parts and measurements of a common bivalve shell. The length is measured from the anterior *a* to *p*, the posterior side. The height is measured from the *umbones*, *u*, to the ventral margin. The umbones turn to the anterior part of the shell. In fig. A, *br br* is the breadth measured through the umbones; *lig.*

the ligament on the posterior side; *a p*, the length. Shells of this division of Acephala are supposed to be placed upon the ventral margin, as it is here, the foot protrudes upon which the animal moves; and hence, *a* being the anterior part of the shell, the valve on the right hand side of the observer is the *right*, and the opposite the *left* valve. The Acephala are divided into three orders. In the first the valves are right and left, and the internal impression of the mantle called the *palleal impression* has a sinus at the posterior part or region. The clam, Venus, belongs to the order.

The second division embraces those bivalves whose palleal impression is entire, and in both orders there are from two to four muscular impressions. The fresh water clam, Unio, and the mytilus belong to the second order.

The third order comprises the shells which are wholly unsymmetrical and irregular, and which lie upon the side, as the oyster. The muscular impressions upon the interior of the valves are strongly marked, and are one or two, and that of the palleal impression is entire.

LAMELLIBRANCHIATA.

1. *Palleal impression with a Sinus.*

Lyonsia (*Turton*).

Modiolopsis and Tellinomya (*Hall*).

Shell thin, long, oval or wedgeform; palleal sinus triangular; ligament internal.

L. NASUTA, *pl.* 17, *fig.* 4.

Tellinomya nasuta (*Hall*).

Fig. 40.

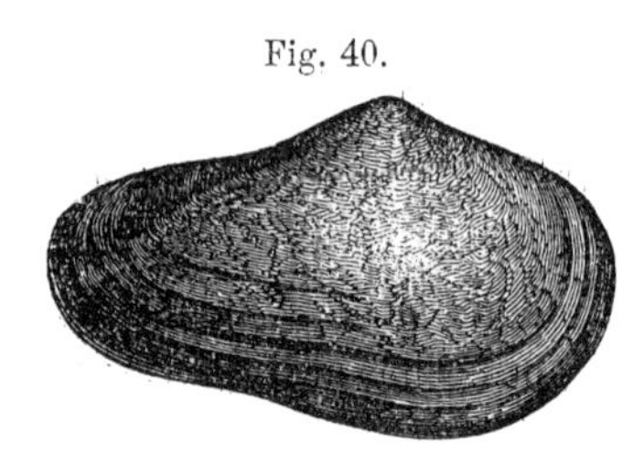

Shell thin, transversely extended, inequilateral, anterior extremity rounded, narrow and more compressed than the posterior, base with a shallow sinus, umbones rather prominent and rounded, surface marked by fine concentric lines. Trenton limestone, Middleville.

L. GIBBOSA.

Pl. 14, *fig.* 3.

Tellinomya gibbosa (*Hall*).

Shell thin, transversely extended; the proportion of height to length as 2 to 3; prominent and thickened at the umbones, sinus of the base quite shallow. Trenton limestone.

L. DUBIA, *pl.* 14, *fig.* 7, 8, 12, 13.

Shell thin, small, inequilateral; extremities subequal, rounded, posterior narrow, umbones subcentral and rather thick. Common at Loraine, Watertown, Middleville, in the Trenton limestone.

L. ANATINIFORMIS.

Shell thin, inequilateral; extremities rounded, anterior narrower than the posterior; the latter subtruncate, oblique and shallow depression on each side of the umbones. Trenton limestone.

L. SANGUINOLAROIDEA, *pl.* 14, *fig.* 2.

Shell thin, transverse, rather ovate, posterior extremity much narrower than the other.

L. TRENTONENSIS.

Modiolopsis Trentonensis, *pl.* 14, *fig.* 4 (*Hall*).

Shell thin, equivalve wide posteriorly, angle line nearly

straight from the umbones to one half the distance towards the posterior extremity; surface marked by fine concentric lines. The shell is rather thick and cylindrical near the anterior extremity. Trenton limestone.

L. SUBSPATULATUS.
Modiolopsis subspatulatus.

This species is distinguished by the broad expanded posterior extremity, the anterior being narrowed and truncate. Southwestern Virginia, northern New York.

L. SUBMODIOLARIS (*d'Orb.*).
Pl. 17, *fig.* 8.
Modiolopsis modiolaris (*Hall*).

Shell thin, equivalve inequilateral nearly oval, wide and expanded posteriorly, rounded and narrowed, anteriorly rounded, about twice as wide as high; surface marked by fine concentric lines of growth. Southwestern Virginia, Loraine shales.

L. SUBTRUNCATA, *pl.* 17, *fig.* 4 (*d'Orb.*).
Modiolopsis truncatus (*Hall*).

Shell trapezoidal oblique, convex breadth one-fourth greater than the height: beaks near the anterior extremity which has the muscular impression. Loraine, southwestern Virginia.

L. CURTA (*Conrad*).
Modiolopsis curta.

Shell orbicular, transverse, breadth about one-fourth greater than the length, convex and base circular; undulations of the surface distinct, but not numerous. Loraine, southwestern Virginia.

L. ANADONTOIDES, *pl.* 17, *fig.* 14.
Cypricardites Sinuata (*Emmons*).
Cypricardites anadontoides (*Conrad*).

Shell long, oval or subelliptical convex; basal line parallel or nearly so with the hinge line; sides rounded, posterior obliquely truncated, concentric striæ strongly marked. The base of many specimens are sinuate on the line beneath the beak.

L. TERMINALIS.

Modiolopsis terminalis.

This species differs from the nasutus by the position of the beaks, which are terminal, and a greater proportional length.

L. NUCULIFORMIS.

Shell small, subelliptic or subtrapezoidal, transverse extremities subacute, beaks elevated, basal margin with a shallow sinus, concentric lines numerous.

L. FABA, *pl.* 14, *fig.* 14, 15.

Modiolopsis faba (*Hall*).

Shell small, subelliptical, constricted near the beak; umbones high and rather narrow, but the posterior side expands; margin is rounded; surface marked by fine concentric lines, and a few oblique ones which run from the beak to the margin. Trenton limestone, Jefferson county, Middleville.

PERIPLOMA.

P. PLANULATA.

Pl. 17, *fig.* 17.

Cleidophorus planulatus (*Hall*).

Shell small, thin, convex, long, elliptical; extremities rounded; anterior marked by a slight groove in front of the beak, which extends half way to the base. Loraine shales, near the bottom.

LEDA (*Schumacher*).

Shell oval and prolonged into a rostrum, or becomes subrostrated, and anteriorly valves closed. Leda resembles the Nucula in its general form.

L. LEVATA (*d'Orb.*).

Pl. 14, *fig.* 10.

Nucula levata (*Hall*).

Small; the general form is oval, but in this respect it is variable; umbones wide or thick, beak incurved; subrostrated

L. PULCHELLA.

Lyrodesma pulchella (*Conrad*).

Shell small, thin, nearly round, and rather angulated; extremities rounded; hinge line curved, and bearing about twelve or thirteen teeth; surface marked by fine concentric striæ.

L. PLANA.

Lyrodesma plana (*Conrad*).

Shell subrhomboidal, compressed; posterior basal; margin rectilinear; extremity rounded; posterior margin truncated.

ORTHONOTA (*Conrad*).

Resemble in their exterior form the Solemya. Upon the cardinal side the shells have teeth and oblique wrinkles or plaits analagous to those seen in the genus Leda. The Orthonota undulata, fig. 41, is the type of the genus, but belongs to the Devonian system.

Fig. 41.

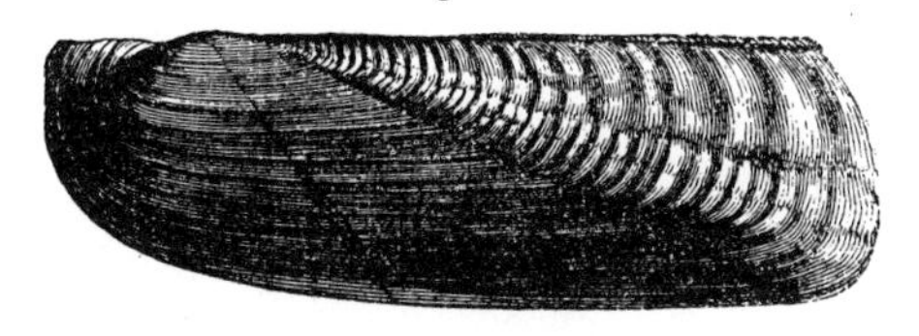

O. PARALLELA.

Pl. 13, *fig.* 14.

Shell extremely elongated, and very narrow; anterior extremity rounded, and contracted just forward of the beaks; cardinal margin straight or gently arched; posterior extremity rounded and broader than the anterior; local margin slightly arcuate; beaks near the anterior extremity; having an obscure carina; surface marked by fine concentric striæ, and a few strong oblique wrinkles along the dorsal ridge.—*Hall*, p. 299, *Pal. Rep.*

O. PHOLADIS (*Conrad*),

Is narrow, and marked by short undulations near the dorsal margin.

O. CONTRACTA,

Is expanded posteriorly, and constricted or sinuate upon the dorsal line.

CYPRICARDIA.

C. AMERICANA, *pl.* 14, *fig.* 11 (*d'Orb.*).

Modiolopsis carinata (*Hall*).

Shell rather small, in the form of a truncate ellipse; it is traversed by a sharp ridge, which extends from the beak to the base, and by a depression which forms a sinus upon the base, giving it the appearance of being constricted; posteriorly it is truncate; surface marked by concentric lines of growth. Trenton limestone.

C. SUBTRUNCATA.

Edmondia subtruncata, *pl.* 13, *fig.* 2.

Shell broadly oval; convex; margins rounded; rather flattened on the posterior slope; a broad ridge leads from the beak to the base; casts show a few concentric undulations.

CARDIOMORPHA.

Contains those Isocardias which are without teeth upon the hinge.

C. VENTRICOSA (*d'Orb.*).

Pl. 14, *fig.* 5, 6.

Edmondia ventricosa.

Broadly ovate; umbones thick and prominent; abruptly cuneate.

C. POSTSTRIATA.

Pl. 17, *fig.* 22.

Nucula poststriata (*Hall*).

Somewhat quadrangular, and traversed by a ridge running from the beak to the base; posterior slope striated. Loraine shales.

AVICULA.

A. DEMISSA, *pl.* 13, *fig.* 10 (*Conrad*).

Shell quadrate, convex; wings extended as wide as the base; umbo rather flat; surface marked by numerous concentric lines. Characterizes the Loraine shales, and is found in south-western Virginia.

A. SUBELLIPTICA.

Avicula elliptica, *pl.* 13, *fig.* 27.

Shell subelliptical; anterior extremity compressed; umbones small; posterior wing triangular. The foregoing comprise the most important characters, as given by Mr. Hall. Being only a cast, it is indistinctly characterized. Trenton limestone.

A. INSUETA, *pl.* 17, *fig.* 15 (*Conrad*).

Shell rhomboidal, depressed; anterior wing short; posterior extended and acute; surface marked by unequal, radiating and concentric, or transverse lines. It is found in the lower part of the Loraine shales (rare).

A. CARINATA.

Pterina carinata *pl.* 17, *fig.* 23 (*Conrad*).

Ambonychia radiata (*Hall*).

Shell ovate, extended into narrow beaks, and not very convex; wing obscure; surface marked with strong radiating striæ; or rather with flat simple plaits; concentric undulations distant.

This species is widely extended, being found in northern New York, Canada, in Ohio, and southwestern Virginia. It is confined to the Loraine shales, and is the most characteristic fossil of the upper part of the Lower Silurian system.

A. TRENTONENSIS (*Conrad*).

Pl. 13, *fig.* 28, 29, 30.

Broadly oval, and wide at base; wing greatly extended; sinuous upon its lower margin; anterior wing small; surface ornamented with longitudinal lines, which are alternately light and strong, concentric lines in their intersection with them produce an elevation, giving the surface a cancellated appearance. Trenton limestone, Middleville and Watertown.

A. SUBARCUATA (*d'Orb.*).

Modiolopsis arcuatus (*Hall*).

Shell, disregarding the wing, ovate, and acute at the beak, with a proportionably wide base, which is produced in part by posterior expansion; below the wing there is a shallow sinus, as usual in this genus; surface marked by undulations; and the anterior slope has obsolete longitudinal lines. The wing of this species is very large in proportion to the body of the shell. The undulations resemble those upon the Posidonomya undata.

POSIDONOMYA (*Bronn*).

P. BELLISTRIATA.

Pl. 13, *fig.* 5, 6.

Ambonychia bellistriata (*Hall*).

Shell long, ovate; length twice as great as the breadth; convex; beaks narrow, elevated, extended, incurved; hinge line straight, oblique; surface ornamented by distinct radiating striæ. Trenton limestone.

P. ORBICULARIS (*d'Orb.*).

Pl. 13, *fig.* 18, 19, 20.

Ambonychia orbicularis (*Hall*).

Shell broadly ovate, regularly convex; expanded behind like an avicula; becomes narrow towards the beak, which is rather incurved; cardinal line straight; surface ornamented by rather waved longitudinal lines, and crossed by lighter or fainter transverse ones; near the beak the portions of the shell show dotted radiating lines. The cast is the form in which it is usually found; it is marked by wrinkles around the beak. Trenton limestone, Middleville.

P. SUBUNDATA.

Pl. 13, *fig.* 23, 25.

Ambonychia undata (*Hall*).

Shell ovate, subquadrate; acute at the beak, with a broad base; length one-fourth greater than the breadth; convex; surface marked by broad concentric undulations, and with faint longitudinal grooves; some specimens have a line extending from the beak to the base; undulations rather irregular, both in height and distance; surface of the shell smooth.

P. AMYGDALINA.

Pl. 13, *fig.* 20, 21.

Ambonychia amygdalina (*Hall*).

Shell long, oval; nearly twice as long as wide; regularly convex; base and anterior margin rounded; umbones elongated; surface of the cast smooth, and with a few obscure undulations.

P. OBTUSA (*d'Orb.*).

Ambonychia obtusa (*Hall*).

General form very much as in the preceding species; but is described by Mr. Hall as shorter, straighter and more obtuse towards the beak, with scarcely any incurvation.

BRACHIOPODA OR PALLIOBRANCHIATA.

§ 23. In many respects the Brachiopoda are the most interesting of all the orders which belong to the subkingdom Molusca. They are the oldest palæozoa of the globe; and being most abundantly distributed and varied in form and clothed with more beauty and elegance, they have become the most attractive objects of study with the palæontologist. They represent the marine fauna of the earliest periods of the earth's history, and as they still exist, connect the past with the present, having survived all the changes which the globe has undergone. Their curious structure has deeply interested the zoologist, and he has sought in the still living representatives of this ancient order, an interpretation of the curious inward organization which many of them still retain in the fossilized state. It is singular that animals so small as the Brachiopoda are, should be provided with an organization so complete, and at the same time so full as to occupy most of the space which is usually allotted to the soft, sensitive parts. We see in their calcified spiral apparatus an immense development of hard matter, disproportionate, it would seem, to the soft, sensitive substance just alluded to, in which the vital functions must reside.

§ 24. The most interesting phase, however, in which we can view them is, as palæozoa, and as representatives of the different stages through which the earth has passed; and practically for the determination and comparison of the geologic stages in different parts of the globe; for, as they are more widely distributed than most orders, they become more applicable to the purposes and objects of the palæontologist. To secure the benefits, however, which have resulted from the study of this order, it is necessary, in the first place, to become familiar with the characteristics of the families and other subdivisions which have been made by the savans of the old world.

§ 25. In this order we find a peculiarity in their attachment or mode by which they are affixed to other bodies. Most of them, for example, are provided with a fleshy process or peduncle, which proceeds from their internal hard parts, and which is attached to other bodies for support, as represented in the annexed fig. 42. It is a representation of the Terebratula, and in other genera, as in this, an orifice may be found, through which a similar organ proceeded for the fulfillment of a like function.

Fig. 42.

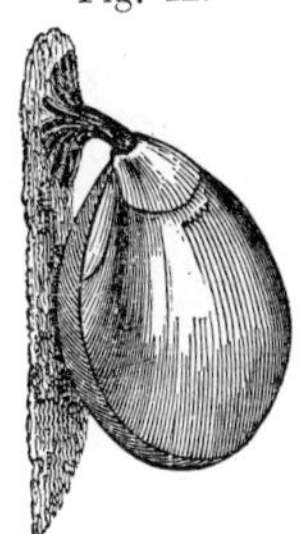

§ 26. In the internal parts, also, this order has another arrangement, which is common to the order. It is in the existence of the organs called the *arms*, which, when at rest, are coiled up compactly within its shell, but which might be unrolled and exposed at the will of the animal, as in fig. 43.

Fig. 43.

§ 27. The mouth is not furnished with teeth; hence, it was not adapted to mastication, and hence, too, their food must consist of minute particles of organic matter floating in the menstruum in which they lived or now live, and which may be brought to the oral orifice by means of the ciliary movements which the arms themselves are fitted to produce.

§ 28. The Brachiopoda preserve not only their distinctive type in the arrangement of the organization of the more vital parts, but also very clearly exists in the structure of the shells. Prof. Carpenter has shown that in this respect the minute structure of their habitations differs essentially from the Lamellibranchiata or Gasteropoda. It is so characteristic that even the smallest particle of the shell of a Brachiopoda can be distinguished by aid of the microscope from the shells of other classes. It shows that they worked on another plan, or followed another model in the construction of their dwellings. This structure can not always be made available to the palæontologist or zoologist, on account of the molecular changes which the shell has undergone.

§ 29. The composition of the shell is also different in this family from those which have been already described. There is more inorganic matter in them, for when submitted to the action of acids, and chemical tests, it is found to consist mostly of carbonate of lime. This fact goes to prove that as a class their rank is lower than the Gasteropoda, whose shells contain less inorganic matter.

§ 30. The position of the muscular system, or that concerned in the opening and closing of the valves, is worthy of a passing notice. In the oyster and pecten, there is one large subcentral muscular impression; in the clam and its allied genera, there are two, situated at the extremeties of the shell. In the Brachiopoda, with the exception of two or three genera, the muscles are arranged just within the beaks of the shell, and it is here that palæontologists find the most important characters for the determination of the genera.

As the Brachiopoda have been worked out in greater detail than any of the preceding classes, I propose to place before the American student a brief statement of most of the subdivisions as proposed by distinguished European palæontologists.*

§ 31. According to the best authority, the BRACHIOPODA are subdivided into ten families, and four subfamilies. The first embraces the TEREBRATULIDÆ. The beak of the larger valve is perforated, the smaller or dorsal valve is supplied with a loop or calcified process for the support of the oral appendages. Fig. 42 shows the perforation and mode of attachment of the family as has been stated. No member of this family is known to have lived until after the Silurian period.

2. The subfamily STRINGOCEPHALIDAE. The labial appendages in this subfamily are supported by an extended circular calcified process or loop, from the internal margin of which numerous rays proceed toward the centre of the shell which is also provided with a mesial septum. The family is confined to the Devonian period.

* Derived from a paper published by the Palæontographical Society, by Thos. Davidson, Esq., vol. I, for 1853.

SPIRIFERIDÆ.

In this family the animal is supposed to have been free, that is, not attached by a fleshy peduncle to other bodies; the oral appendages were largely developed and were supported by

Fig. 44.

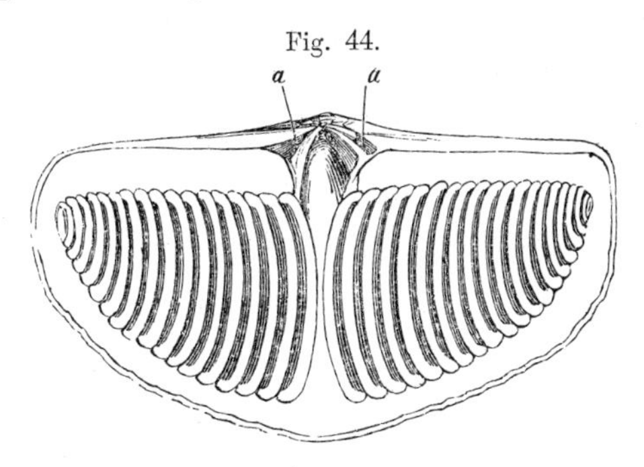

spirally rolled lamella. Its five genera which are embraced in this family are arranged in two subsections.

Genus SPIRIFER. Shell in its normal shape has an elongated trigonal form, inequivalve with or without a mesial fold. Fig. 45 exhibits the elongated form as it occurs in the Spirifer

Fig. 45.

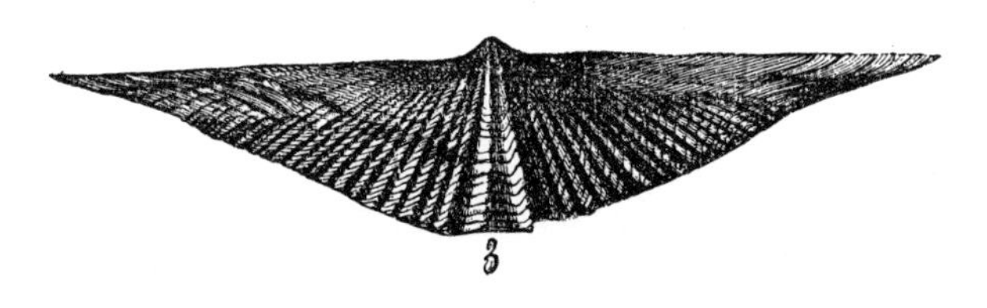

mucronatus. This Spirifer belongs to the Devonian period. These beautiful external forms are accompanied with internal arrangements and apparatus equally beautiful and more remarkable Thus, fig. 44 exhibits the spiral appendages as they are developed in the Spirifer striatus, the type of the genus.

The Spirifers make their appearance in the Trenton limestone and continue through the whole of the Palæozoic period.

Passing over Athyris and Uncites, we come to the genus Atrypa, one which makes its appearance in the lower Silurian and becomes one of the most common forms among the Palæozoa of this period.

The form of the Atrypa is circular, and may be transverse or elongated; the beak is incurved and has frequently a small round orifice, the valves are articulated by teeth and sockets; the dental valve is convex or may be nearly flat; the socket valve is convex; the mesial folds may, or may not exist. The arrangement of the spiral appendages* are the reverse of the Spirifer, being conical, but with the apex directed inward, as exhibited in fig. 46.

Fig. 46.

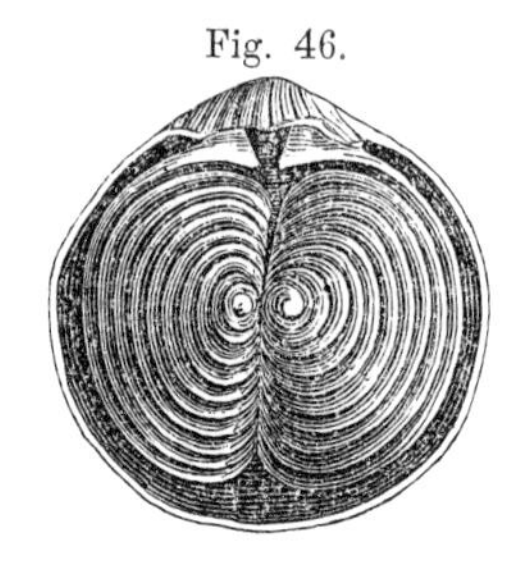

The outline of the Atrypa is shown in fig. 47, though it deviates in many species from this form. Thus, the A. lævis, fig. 47, is slightly elongated, it is an upper Silurian form; the A. elongata, is an extremely elongated one, fig. 48, and belongs to base of the Devonian system. In regard to the existence of spires in this and the genus Spirifer, it seems to require a favorable combination of circumstances, to preserve them, for it is extremely rare to find them remaining in the fossil.

Fig. 47

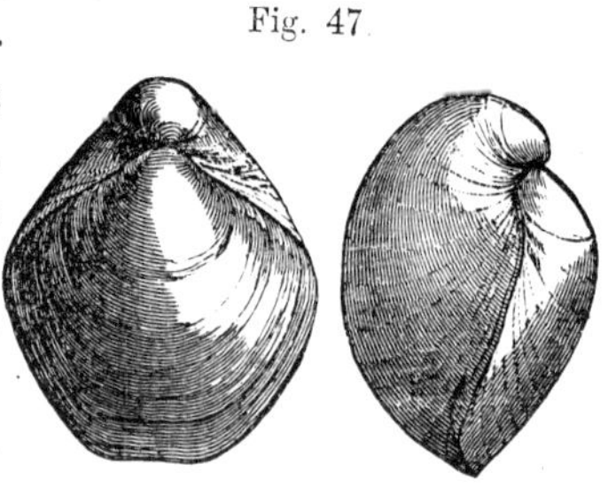

* From Mr. Davidson's paper, published by the Palæontographical Society.

Fig. 48.

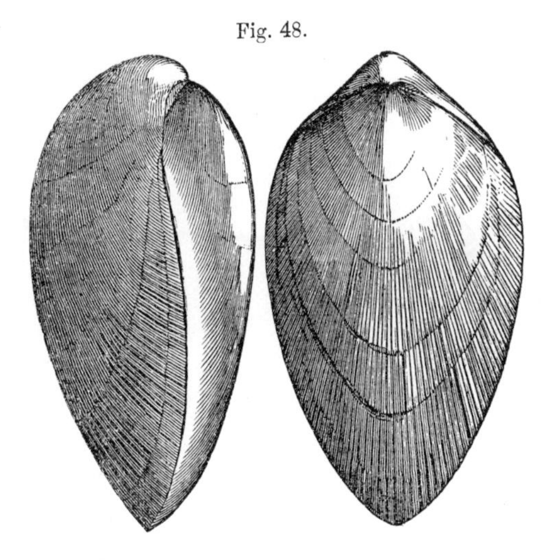

SPIRIGERA (*d'Orb.*).

The form of the Spirigera is variable; it may be circular, subquadrate, transverse and globose. It has the internal spiral apparatus and valves articulating by teeth and sockets, a short incurved beak lying contiguous to the umbo of the socket valve, or separated by a deltidium in two pieces; no true area; valves convex and divided or not by a mesial fold or sinus; surface smooth, striated or costated and marked by numerous lines of growth. There are four pits, or muscular impressions in the smaller valve and a small circular aperture close to the extremity of the umbo, the extremities of the spiral are directed to the lateral margins of the shell. In the ventral valve the dental plates extend downwards half across the shell running parallel with each other, showing within the latter elongated muscular impressions. Davidson, Spirigera concentrica, fig. 49, belongs to the Devonian system.

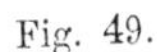

Fig. 49.

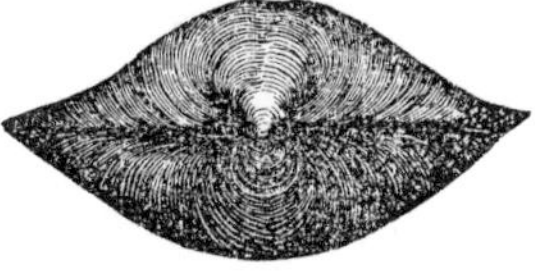

RHYNCHONELIDÆ.

Fig. 50.

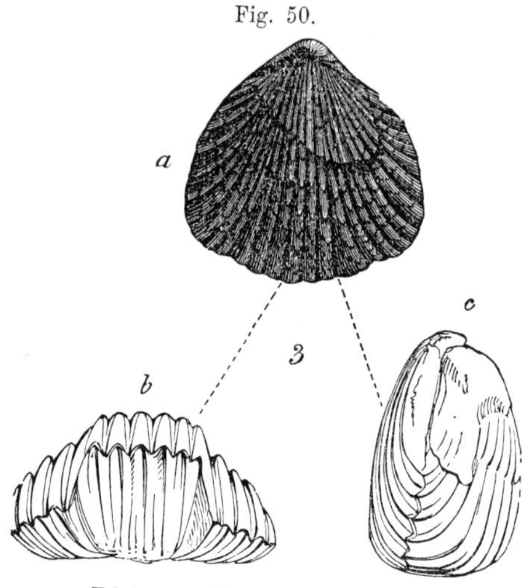

"The animals belonging to this order were free or fixed by a muscular pedicle issuing from an aperture under the curved beak of the ventral valve. They are supplied with oral appendages which are spiraly rolled, supported by a pair of short curved processes."

Fig. 50 exhibits the general form of the genus Rhinconella.

The family at present contains only three genera. The genus Rhinconella existed in the Palæozoic seas and has come down to the present time.

Fig. 51.

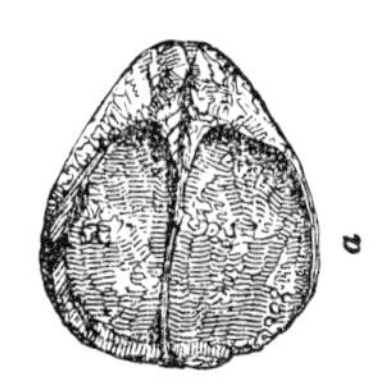

The Camaromorpha, of which none have been discovered below the carboniferous system, and the Pentamerus which appeared for the first time in the lower part of the upper Silurian, fig. 51, is of an oval elongated form with both valves convex; the ventral, has an entire incurved beak, and also is more convex than the dorsal. In the young state there is a triangular fissure, but no deltidium. The dorsal valve has two distinct longitudinal septa, while the ventral has two contiguous septa coalescing into one median plate. Fig. 51, *a*, Pentamerus galeatus, shows the inside plates of the dorsal valve, and fig. *b* is a side view of the incurvation of the beak; *a* shows the septa of the larger valve.

STROPHOMENIDÆ.

§ 32. The shell in this family is convex, plano-convex, or with one valve convex and the other concave, following the curvature of the valve. The cardinal area belongs to both valves and they have a straight hinge. The family forms a large proportion of the Palæozoa of the lower Silurian system. It contains, as constituted by Mr. Davidson, four genera: *Orthis*, *Orthisina*, *Strophomena*, and *Leptæna*.

The genus *Orthis*, as now constituted, contains those fossils whose valves are usually unequally convex, but variable in shape, being subcircular or quadrate; the socket valve is sometimes flat, or even slightly concave; hinge line straight and usually shorter than the width of the shell. Both valves are furnished with an area divided by an open triangular fissure; beaks incurved, but in the ventral valve it is longest. Fig. 52

Fig. 52.

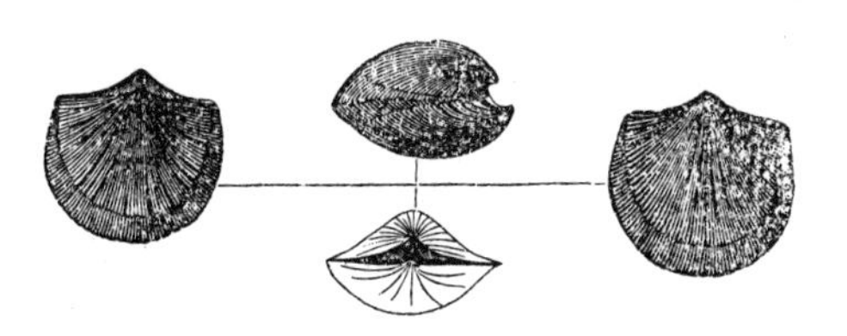

shows the most common form in the Orthis testudinaria. The internal parts of the valve furnish important characteristics also of this genus.

Fig. 53, *a b*, exhibits the teeth and muscular impressions of

Fig. 53.

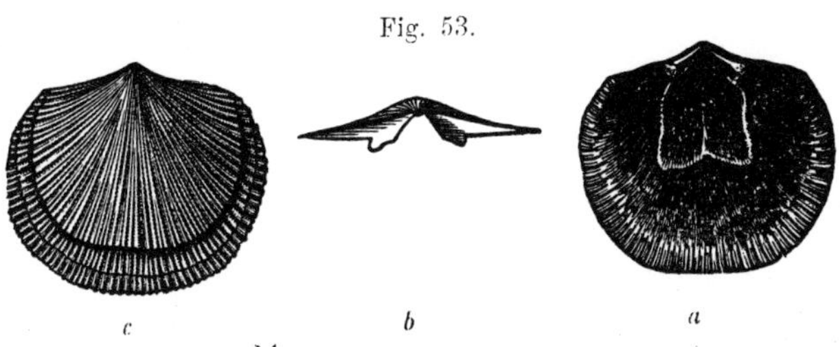

the ventral and dorsal valves; *c*, the most common circular form of the valves in the Orthis quadrata.

ORTHISINA.

This genus has the general form of the Orthis. The cardinal area is double, hinge line straight, less than the width of the shell; fissure is concealed by a convex deltidium, which is perforated in a few species. The muscular impressions of the ventral valve are situated just within the beak in an area which is divided by a mesial ridge.

STROPHOMENA.

This genus may be known by its flat or depressed form, its great expanse and diminutive thickness, its straight hinge line which equals the width of the shell and the conformity of the valves to each other, whether convex or concave. The cardinal area is double and crenulated at its inner edges. The fissure in the larger valve is partly covered with a convex deltidium; a small circular orifice exists in the young shell. The dental valve has two divergent teeth which are fitted to sockets in the dorsal valve. The muscular impressions are bounded by a circular ridge, open below, through which a low mesial ridge runs in the direction of the basal margin.

Fig. 54, Strophomena planumbona, shows the general form of

Fig. 54. Fig. 55. Fig. 56.

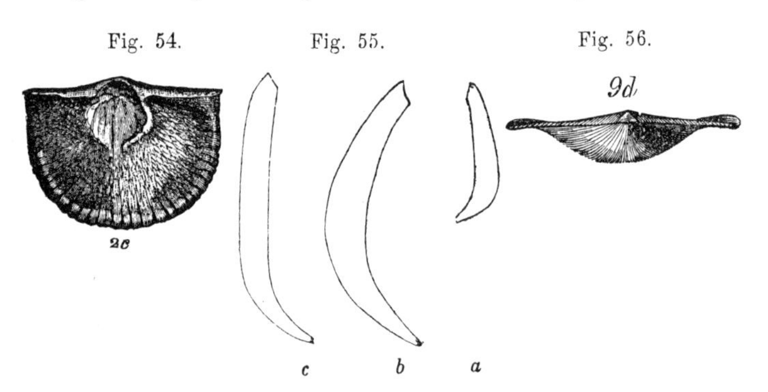

the shell and cardinal area; the semicircular ridge which limits the muscular impressions; and fig. 55, *b*, is a section of the valves showing the mode in which the margin is frequently bent. Fig. 56, *a b c*, the cardinal area.

LEPTÆNA.

Shell involute, semicircular; ventral valve regularly convex; dorsal valve concave conforming in its curvature to the other valve; area double, fissure in the ventral valve partly covered by a deltidium. In the dorsal valve the muscular impressions are large and bordered by a low ridge, and which extend to two-thirds the length of the valve, while they are small in the ventral valve, and without a margin. Fig. 57 shows the form of the shells belonging to the genus.

Fig. 57.

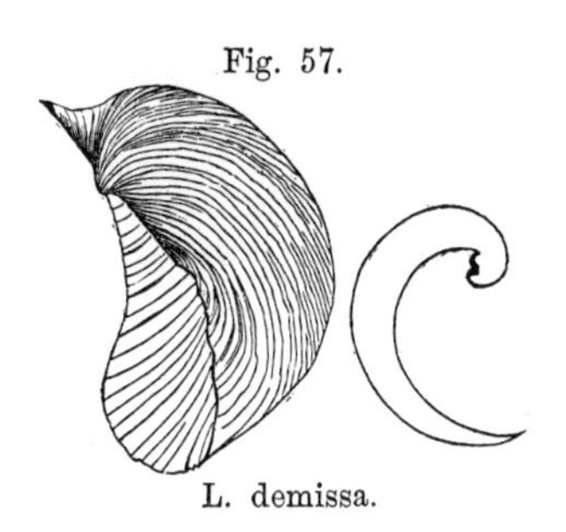

L. demissa.

The genera Davidsonia, Productus and Calceola are found in the Devonian system, and, as yet, being unknown in the Silurian system, will be passed over.

3. Shell unarticulated, texture horny, or calcareo-corneous, depressed, circular, and only slightly transverse; muscular peduncle, passing through a slit or foramen excavated in the lower valve.

CRANIADÆ.

The shell fixed to stones and other bodies by the substance of the ventral valve. The animal is provided with spiral arms. The upper valve is patelliform, and without articulating processes; the valves and their attached muscles being so constructed and arranged as to slide upon each other.

DISCINIDÆ.

Shell circular or oval, patelliform or conical, with the apex central or subcentral, inclining to the posterior margin; lower valve opercular, flat or subconvex, and perforated by a narrow, oval or longitudinal slit; valves unarticulated. It includes the common Orbicula of authors.

Discina.

Shell circular, oval, patelliform; the upper valve conical; the the lower, flat or convex; unarticulated; the surface is marked by lamellose lines of growth, and radiating ones proceeding from the apex. The shell structure is horny, differing in this respect from the preceding, which are calcareous.

Fig. 54.

a *b*

Fig. 54 shows the form of the valves of Discina. *a* shows the lamellose structure of the upper valve, and *b*, the lower or perforated valve.

Trematis (*Shap*).

Both valves convex; the lower with a greater convexity than the upper; the umbo on the former is subcerted in the latter; submarginal fissure in the lower valve, oblong, and originating just beneath the umbo, and extending to the margin. Includes the Orbicula in part of authors.

Fig. 55.

Fig. 55 shows the form of the two valves, and the aperture, *a*, in the lower valve for the transmission of the pedicle.

Lingulidæ.

Shell horny, and fixed by a muscular peduncle, as represented in fig. 56 of the recent Lingula; shell unarticulated, retained in position by muscles only. It contains two genera, Lingula and Obolus.

Fig. 56.

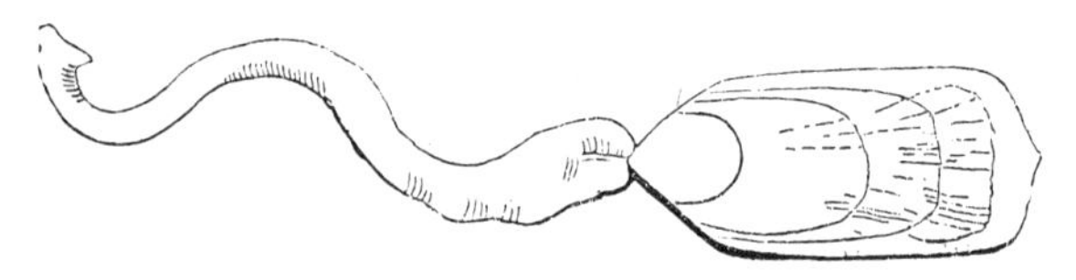

Lingula.

Shell thin, equivalve, subelongate; edges parallel, or nearly so; apex acute; the peduncle passes out between the shell.

Obolus.

Shell circular, and orbicular, equilateral, inequivalve; slightly transverse; depressed; unarticulated; larger valve most convex, and with a short, obtuse beak, and a flattened cardinal edge, or a false deltidium produced by lines of growth, in which the groove for the transmission of pedicle is excavated.

SPIRIFERIDÆ.

Atrypa.

Synopsis of the genus:

A. *Shell smooth, with a mesial fold or not, marked with concentric lines.*

B. *Shell plicated; plaits simple.*

C. *Shell plicated; plaits imbricated.*

A. Shell smooth, with or without a mesial fold, with or without concentric or radiating lines.

A. extans (*Conrad*).

Pl. 10, *fig.* 1 *and* 11.

Shell ovate, transverse, subglobose; dorsal valve with a broad, deep sinus; ventral valve with a corresponding rounded elevation, and considerably elongated in front; beaks small, surface marked by concentric lines, longitudinal striæ less distinct. Trenton limestone, Watertown.

A. nucleus, *pl.* 10, *fig.* 2.

This specie has the general form of the A. extans, but smaller, beak of the ventral valve incurved and compressed over the dorsal valve. Its characters do not clearly separate it from the A. extans.

A. BISULCATA, *pl.* 10, *fig.* 3, *a b c d e.*

Shell small, round, oval, unequally convex; dorsal valve bisculated; beak of the ventral valve incurved and close pressed upon the other; surface smooth, concentrically striated, ventral valve* with a single sulcus. Trenton limestone, Middleville.

A. CUSPIDATA.

With the mesial fold in front, the shell triangular, laterally it is subglobose.

A. EXIGUA, *pl.* 10, *fig.* 6.

Shell small, ovate valves unequally convex, beaks of the ventral valve acute, valve itself with a ridge in the middle, dorsal depressed surface marked with fine concentric lines, varying from a line to one-fourth of an inch in length.

A. CIRCULUS, *pl.* 10, *fig.* 7.

Shell small, with a suborbicular form; outline circular convexity of the valves subequal; beaks subequal; surface marked by fine concentric lines.

A. AMBIGUA, *pl.* 10, *fig.* 8, 9, 8*c.*

Form of the shell suborbicular, outline trigonal or subtromboidal, and of a medium size; beaks subequal; valves unequally convex, ventral valve regularly convex, rounded near the beak, sinus extending to the front margin; dorsal valve has a corresponding fold; surface marked by fine concentric lines, obscurely plicated; ribs plain or unimbricated.

A. HEMIPLICATA (*Hall*).

Pl. 10, *fig.* 10.

Shell wider than long or transverse, unequally convex, beaks small, ventral valve with a broad fold, plicated below and

---

* I apply the term ventral valve to the one whose beak is curved over and upon the other; dorsal valve is therefore the small valve.

with lateral plications, none of which extend to the beak, mesial fold in the dorsal valve plicated, broad not deep; plications usually two; it is also laterally and coarsely plicated upon the margin; surface usually smooth. There are some variations in form, some having wide frontal margins; in others the plications are variable in extent, and in some the ridges bounding the sinus are more angulated than in others; the ventral valve is usually quite prominent, giving a gilbous form to the fossil. Trenton limestone, to which it is confined.

A. SUBTRIGONALIS, *pl.* 10, *fig.* 12.

Form of the shell trigonal, rounded in front, which is broad and from which it tapers rapidly to a small beak, convexity of the valves subequal; ventral valve regularly convex with a small triangular foramen, under a slightly incurved beak; valves with about twenty plications of which the mesial sinus has three, and the corresponding fold has four.

A. RECURVIROSTRA, *pl.* 10, *fig.* 5.

Shell small, ovoid; ventral valve regularly convex; beak incurved, sinus shallow and obscure, plications small and about twenty-four, and extending from the beak to the base.

A. PLICIFERA, *pl.* 3, *fig.* 29.

Shell somewhat trigonal, small, incurved and of a medium size; convex, but rather depressed or flattened; marginal line undulating; sinus shallow and not extending farther than the middle of the shell; plications about twenty rounded and equal, five or six belong to the sinus. Chazy limestone.

A. PLENA (*Hall*).
Pl. 3, *fig.* 28.

Shell of a medium size, rounded and convex, rather ventricose; ventral valve with the short, small beak only slightly incurved and close pressed upon the other; plications undivided and from fifteen to twenty and plane, with about four or five belonging to the sinus and opposite fold. Common and in beds of the upper part of the calciferous sandstone.

A. ALTILIS, *pl.* 3. *fig.* 30.

Shell of a medium size, rotund or convex; dorsal valve rather more convex than the ventral; beak small acute, and the ventral one incurved over the other. Plications equal, rounded and about twenty-four, eight of which belong to the mesal sinus and mesial fold; marginal line waving; occurs in the Chazy limestone and upper part of the calciferous sandstone.

A. ACUTIROSTRA.

"Shell small or minute trigonal; plications equal, and about twelve or fourteen; the central one on the ventral valve extending half way to the beak."

A. DUBIA, *pl.* 3, *fig.* 23.

Small, wide in the middle and converging rapidly to the beak; beak of the ventral valve small, incurved; plications about thirty. Chazy limestone.

A. SORDIDA, *pl.* 10, *fig.* 16.

Small circular convex, ribs undivided, rendered rugose by intersecting lines, without a mesial fold, supposed to be a young shell.

A. MODESTA (*Say*).

Shell small, circular, unequally convex; dorsal valve broadly oval, convexity much less than the ventral valve; beak of the ventral valve prominent, incurved, perforated, plications about eighteen, four of which belong to the mesial fold. Trenton limestone, extending into the shales above.

C. Ribs or plaits imbricated.

A. INCREBESCENS, *pl.* 10, *fig.* 13, *from a to x.*

Shell of the medium size, full or inflated convex, length and breadth subequal, variable in form and contour; beak of the ventral valve acute, perforated when young with a broad sinus undivided and imbricated and plaits about sixteen, with three in the sinus and four upon the mesial fold.

A. DENTATA (*Hall*).

Pl. 10, *fig.* 14.

Shell small, form somewhat triangular; outline cordiform; valves unusually convex; plications undivided, angular, strong, two of which form the mesial fold; surface crossed by numerous imbricating lines. Trenton limestone.

STROPHOMENIDÆ.

ORTHIS.

Synopsis of the genus:

A. *Ribs or plaits simple or undivided, and not imbricated.*

B. *Ribs or plaits fasciculated, or forming clusters having a common origin, and not imbricated.*

C. *Ribs or plaits imbricated.*

A. ORTHIS TRICENARIA (*Conrad*).

Shell semicircular; cardinal area high; foramen narrow; cardinal line equal to the width of the shell; apex of the central valve elevated and pointed; beak scarcely incurved; dorsal valve flat circular; cardinal angles rather rounded, inside with a mesial ridge, extending from the socket to the opposite margin; plaits numerous, variable, and undivided, from twenty-five to thirty; the points of the ribs inside have linear indentions.

O. PLICATELLA (*Hall*).

Pl. 9, *fig.* 9, *a*, *b*.

Shell rather small, transverse or broadly ovate; convexity of the valves subequal; plications merely simple, or about twenty-four in each valve, number variable. Trenton limestone.

O. PECTINELLA (*Conrad*).

Pl. 9, *fig.* 10, 11, *a*, *b*.

Rather large, circular, sometimes transverse, obliquely convex; cardinal line rather less than the width of the shell, and largest upon the less convex valve; plications equal, about twenty-eight, and mostly simple, a few do not extend to the beak; a few elevated concentric lines sometimes exist near the margin. It appears to be confined to the Trenton limestone.

B. *Ribs or plaits divided, sometimes fasciculaled.*

ORTHIS TESTUDINARIA (*Dalman*).

Pl. 9, *fig.* 1, *a* to *i*.

Circular, rather transverse, planoconvex; cardinal line straight, less than the width of the shell; angles rounded; cardinal area rather small; beak of the ventral valve prominent, slightly incurved; foramen rather wide; plaits numerous, in fascicles of about six in each, but not constant; middle of the ventral valve prominent upon the umbo, against which there is a corresponding shallow sinus in the opposite valve; inside of the dorsal valve there is a mesial ridge, which disappears about the middle of the valve, or becomes less prominent. The O. testudinaria is the most common Orthis of the Trenton limestone; it extends into the Loraine shales, where it is less abundant.

O. SUBÆQUATA (*Conrad*).

Pl. 9, *fig.* 2.

Differs from the O. testudinaria in the nearly equal convexity of the valves.

O. DISPARILIS (*Conrad*).

Pl. 9, *fig.* 4, *a*, *c*, *b*.

Small, semicircular, concave or planoconvex; area prominent, or the beak of the ventral valve elevated; foramen narrow; plaits about twenty-eight, one-half commencing upon the umbo, traversed by numerous concentric lines. Trenton limestone.

O. PERVERTA (*Conrad*).

Small, circular; convexity of the valves unequal; area of the valves nearly equal; surface marked by numerous radiating lines, which divide upon the umbo.

O. FISSICOSTA (*Hall*).

Pl. 9, *fig.* 7, *a*, *b*.

Circular and transverse; cardinal line less than the width of the shell; beaks incurved; valves unequally convex; plaits about twenty; angular and divided twice or three times near the middle.

O. INSCULPTA, *pl.* 9, *fig.* 12.

This fossil is described by Mr. Hall as resupinate; dorsal valve depressed, convex; beak elevated, not incurved; cardinal line less than the width of the shell; area short; surface marked by fine elevated radii; bifid upon the umbo, and again bifid or trifid towards the margin of the shell; and transversely marked by strong concentric lines. The interior vascular impressions of the valves renders it quite doubtful whether this is an Orthis; the muscular impressions, and those of the vessels are quite different from an Orthis.

O. DICHOTOMA (*Hall*).

Pl. 9, *fig.* 13.

Convexity of the valves unequal; plications bifurcate midway between the beak and the base; crenulations indistinct. Blue limestone of Ohio.

O. SUBQUADRATA (*Hall*).

Form of the shell subquadrate; valves unequally convex; cardinal line much shorter than the width of the shell; angles rounded; foramen small; ventral valve flattened near the margin; plications numerous, sharp, crossed by numerous concentric lines, which give the appearance of having an imbricated structure, yet, when worn, these lines are obliterated. Blue limestone of Ohio, and unknown in New York.

O. OCCIDENTALIS (*Hall*).

Shell transverse, or wider than high; beak unequally convex; hinge line shorter than the width of the shell; beak of the ventral valve large, incurved, with an elevation extending from the middle of the valve to the margin; mesial sinus wide and shallow in the dorsal valve: surface marked by numerous, sharp, bifurcating radii. Blue limestone of the West.

O. SINUATA (*Hall*).

Rather large; wider than high, semicircular; umbo rounded, prominent; beak incurved; area narrow; valves unequally convex, ventral, with a broad rounded elevation, to which the sinus of the dorsal valve corresponds; surface marked by numerous radiating striæ, which are somewhat crenulated.

O. SUBJUGATA (*Hall*).

The O. subjugata is described by Mr. Hall as semioval, with a point somewhat produced, dorsal area large; ventral area narrow; dorsal (ventral) having its greatest elevation just below the beak; sides somewhat depressed, often flattened with a deep broad mesial depression; ventral valve convex, gibbous; surface marked by even rounded striæ, which are dichotomous and trichotomous toward the margin. Blue limestone of Ohio and the western states.

O. ÆQUIVALVIS.

Pl. 9, *fig.* 6. *a*, *b*, *c*.

Shell symmetrical; valve subequal; hinge line less than the breadth of the shell; plications above thirty, bifurcating once or twice. Trenton limestone.

C. Ribs or plaits imbricated.

ORTHIS LYNX, *pl.* 14, *fig.* 9, *a p.*

Shell thick and comparatively massive in the old shells, form variable, but usually transverse, or it may be subquadrate and globose, frequently very thick, cardinal line usually less than the width of the shell; area of the valves unequal; plaits undivided and twenty, twenty-four thick, angular and imbricated, forming with each other a crenulated margin; mesial fold has four, and the sinus three ribs, this number is variable with age.

Trenton limestone. In New York the O. lynx is much smaller than at the west in the blue limestone. It is widely distributed, being found at most, if not all the localities where the Trenton limestone occurs.

O. BELLA-RUGOSA (*Conrad*).

Pl. 9, *fig.* 3, *a b c d.*

Shell rather small, semicircular or semioval, convexity of the valves subequal, hinge line rather less than the width of the shell, dorsal valve with a narrow depression in the middle, the ventral valve has a rather elevated beak, and slightly incurved; plaits numerous, bifurcating unequally and imbricated. Trenton limestone.

STROPHOMENA.

Synopsis of the genus:

A. *Longitudinally striated.*
B. *Striated and transversely wrinkled.*

S. ALTERNATA (*Conrad*).

Pl. 11, *figs.* 1, *a b c; and fig.* 3, 5, *b c.*

Large, circular or semicircular; area very narrow, hinge line straight, sometimes the angles are bent towards the convex or ventral valve, where the angles of the valves become very thin; ventral valve perforated and bent near the middle, disc flattened or only slightly convex; dorsal valve with a convex deltidium or boss projecting into the fissure of the ventral; surface firmly striated, every third or fourth being stronger and more distinct. Cabinet of Williams College.

S. CAMERATA (*Conrad*).

Resembles the S. deltoidea, having very nearly its proportions, striæ or surface markings, but its convexity is greater than that of the deltoidea.

S. DELTOIDEA.

Shell higher than wide, somewhat deltoid, hinge line longer

than the shell is wide. Ventral valve plano convex upon the disc and becoming convex below and with the middle of the marginal part produced in front, while the lateral parts are bent down. Cabinet of Williams College.

S. ALTERNISTRIATA (*Hall*).

Large, circular, somewhat wider than long; hinge line longer than the shell is wide; ventral valve with a large subconvex disk, bent at one-fourth the distance from the basal margin to the cardinal area; striæ rather coarse, subequal.

In this species the striæ are coarser than those of the alternata and more rugose; the ventral valve has punctures between the striæ, while in the dorsal valve the surface appears to be marked by obscure transverse lines. Cabinet of Williams College.

S. FILITEXTA (*Hall*).

Pl. 11, *fig*, 9, *c*.

Shell large, circular, height and breadth subequal, hinge line prolonged beyond the margin of the shell along a projecting angle; dorsal valve, the muscular impressions strong, the saucer form area formed of a distinct layer of shelly matter which is coarsely plicated; shell punctated throughout. Cabinet of Williams College.

S. PLANUMBONA (*Hall*).

Semicircular hinge line extended beyond the margin of the shell, whole margin bent; ventral convex, dorsal valve, the fissure partly occupied by a convex deltidium; muscular impression bordered on each side by a semicircular ridge of shelly matter; which is wanting in front and permits the passage of a mesial ridge; internal margin indented with furrows alternating with ridges along the bent border; surface striated with raised lines variable in strength and in thickness. Cabinet of Williams college.

S. DEFLECTA (*Conrad*).

It is described by Mr. Hall as resupinate, semioval; dorsal

valve slightly concave, elevated towards the beak, and deflected at the angles; ventral valve moderately convex, cardinal extremities deflected, cardinal area wide, and the surface has a crenulated appearance, from the crossing of the longitudinal and concentric lines.

S. recta (*Conrad*).

Shell small, semicircular, as wide again as high; cardinal line extended into the acute angle beyond the margin of shell; cardinal area narrow; striæ bifurcating; surface crenulated. Cabinet of Williams College.

S. sericea, *pl.* 11, *fig.* 6. *a b c d.*

Shell small, semicircular, depressed; hinge line longer than shell is wide; base line subparallel with the hinge line, or forms a true semicircle; fissure of the ventral often closed by the deltidium of the dorsal valve; perforation generally closed, greatest convexity immediately below the hinge line, or near the umbo; muscular impressions of the dorsal valve large, subquadrangular and divided by a mesial groove. Extends from the Trenton limestone to the Loraine shales. Most widely distributed of the Branchiopoda. Cabinet of Williams College.

S. sinuata (n. s.).

Fig. 61.

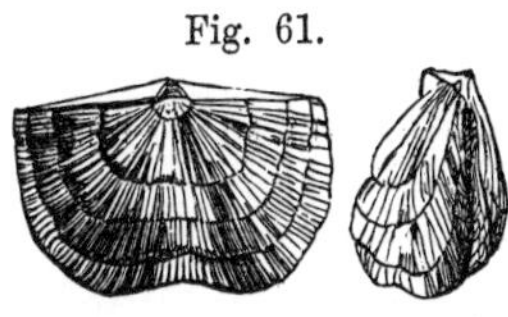

Shell rather small, thick, stout for its size, semicircular, cardinal line equal in length and width; aperture open at the apex of the deltidium; umbro slightly elevated; ventral valve bent in the middle of the margin, producing a deep sinus and a projection or fold of the opposite valve. Concentric lines of growth strong, striæ bifurcate at different distances be-between the margins. The species is one-third larger than the S. sericea. The sinus resembles that of the S. nasuta, or the Orthis sinuata. Cincenati. It belongs to the blue limestone.

B. *Surface transversely wrinkled.*

S. TENUISTRIATA (*Sowerby*).

Semicircular, hinge line longer than the width of the shell; angles produced; ventral valve rather flat, margins bent down; transverse wrinkles numerous and narrow; longitudinal lines numerous, equal.

DISCINIDÆ.

Discina (*Lamark*).

The genus Discina, Orbicula lamellosa, is supposed by M. d'Orbiny not to have made its appearance prior to the Tertiary period.

ORBICULOIDEA (*d'Orb.*).

It is characterized by Mr. Davidson as suborbicular, patelliform, transversely or longitudinally oval; lower valve conical or concave; no pedicle disk, but a narrow, oval or circular aperture, situated in a furrow.

O. LAMELLOSA, *pl.* 8, *fig.* 10.

Orbicula lamellosa (*Hall*).

Rather small, circular; apex of the upper shell small, and situated about one-third of the breadth of the shell from the margin; squarmose concentric lines mark the surface.

O. FILOSA, *pl.* 8, *fig.* 9.

Orbicula filosa (*Hall*).

Mr. Hall hesitates in referring this shell, figured in the Palæontology of New York, pl. 30, fig. 9, to the Orbicula. I have not seen this fossil, and therefore shall merely refer to the figure.

O. TRUNCATA (n. s.).

Fig. 62.

Shell corneous, circular; apex eccentric, and marked with a shallow furrow, which extends one half the distance towards the margin: posterior margin truncate; cuticular surface marked by fine radiating lines, and rather strong concentric ones of growth. Trenton limestone, Middleville. Cabinet of Williams College.

TREMATIS (*Sharp*).

Orbicella (*d'Orb.*).

Shell depressed; valves unequally convex, greatest in the lower valve; umbo of the lower valve subcentral, of the upper, marginal; lower furnished with an oblong aperture for the transmission of the pedicle.

T. TERMINALIS (*Emmons*).

fig. 55.

Orbicula terminalis.

Orbicella terminalis (*d'Orb.*).

Fig. 63.

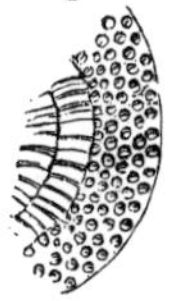

Shell circular; length and breadth of the upper valve equal or subequal; cuticle quincuntially punctate; punctures beneath the cuticle less regular and more distant than upon the cuticle, but arranged somewhat in concentric lines; shell inflected about the aperture, which opens just beneath an elevated umbo; umbo of the upper valve marked by three longitudinal lines. Fig. 63 shows the surface markings. Trenton limestone (rare). Cabinet of Williams College.

LINGULIDÆ.

LINGULA QUADRATA.

Pl. 8, *fig.* 4, *a*, *b*.

Shell large, quadrilateral; sides rounded; extremities subequal; base circular or rounded; cardinal extremity subacute; beak submarginal with a false area; marginal line circular; ventral valve rather more convex than the dorsal; surface finely marked by concentric lines, the longitudinal ones are more distinct upon the cast than upon the shell. In all my specimens there is a longitudinal groove in the middle of the shell or upon the ventral valve. The sides being thin and brittle, the margins are often broken, so as to appear parallel. Trenton limestone. Cabinet of Williams College.

L. ELONGATA (*Hall*).

Pl. 8, *fig.* 5.

Shell oval; ends subequal, cardinal one narrower than the basal; elevated in the centre; surface marked by fine concentric striæ.—*Hall, Pal. Rep.*

S. PAPILLOSA (n. s.).

fig. 64.

Fig. 64.

Umbo marginal, obtuse; surface faintly striated longitudinally; surface finely but irregularly papillose; middle marked by two smooth furrows. Trenton limestone.

L. OBTUSA, *pl.* 8, *fig.* 7.

Shell ovate; sides rounded, and curving towards an obtuse beak; beaks produced beyond the cardinal area; concentric and longitudinal lines extremely fine.

L. PRIMA (*Emmons*).

Shell small, oval, obtuse, rounded, nearly straight at the base; beak scarcely elevated above the surface of the wider part of the shell; acute; surface faintly marked with concentric lines of growth. Potsdam sandstone, Keeseville, High bridge, where it is found through a thick mass of rock for sixty or seventy feet. Cabinet of Williams College.

L. ANTIQUATA (*Emmons*).

Pl. 4, *fig.* 7.

Shell ovate, rather wide near the base, begins to taper above the middle and terminates in a rather acute beak; the shell is made up of lamellose plates, somewhat longitudinally wrinkled. Occurs at French creek, Jefferson county, N. Y., in a friable variety of Potsdam sandstone, near the top but below the Fucoidal layers.

Probably the L. prima and antiqua should constitute but one species, as both vary in form and size. The L. prima is however much smaller, upon the whole, than the antiqua. Mr.

Hall remarks that L. antiqua, figured in my Geological Report of the Second Geological District of New York, is the L. acuminata of Conrad; the figures, however, were drawn from specimens I obtained at the High bridge, and were taken from the Potsdam sandstone, and though enlarged, my shell represents the form of this species and not that of the Acuminata.

L. ACUMINATA (*Conrad*).

Pl. 4, *fig.* 9.

Shell elongated, rounded at base; sides gradually converge to an acute and subacuminated apex; umbo rather elevated, forming a ridge. Calciferous sandstone.

L. RICINIFORMIS.

Pl. 8, *fig.* 2, *a*, *b*, *c*.

This species is described as oval, convex, slightly attenuated towards an obtuse beak, with a smooth surface and obsolete concentric lines, and not exceeding one-third of an inch in length. Trenton limestone.

L. ÆQUALIS (*Hall*).

Pl. 8, *fig.* 3.

It has a very close resemblance to the L. riciniformis, but is a little larger.

L. CRASSA (*Hall*).

Pl. 8, *fig.* 8.

Shell thick, broadly ovate, subacute at the beak, and wide at the base. The unequal convexity of the valves of this species can not form a distinguishing character, as it is rather common to many others. Trenton limestone, to which it is confined.

OBOLUS.

This genus has not as yet been discovered in the Lower Silurian rocks of New York. Dr. Owen has given several figures of fossils which he refers to this genus. It appears to me that there is still some doubt respecting the identity of the fossils which have been thus referred to, inasmuch as the lower valves can not, as yet, be identified among the numerous forms which have been illustrated in Dr. Owen's valuable work.

MOLUSCOIDES.

The covering of the Tunicata not being composed of matter capable of being preserved, no animals of the class have been discovered in any formations.

BRYOZOA.

The Bryozoa resemble the polypi in their general appearance, form, distribution of the organs for conveying food to the mouth, and in being fixed to a common calcareous or corneo-calcareous support, but are still more closely related to the Molusca than to the Zoophyte. They are supplied with ciliated tentaculæ which surround the mouth, a digestive apparatus like the Molusca. They are enveloped in a mantle and pròtected in a cell often composed of carbonate of lime. This cell has an operculum, and contains a single individual. But these cells, though they adhere to each other and are placed upon a common axis, are never divided by septa radiating from the outer wall, and this fact will enable us to distinguish a Bryozoon from a coral, or Zoophyte. A coral has a single orifice for the reception of food and the rejection of excrementitious matter: a Bryozoon has two, and besides, the cell of a coral never has an operculum. When the septa or partitions of a coral cell are broken down, or dissolved, as they may be in a fossilized state, it is then difficult to distinguish them from each other, but we never see the cellule of a Bryozoon with septate lamina.

The number of genera belonging to the Bryozaries is not numerous, I shall however describe those only which are known in the lower Silurian system

Genus Ptilodictya (*Lonsdale*).

The cells are either round or rhomboidal, growing upon two sides of a lamellous support and arranged in a quincunx order.

P. ramosa (*d'Orb.*).

Pl. 4, *fig.* 1, 1 *a.*

Stictopora ramosa (*Hall*).

Cells on both sides, of a branching calcareous support, forming a crust upon which they are arranged in parallel lines, so as to form a quincunx order; mouths oval opening obliquely upward.

P. labyrinthica (*d'Orb.*).

Pl. 4, *fig*, 14, 18, 22 *and* 3.

Stictopora labyrinthica (*Hall*).

The branching stems are flattened and flexuous, branches forming an axis which bear upon both sides oval cells arranged in quincunx order. This Bryozoon is abundant on the weathered surface of the Birdseye limestone at Chazy.

P. elegantula (*d'Orb.*).

Pl. 7, *fig.* 4, *a b c d.*

Stictopora elegantula (*Hall.*)

The branching celluliferous axes are somewhat flexuose supporting oval cells arranged in close lines, upon which they alternate. Trenton limestone.

P. recta (*d'Orb.*).

Escaropora recta (*Hall*).

Axis cylindrical, simply tapering, the tubes with open mouths and oval, margined with or inclosed in rhomboidal spaces, and arranged in regular lines.

Subretepora (*d'Orb.*).

Intricaria (*Hall.*)

Cellules large, and arranged in single lines upon anastomosing or dichotomous branches.

S. RETICULATA, *pl. 7, fig.* 8, *a b.*

Intricaria reticulata (*Hall*).

Axes branching and anastomosing, sometimes regularly and forming thereby pentagonal or hexagonal meshes; mouths circular and in single rows. Trenton limestone, Bridport, Vt.

SULCOPORA (*d'Orb.*).

It differs from Ptilodictya in the cells being arranged in lines with a furrow between them.

S. FENESTRATA (*d'Orb.*).

Stictopora fenestrata (*Hall*).

Pl. 3, *fig.* 5, *a b.*

Cellules in rows separated by a furrow running in the direction of the axis of support, mouths oval or rounded. Chazy limestone.

ENALLOPORA (*d'Orb.*).

Cellules projecting upon the sides of a compressed branching axis, and never anastomosing; cellules alternating; mouths circular or nearly round.

E. PERANTIQUA (*d'Orb.*)

Gorgonia perantiqua (*Hall*).

Pl. 7, *fig.* 5. *a b.*

M. d'Orbiny refers to the Gorgonia perantiqua of Hall, the foregoing fossil. The celluliferous branches are subdivided and appear to proceed from a common central support. Bridport landing, Vt.

STELLIPORA (*Hall.*)

This genus is characterized by Mr. H. as being formed of a thick expanding crust having star shaped elevations composed of from four to nine rays, the upper edges of which present numerous elevated pores. Fig. 19, pl. 7, is referred to this genus under the name of S. anthoidea. Trenton limestone.

PTEROPODA.

Conularia (*Miller*).

Shell straight pyramidal; opposite sides similar and equal; angles grooved; texture delicate and somewhat like a woven fabric; apex solid, and separated from the open shell above by a simple imperforate, very convex septum.*

C. Trentonensis (*Conrad*).

Pl. 16, *fig.* 4.

Shell pyramidal, four sided, equal, and separated by longitudinal furrows, angles of divergence of two opposite sides, about 28°; sides ornamented by numerous parallel ridges meeting in the middle of each face at an angle, and forming shallow grooves; spaces between the ridges concave and traversed by other sharp ridges and interrupted by the transverse ones. Trenton limestone, cabinet of Williams College.

C. granulata (*Hall*).

Pl. 16, *fig.* 5.

Angle marked by linear grooves and surface by striæ, which are crossed by finer longitudinal ones, giving a granulated appearance to the shell under the microscope.—*Hall.* Trenton limestone.

C. papillata (*Hall*).

Pl. 16, *fig.* 6.

It is described by Mr. Hall as gradually diminishing towards the apex, and as marked by regular lines of granulations, the spaces between which are elevated.

C. gracile (*Hall*).

Pl. 16, *fig.* 7.

Slightly arcuate; surface marked by sharp undulating transverse striæ, longitudinal ones rather indistinct.

---

* In the Palæontology of New York, it is stated that the Septa are perforate; this statement, however, requires confirmation. There is no perforation in a septum which is well exposed in a specimen in my collection.

C. HUDSONIA (n. s.).

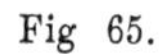
Fig 65.

Pyramid elongated; sides equal or subequal; transverse furrows meet in the middle of a face at an angle of 130°. The edges of a side diverge at an an angle of about 25°. Both the transverse and longitudinal striæ are stronger than those of the C. trentonensis; there being more than twice as many in the latter as in the species under consideration. The markings have a general resemblance to the C. trentonensis, excepting that they are much coarser and the fossil is much larger. Loraine shales, Loraine, Jefferson county. Cabinet of Williams College.

CRUSTACEA.

It embraces those animals which belong to the subkingdom Articulata. The more perfect orders of this subkingdom have a true vascular circulation and a branchial respiration. Their thorax which is covered with a shield, conceals more or less the the head. The abdomen is usually composed of many joints or rings, but in this respect the number varies from five to seven pairs. There are types of Crustaceans, however, in which some of the foregoing characters are effaced.

The crabs and lobsters represent this class or subkingdom.

Those crustacea, however, in which the palæontologist is most interested, are the Trilobites. This order of crustacea deviates somewhat from the true type representing the class, rather in its embryo than in its mature state. They form an order, the species of which are closely related. They are all extinct, and indeed they belong to the oldest palæozoa of the globe. None are found in the Mesozoic period. The normal type of the

Trilobite is expressed in the ordinal name which has been universely applied to them; thus the shield or integument which covers them is divided usually into three lobes by mesial furrows which traverse them longitudinally and then the anterior part the head, the middle or abdomen, and the posterior extremity are clearly indicated by marks which are easily recognized. Thus in the Conocephalus striatus, fig. 66, A marks the head; B, the abdomen, and C the tail; the latter is usually called *pygidium*. The proportional extent of the foregoing parts is variable in different genera. The form of the head is more or less expanded into a border, which is frequently ornamented by granulations, spines or perforations. Behind, the head is limited by a kind of border which is called an occipital ring. The middle part of the head, which is really one of its lobes, is called a glabella; it is not always distinct, for in the Illaenus crassicauda, pl. 18, fig. 5, the longitudinal grooves are very indistinct. The parts of the head are united by suture. The principal suture which is seen in fig 67, forms a double curved

Fig. 66.

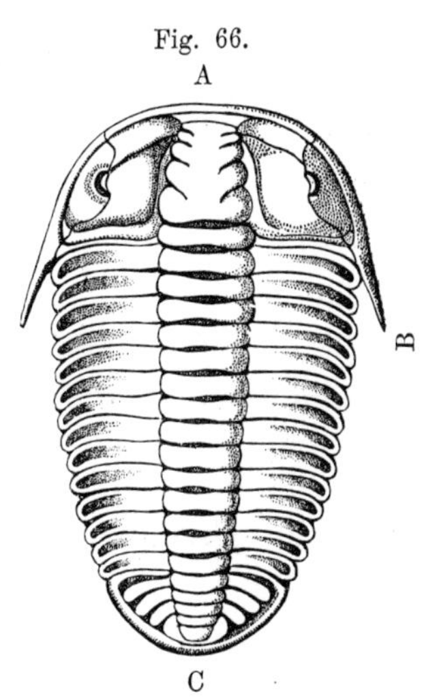

Fig. 67.

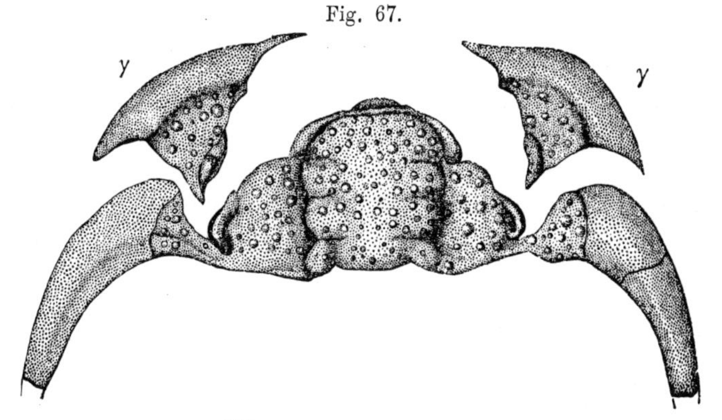

line, beginning before, near the front, it extends back to the posterior angle of the buckler, thus *y y*, cheek pieces separated from the head through the principal or great suture in the ceraurus pleurexanthemus. The eyes are apparently formed upon the type of the insects of the present day, consisting of fixed sperical lenses, which are sometimes flattened, they are large and nnmerous in the genus Phacops and in the Isotelus.

Fig. 68.

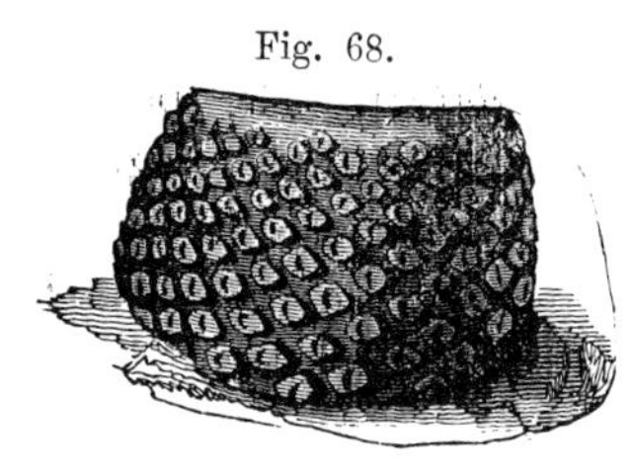

This figure is an enlarged view of the compound eye of the Phacops macrocephalus which belongs to the Devonian period, but the eye is not more perect, however, than those of the Phacops calicephalus and Isotelus gigas of the lower Silurian period. But many of the genera seem to be deprived of eyes, as the Atops and Eliptocephalus of the Taconic system, and the Triarthrus, Beckii of the lower Silurian. The number of lenses is also variable in different genera. They are placed upon the great or principal suture of the head, or near it.

The abdomen is composed of rings which are variable in number in the different genera, and, indeed, in consequence of this metamorphosis, the number is variable in certain species at different periods of their existence. In the adult state the number is constant in the same species.

The body of the trilobite being composed of separate parts arranged transversely, enables it to roll itself up on its longitudinal axis, as represented in *pl.* 18, *figs.* 6, 7, 8, 9. Certain trilobites, however, are incapable of performing this peculiar movement, and of assuming the globular or spheroidal form.

The structure of the pygidium is as variable as that of the abdomen; the number of rings is inconstant; in the oldest trilobites, or those of the palæozoic age, the organ bears generally a smaller proportion to the rest of the body than the more recent; fig. 68 and 69. This is seen in the Atops and Paradoxides, where the pygidium is reduced to the smallest number of elements, and in the Asaphus expansus, *pl.* 18, *fig.* 4, of the Upper Silurian.

Pictet, in his most valuable work, "Traité de Paléontologie," proposes the following modified classification of Barrande:

Fig. 69.

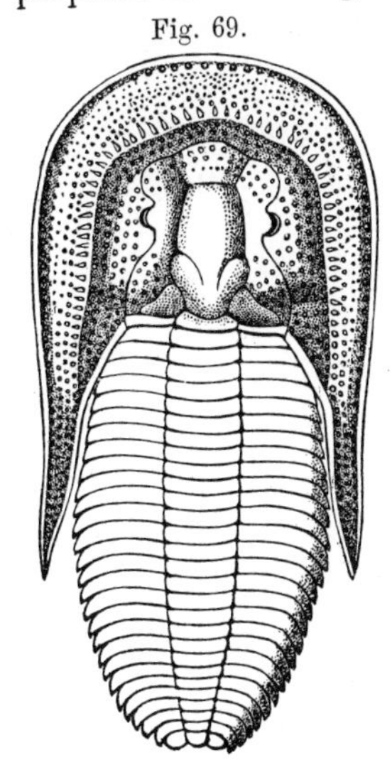

I. *Head very distinct in its conformation from the pygidium, which is very small, and the thorax large. It contains two genera, viz:* Harpes, *fig.* 66, *and* Paradoxides, *p.* 115.

The Harpes belong to the Lower Silurian. It has been found in Ireland and Bohemia, but has not as yet been discovered in this country. It is remarkable for its wide, perforated border of the cephalic shield. It is also doubtful whether the Paradoxides has as yet been discovered in our rocks.

II. *Pygidium and thorax subequal.*

It contains two families: the Calymenides, and the Lychasides. A perfect example of the latter family is known in the Lichas boltoni of the upper Silurian, and the Phacopiens, in fig. 67, Phacops limulurus.

III. *Pygidium large, thorax small.*

It contains four families; the Trinuclides, *pl.* 17, *fig.* 1, the Asaphides, the Æglinides and the Illænides, *pl.* 18, *fig.* 5.

Fig. 70.

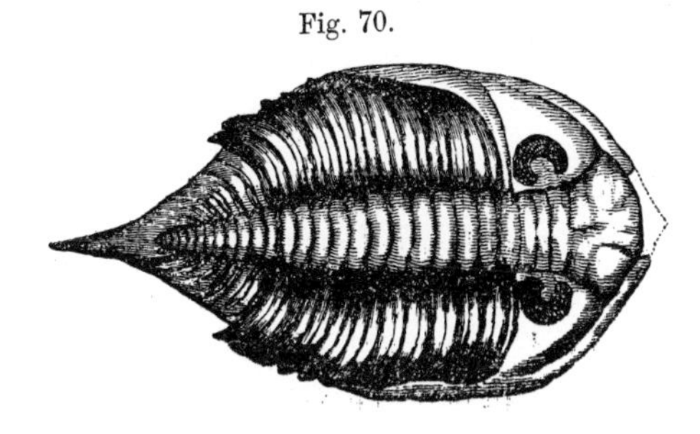

IV. The division contains three families.

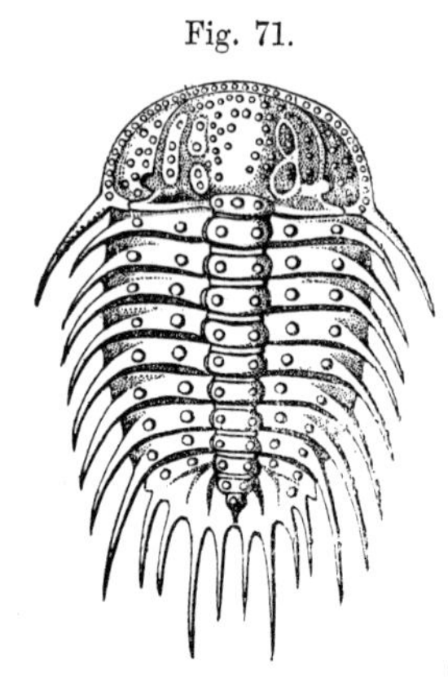

Fig. 71.

1. Pygidium from two to five segments, terminating in points, as Acidapsis, fig. 71.

The second family, under this section, has the pygidium armed with points; but the segments are more numerous than in the preceding, as the Amphionides.

The third family, the Brontides, the pygidium has a very short axis, and a wide lateral extension, and marked by radiating furrows extending from the centre to the circumference.

V. *Head and pygidium very nearly of the same form, as* Agnostus, fig. 74, *d*, Beyrichia, fig. 74, *a*, *b;* p. 218

I. *Head and pygidium distinct.*

TRINUCLIDES.

II. *Pygidium large; thorax small.*

TRINUCLEUS.

Form a short ellipse; smooth surface; head largely developed, and bordered with a perforated limb, which terminates in long spines; cephalic shield composed of three prominent convex lobes; the furrows of the glabella converging towards the axis; thorax with six rings, those of the pygidium variable.

TRINUCLEUS CONCENTRICUS.

Pl. 17, *fig.* 1.

Border with three to five rows of punctures; glabella high and projecting; maxillary spine equals the length of the thorax and pygidium; spine of the glabella short, pointed; rings of the thorax grooved, those of the pygidium soldered and terminating in a border; rings of the axis twice as many as the sides

III. *Pygidium and thorax subequal.*

CALYMENIDES.

Calymene. Triarthrus.

This family has about thirteen joints or segments in the thorax; the large suture of the well developed head terminates in the angle of the cheek.

CALYMENE.

The form of the shield is oval; the head larger than the pygidium; and the latter is about one-half as large as the thorax; the crust is granulated; the front lobe of the head projecting, the lateral lobes of the glabella are globular and separated from the middle part by deep winding furrows, which nearly isolate them, the main furrows being divergent from the the front; the eyes are reticulated, but not well developed; the thirteen segments of the thorax are angulated, and the ribs or lateral lobes are bent and rounded at the extremeties; the pygidium is convex and rounded, and its centre lobe distinct and narrower than the lateral lobes. Tbe Calymene senaria belongs to the Lower Silurian, and the C. blumenbachii to the Upper Silurian.

C. SENARIA (*Conrad*).

Pl. 15, *fig.* 16, *and p.* 216, *fig.* 9.

Shield ovate, and uniformly granulated; lobes very convex and prominent; posterior tubercle of the glabella very large and globular; the anterior very small; ribs deeply furrowed; pygidium with seven or eight segments in the middle lobe, and five in the lateral; crust continuous around the middle lobe, and each segment is so divided by a furrow as to form a subsegment to each ring. If we count the apex of the middle lobe of the pygidium, which is really divided from the others by a shallow groove, there are eight, instead of seven segments. Trenton limestone in New York, the Blue limestone of Ohio and the equivalent rock in the southwest, Virginia and Tennessee.

TRIARTHRUS (*Green*).

Shield or crust an elongated ellipse, with the posterior extremity narrower than the anterior; crust comparatively smooth, with a single row of tubercles in the middle in the young, but often obsolete in the old; furrows of the cephalic shield parallel, straight, and in a line with those of the thorax; eyes none; axis wider than the lateral lobes; thoracic rings, fourteen; rings of the pygidium, six in the middle lobe, and five in the lateral.

T. BECKII (*Green*).

Calymene beckii (*Hall*).

Pl. 15, *fig.* 12.

Glabella nearly twice as wide as the lateral lobes, and marked by two equal, oblique, impressed lines upon the posterior half, which begin in the furrow and run obliquely downwards and inwards, nearly to its middle; cephalic ring distinct, with the extremities of the furrow parallel with the oblique impressed lines; surface rather finely sculptured, as it appears under the microscope, ribs grooved, but often appear forked and rounded; showing that the groove separates them into two parts, and that they are movable on each other. In specimens less than an inch in length there are fourteen or fifteen thoracic rings.

PHACOPS.

Form an elongated ellipse broad in front, broad or acute behind. Surface pustulose, or coarsely granulated; cephalic shield lunate from by the extension behind of the cheeks. Glabella separated from the eyes and cheeks by deep furrows convergent behind; eyes large and lunate; thoracic rings, eleven; rings of the pygidium variable, but terminating in a border.

P. CALLICEPHALUS, *pl.* 15, *fig.* 7 *a and* 7.

Form an elongated ellipse, rounded and broad in front, narrow and subacute behind; glabella, with rather deep winding furrows separating from its body a lobe directly behind the eye, which is marked by a furrow on its inner margin; front lobe large and rounded; posterior small, and somewhat in the form of a ring; rings of the lateral lobes of the thorax, grooved; rings in the axis of the pygidium, fourteen to fifteen, and nine in the lateral, terminating in a smooth border.

IV. Illenides.

Large smooth convex; head semielliptical or globose, usually wider than long, or transverse; eyes distant and lateral; rings of the thorax simple, eight or ten; pygidium very convex and trilobation rudimentary.

Illaenus crassicauda, *pl.* 18, *fig.* 5.

Semielliptic; sides parallel, and extremities very convex and rounded; anterior extremity rather larger than the posterior; eyes prominent and placed on a line of the middle of the pleuræ; thoracic rings, ten. Extends from the upper part of the calciferous sandstone to the trenton limestone.

I. trentonensis, *pl.* 15, *fig.* 13.

Form semielliptical, convex and globose; eyes distant, lateral; crust marked with imbricating striæ, similar to those of the Isotelus; ribs of the thorax linear simple or without furrows; trilobation none, or only rudimentary.

Isotelus (*DeKay*).

Form ellipsoidal; extremities subequal, smooth and finely pustular, convex; facial suture running at the inner base of the prominent eyes, and terminating within the angles of the cheeks; eyes placed on a line with the furrows of the thoracic rings; thoracic rings, eight; axes wider than the pleuræ; pygidium similar to the cephalic shield, but more obtuse; articulations soldered; margin broad.

Isotelus gigas, *pl.* 16, *fig.* 12.

Large extremities subtrilobate, more rounded behind; becomes obtuse by age; eye piece subglobose or spherical, supported on the inner margin of the cheeks; rings of the thorax strongly furrowed upon the axial border; surface of the pygidium marked with many very shallow or obsolete transverse furrows.

This trilobite is the characteristic fossil of the Trenton limestone, being widely distributed in this or its equivalent rock throughout the United States.

V. Asaphides.

Ogygia.

The large caudal shield, fig. 72, has been referred to Ogygia

Fig. 72.

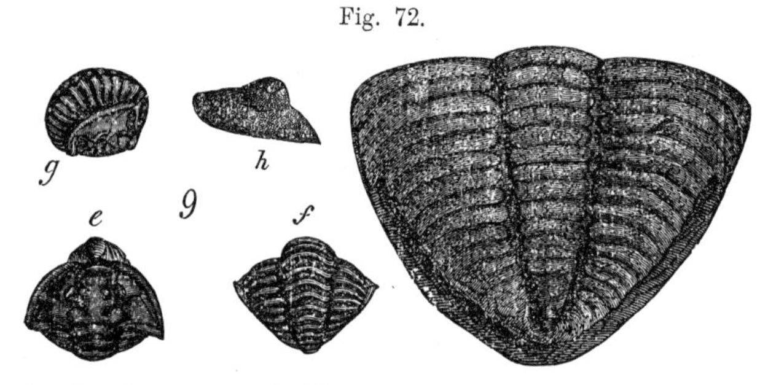

vetusta, but is more probably an Asaphus. It was found in the Birdseye limestone of the Mohawk valley; *e f g h* 9, small variety of Calymene senaria; *h*, the eye, showing the lenses; 9, the folding of the thoracic shield; *e*, the cephalic shield; *f*, pygidium.

VI. Odontopleurides.

Acidaspis trentonensis.

Fig. 73.

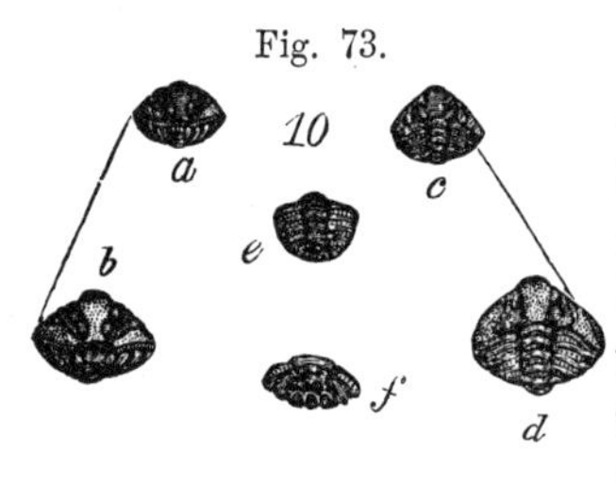

This species has a shield subcrescentiform, rounded and dentated in front with the cheek angles produced into spines; glabella quadrate, with a spine at the posterior margin; thorax with ten rings, and ornamented with a row of small tubercles upon the lateral lobes; pygidium with two rings and margin spinous.

ODONTOPLEURIDÆ.

CERAURUS.

Large or small; form somewhat triangular; pustulated, granulated and spinous; cephalic shield crescentiform; thoracic rings eleven; glabella quadrangular and four lobed; furrows forming the lobes extend transversely one-fourth across it; main furrows, separating it from the cheeks, nearly in a line with those on each side of the axis; eyes distant, granulated; cheeks triangular.

C. PLEUREXANTHEMUS.

Pl. 15, *fig.* 1, *a to k.*

Large; cephalic shield terminating behind in long, robust, curved spines; a shallow, or wide sinus in front; thoracic rings with two rows of tubercles on the axis, and three upon the pleuræ, which are sinuously furrowed; pygidium with about four rings, and its lateral lobes armed with long, stout, curved spines, and wider than the axis; rings very prominent. Trenton limestone in Northern New York, in Ohio, and in Wythe Co., Virginia.

C. VIGILANS.

Pl. 15, *fig.* 2 *a to c.*

Small, subtriangular; cephalic shield granulated, crescentiform, and terminating in long, flat spines; glabella quadrangular; entire; thoracic rings eleven; axis wider than the pleuræ; pygidium triangular; axis with many rings, with every third tuberculated, terminating in a kind of spine; lateral lobes with nine ribs. Trenton limestone, Middleville; this species is quite common.

ACIDASPIS.

Odontopleura.

Ellipsoidal, remarkably spinous, with rows of tubercles; cephalic shield subcrescentiform, rounded in front; thorax with from eight to ten rings armed with spines; pygidium small, circular, and with few rings and spinous.

A. TRENTONENSIS.

Cephalic shield dentated in front and produced into lateral spines; glabella subquadrate; lateral furrows in a line with those of the axis; pleuræ with a row of tubercles; eyes smooth; thorax with ten rings; pygidium with two rings; margins pinous. Trenton limestone.

LYCHASIDES.

LYCHAS.

Fragments of the genus Lychas have been found in the Trenton limestone, *pl.* 15. *fig.* 2*a*; and probably an asaphus, *pl.* 15, *fig.* 11. Also the genus Thaleops, which is closely related to Illænus.

AGNOSTIDES.

Cephalic shield and pygidium nearly alike.

The agnostides are small or even minute crustaceans, having two shields very much alike; the number of thoracic rings being very small, never more than two.

BEYRICHIA SIMPLEX* (*Jones*).

Fig. 70, *a*.

It is described as broadly ovate, globose, smooth; ventral border rounded; dorsal border somewhat angular; hinge line oblique, about two thirds the length of the valve; dorsal sulcus faintly marked on the anterior half of the valve. The description does not agree with the fossil, fig. 74, *a*; both borders are rounded, and the sulcus is variable in its depth and position. Reference is made to B. logani, Canada, and it is probable, from that reference, that it is the Canada species. Abundant in the Blue limestone of Ohio.

---

* Journal of the Geol. Soc. of London; vol. ix, p. 161.

B. REGULARIS (n. s.).

Fig. 74, *b.*

Elliptic, smooth; dorsal border straight, or nearly straight; ventral regularly rounded, with three distinct ribs; one of the angles is more rounded than the other, and a slight obliquity is observable in respect to the direction of the ribs. Blue limestone of Ohio.

Fig. 74.

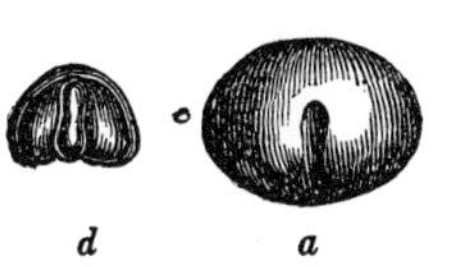

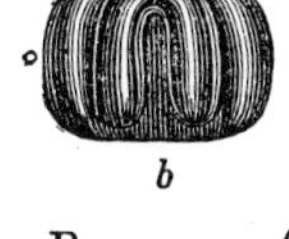

B. CILIATA (n. s.).

Fig. 74, *c.*

Form, an elongated ellipse, smooth, with a straight dorsal border; ventral border strongly rugose, ciliate; ribs three, and oblique with respect to the straight border. The ciliæ scarcely come under that denomination, as they appear too rugose when seen under the microscope. Blue limestone of Ohio.

AGNOSTUS LOBATUS (*Hall*).

Fig. 74, *d.*

Small, trilobate; base and sides furnished with a narrow border; a small tubercle often exists near its larger extremity. —*Hall, Pal. of New York, Vol. i.*

This species, which is referred by Mr. Hall to the Hudson river group, is really confined to the Calciferous sandstone; the mass of limestone to which he refers is that rock, and contains numerous fragments of Illænus, similar to one in the same rock at Chazy; and also found in fragments in the Trenton limestone and calciferous sandstone at Greenbush.

CYPROIDES.

The animal is enclosed in a bivalve ovoid carapace, or of a reniform shape, supplied with a dorsal hinge, and which is completely shut, but when open allows the extrusion of its feet and antennæ. They are small crustaceans, some of which are microscopic.

CYTHERINA SUBELLIPTICA (n. s.).

Form subelliptical, smooth; dorsum nearly straight, sometimes sinuate; anterior edge with an oblique truncation extending from the hinge to the anterior edge, and forming a kind of short beak; posterior and anterior edges rounded; it is about three-fourths of an inch long. It is abundant in the upper part of the calciferous sandstone, about one mile from Watertown upon the Black river. Fig. 75, *a*.

C. CRENULATA (n. s.).

Fig. 75.

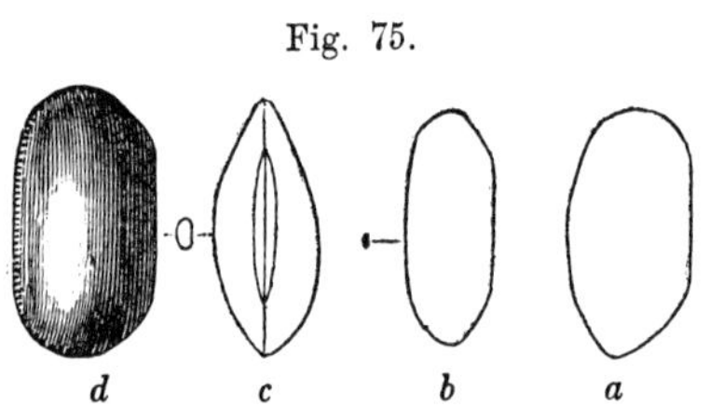

Obtusely oval, smooth, inflated; hinge straight with the valves extended back, and forming, apparently, a groove; extremities rounded; anterior edge crenate. Fig. 75, *d*, *c*, hinge or dorsal side. Trenton limestone, Middleville.

C. SUBCYLINDRICA (n. s.).

It is smaller than the preceding, and of a cylindric or sub-quadrate form; smooth; extremities dissimilar, and without crenulations upon the edge. Fig. 75, *b*. Middleville.

DIKELOCEPHALUS (*Owen*).

This new genus is described by its discoverer as having a semicircular, rather flat, cephalic shield; a moderately convex glabella, separated from the cheeks by parallel furrows, giving it a quadrate form, and marked by two curved furrows extending from side to side, and another impressed faint one on the outer margin; facial sutures distinct, and somewhat sigmoid. The cheek pieces are produced into spines of moderate length; rings of the thorax supposed to be eight, with an axis narrower than the lateral lobes; lateral lobes wide, plain, and without grooves; pygidium with seven or eight rings in the axis, and the lateral lobes expanded widely into a flat, grooved border, with the posterior corners produced.

The large trilobite belongs to the lowest sandstones of Iowa, which are equivalent to the Potsdam sandstone.

Prof. Owen describes five species of Dikelocephalus, viz: D. minnesotensis, D. pepinensis, D. miniscaensis, D. Iowaensis, D. granulosus. They all belong to the lower sandstone, No. 1; but some pass up into the lower magnesian limestone. Dr. Owen has also discovered other forms of trilobites, some of which constitute new genera, and a new species closely allied to the Isotelus gigas, having a facial suture, eyes, extremities and ribs similar to this common form of Trenton species; the eyes, too, are placed on longitudinal furrows as in Isotelus. It differs from the Isotelus in having its cheek produced into a long spine.

RADIATA.

The kingdom contains those animals whose constituent parts are arranged about a centre in a radiate form. This radiate form is characteristic of one of the four great types upon which animals are created. Radiated animals are divided into three classes. The Echinodermata or Starfishes, the Acalepha or Jelly-fishes, and the Polypi or Corals. The Jellyfishes being composed of soft, perishable matter, are never found in a fossilized condition. In the Lower Silurian rocks the animals belonging to the other two divisions occur.

The first which I shall describe belong to the *Echinodermata*, or *Echinodermis*. The name which has been applied to this division means skin with spines. These appear under three forms. 1. the starlike form, as the starfish of our coast. 2. The Echinidæ, the egglike form, which are called sometimes sea-eggs; and 3. The Holothuridæ, which are mostly sack-like, and contain only a few, internal, semi-ossified supports, but do not occur in a fossilized state.

Only two forms of the Echinoderms are known to belong to the Lower Silurian period. One which is like the starfish, and the other which is supported upon a jointed stem, which was attached to the sea bottom. To the latter form has been applied the name *Encrinite.* The general form of the Encrinite is exhibited in the annexed wood cut of the Eucalyptocrinus decorus of the Niagara group, or Upper Silurian.

Fig. 76.

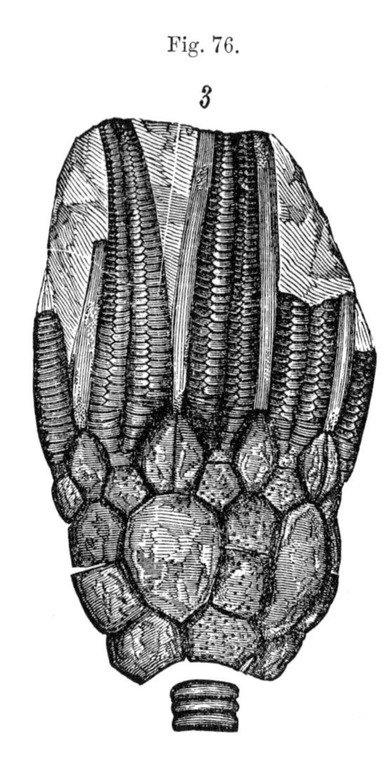

It is composed of a head of a globose form, supported on a stem composed of joints of which a few remain below; above, the head is encircled by arms, composed of numerous plates, which diminish in size; within this circle of arms there is an oral aperture or mouth; the head is made up of numerous plates of a hexagonal or pentagonal form, sometimes heptagonal, which are fitted together by suture, and which, on the death of the animal, may separate from each other. The plates are variable in number; at the base some begin in threes, others in fives, and are sometimes arranged in as many as nine series, varying from three to nine in the series. These are arranged in rows, encasing within the hard, dermic walls the soft parts necessary for the performance of the vital functions of this peculiar organic type. A horizontal view of the elements composing the head is shown in *pl.* 18, *fig.* 1, 2. A detailed explanation will be given hereafter.

Encrinites appear to have been among the oldest fossils of the Silurian rocks. None occur in the Potsdam sandstone, but in the lowest layers of the Calciferous sandstone, not only joints and fragments of stems occur, disseminated through the layers of the rock, but entire beds of columns occur from two to fifteen feet thick. Some of the plates are figured, *pl.* 3, *fig.* 9.

As these are fragments more or less broken and no well defined combination of plates occur, or have yet been found, it is difficult to determine the genera and species to which they belong. Some of the elements of the columns belong to the Cystidea. Some of the plates preserve the ovarian aperture, and exhibit the saucerform joints of the stem.

Of the Echinoderms, two species have been discovered by different individuals, and which have been described and figured in the Palæontology of New York. The most distinct and perfect resemble a small star-fish. It belongs to the genus *Cœlaster* of Agassiz. Two species are described the C. matutina, and the C. tenuiradiata. The five arms of the C. matutina have each three rows of plates to the pelvis. It belongs to the Trenton limestone. The latter, the C. tenuiradiatus is known by a single plate, or the madreporiform tubercle, of the back of an asterias. Pl. 3, fig. 11, is a figure of the plate.

GLYPTOCRINUS (*Hall*).

Head conical; pelvis composed of five plates, the remainder are hexagonal and six-rayed, and the entire series consists of nine rows. The peculiar sculpturing of the plates form an arrangement which will serve to identify this beautiful fossil of the Blue limestone of the west. The elements of the upper part of the column are round with fine and stellate sculpturing upon the disc, the lower and larger plates are crenulated upon their margins.

HETEROCRINUS (*Hall*).

The number of pelvic plates is five, the series is composed of this number throughout, but the rows vary from four to nine. The head is small, short, and tapers from the middle.

H. DECADACTYLUS.

The column is composed of pentagonal plates with stellate disks and a central tubercle or eminence; sides sinuous and margins of the disc sculptured with elevated stellate lines or ridges.

It occurs in the Loraine shales. The small cylindrical heads are not uncommon in the fine soft slates of this formation.

SCYPHOCRINUS (*Zenker*).

Schizocrinus (*Hall.*)

The base of the series of plates, five; the head cup form, and the series in six rows; hand and fingers bifurcate.

S. NODOSUS.

Column long, round, composed of elements of variable forms or patterns, according to their distance from the head. The most obvious character consists in the regular occurrence of projecting nodulose plates, or rather, their margins are somewhat moniliform. Their discs are finely stellated. Trenton limestone.

POTERIOCRINUS (*Miller*).

The cupform calyce is composed of a pelvis with five plates in the base of the series. There are five additional series, one composed of rather large hexagonal elements, and the last heptagonal ones which are connected with the arm plates.

P. ALTERNATUS.

Fig. 77.

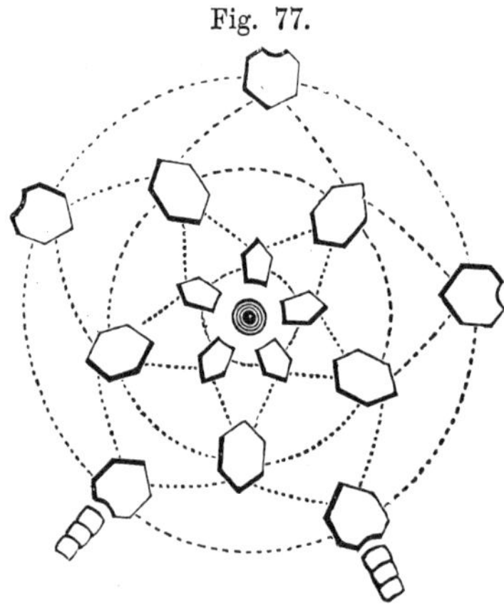

The tentaculæ are fimbriated the pelvic plates are narrow; elements of the upper part of the column, round and small making a slender support. Fig. 77, arrangement of the plates in horizontal projection.

CUPULOCRINUS (*d'Orb.*).

Scyphocrinus (*Hall*).

The saucer shaped pelvis is composed of five pentagonal pieces, and the head is made up of five series of elements. The five arms bifurcate.

C. HETEROCOSTALIS (*Hall*).

The column is round. The alternation of the plates is somewhat peculiar as described by Mr. Hall: the first from the pelvis is a plate with crenulated edges; second, a thicker joint with a round, smooth edge; and third, a thin one with a fimbriated edge. Trenton limestone.

TENTACULITES FLEXUOSUS.

Shell small, free and regularly tapering, section circular, surface annulated, annulations sharp and irregular in their distance from each other; longitudinally striated and somewhat bent or flexuous. This species is about one inch long and about a line in its greatest diameter, from which it tapers to a point; it has sharp, raised annulations or sharp ridges with striæ between.

In the Blue limestone of Ohio, there is a perfectly straight one, and smaller than the flexuous one. It may be a variety of the latter. Trenton limestone, Loraine shales.

ZOOPHYTES,

Or Polypi, possess a structure which serves at once to distinguish them from the Bryozoa and Sponges. The character of the former has been already stated, while of the latter, it is necessary to say that their tissues are reticulated, and at the same time traversed by wide and winding aquiferous canals, and that their hard parts consist of simple spicula which are peculiar to this class of animals.

The hard or stony tissue of polypi or corals, has been technically called *sclerenchyma.* It belongs to the tegumentary system, or rather, it is a product of this system. The organization of the Zoophite being quite different from that of the Molusca, its hard, imperishable parts are necessarily quite different. The former has but a single orifice for the double purpose of the reception of food and the rejection of the excrementitious matter. Its digestive cavity is a sack, and what is exterior to the digestive surface has the property of secreting the sclerenchyma which begins to be formed at a central point at the inferior part of the stomach. It is here the process of calcification begins, and from this it extends upward inclosing the gastric cavity in a cup, or cell whose form will vary according to the character of the individual which occupies it or is its maker. The cup or cavity may be cylindrical, or broad stellate open throughout or closed at the bottom. This cell or cavity in the stone corals is never perfectly simple, as the wall of calcified tissue grows upward, vertical lamina forming partitions grow from the inner surface towards the centre. Their number is always the same in the same species. These lamina or vertical plates are called *septa*. If they reach the centre and unite, they form a kind of vertical axis to the cell, which is the *columella*. The spaces between the septa are called *loculi*. When the outer surface of calcified wall sends out ridges or spines, they form *costæ*. The whole cell with its modifications forms a *calice*. In certain families the calice is open at both ends; in others it is closed at the bottom; and in others still, there is a repetition of transverse bottoms, tabula, or floors, which divide the calice into many small cells placed vertically one above the other. Fig. 78 shows the calice with its septa of the Cyathophylhum turbinatum extending from the wall towards the centre; the depression or the cup being produced by the discontinuance of the septa inwards.

Fig. 78.

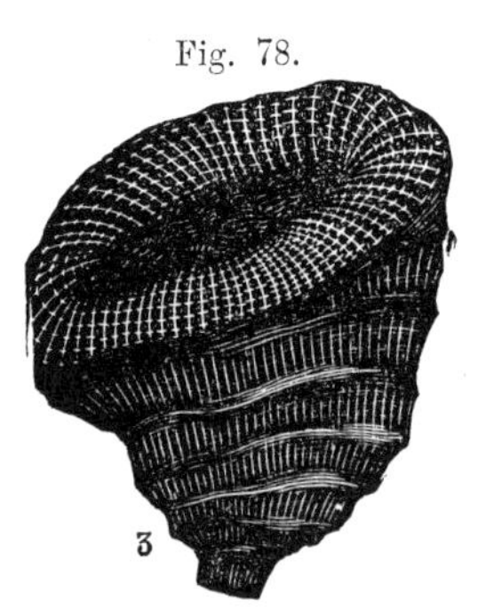

The number of chambers formed by the septa is variable. They increase regularly in successive circles. The regular primary number being six; in each chamber there will be formed other septa in succession, constituting another circle. To this may be added two more, forming a third order of circles. The Polypiers increase or multiply in three different ways, by ova, by buds and by division. The simple polype is produced from ova, it is an isolated individual either free or fixed, the transverse section is usually round, as the Turbinolia. The second mode is by buds which results in the aggregation of individuals, but each individual is perfectly distinct from the others; the calice is generally circular, as in fig. 79, the Favastræ rugosa. The last mode, by division or fissiparity, the individual is divided and forms two individuals, this form also results in the aggregation of individuals; the calice is irregular, or oval, the elongation being the first step towards a division of the individual. The form therefore, of the calice, together with its mode of aggregation are important generic characters. The Polypiers are divided by M. Milne Edwards into three orders, the *Zoantharia*, the *Alcyonaria* and the *Hydraria;* the two former only, are represented in a fossilized state.

Fig. 79.

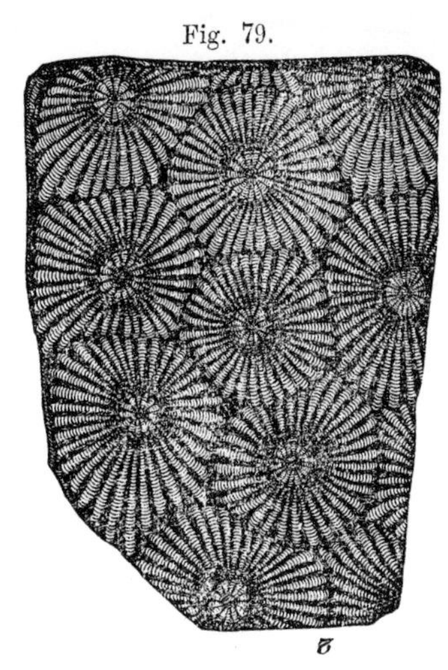

The general characteristics of the subkingdom Zoophita are, animals formed for a sedentary mode of life, being provided with a circle of retractile tentaculae around the mouth, and a central gastric cavity with only a single orifice, and in which are lodged the reproductive organs.

The Corallaria which forms a subclass under Zoophyta, embrace the calcareous corals which in form may be tubular, cyathoid or discoidal, but exclude all the cylindrical, tubular or horny sprigs, which bear bell shaped cells for the reception of contracted tentaculæ.

The Zoantheria embrace most of the known fossil stony corals. They have either a conical, tubular, simple or arborescent forms, and an internal gastric cavity divided vertically by a circle of septa, radiating from the internal wall of the corallum.

Family CYATHOPHYPLIDÆ.

The family embraces those corals whose septa are incomplete, or septa which do not extend from the bottom to the top of the visceral chamber, in the form of uninterrupted lamina.

STREPTOPLASMA (*Hall*).

"Corallum simple, and differing from Cyathophyllum by the structure of the wall, which is destitute of Epitheca, and covered with sublamellar costæ."—*Ml. Edwards.*

S. EXPANSA, *pl.* 3, *fig.* 6, 7, *a*, *b*.

Small, turbinate; cup deep; septa numerous. It is often worn down so as to present a forked, triangular form. Chazy limestone.

Mr. Hall has described five other species belonging mostly to the Trenton limestone, viz: *profunda*, *corniculum*, *crassa*, *multi-lamellosa*, *parvula*.

Family FAVOSITIDÆ.

"The coral is formed of lamellar walls, with little or no cænenchyma; visceral chambers divided by numerous and well developed complete tabulæ."

The family is divided into two tribes: 1. favositidæ, the corallum massive, with its walls perforated and its septa rudimentary, no cœnenchymæ. 2. Chætetinæ, corallum massive, walls not perforated, neither septa nor cœnenchymæ.

CHÆTETES.

The genus Chætetes is now regarded as a fossil of Carboniferous limestone; and the fossils which had been referred to Fischer's genus, are now, in part, placed in Monticulipora, (*d'Orb.*). In the Chætetes the *fissiparous* mode of reproduction prevails, while in the Monticuliparous, the mode of reproduction is *gemmiparous.*

---

* Cœnenchyma; it is the cellular mass formed upon the outside of the wall of the corallum.

MONTICULIPORA (*d'Orb.*).

Monticulipora petropolitana.

Favosites lycopodites (*Vanuxem*).

Chætetes lycoperdon of various authors, in part.

Fig. 80.

The species is thus defined, by MM. Edwards and J. Haimes. Corallum in general, free; basal plate flat or concave, and completely covered with concentrically wrinkled Epitheca; upper surface regularly convex; tabulæ complete and horizontal. The calices are unequal in size, polygonal, sometimes round; the largest one-fifth of a line in diameter; the floors and tabulæ are about one-twelfth of a line apart. The fossil is one of the most common in the Trenton limestone.

FAVISTELLA (*Hall*).

Fig. 81.

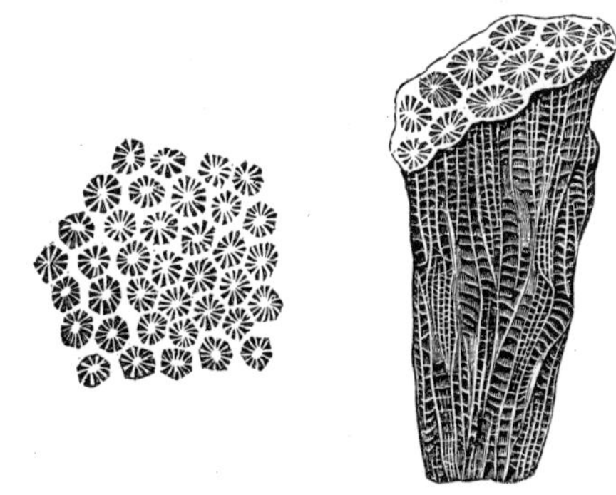

Coral massive, cellular, hexagonal; septa about twelve; cell walls with interposed cœnenchyma.

F. STELLATA.

Corallum hemispheric, spheroidal; the cell walls soldered, and the and are not separable, like the Favosite; Tabulæ nearly direct. Loraine shales. Fig. 81.

COLUMNARIA (*Goldfuss*).

C. ALVEOLATA.

Pl. 4, *fig*. 10.

Corallum massive, convex; calices arranged in parallel radiating columns; hexagonal; walls furrowed and separable, and without cœnenchyma; septa rarely if ever meet and form a columella, often obsolete. Chazy and Birdseye limestone.

RECEPTACULITES NEPTUNI.

Pl. 14, *fig.* 1.

This fossil is described by Mr. Hall as suborbicular, hemispherical; depressed in the centre; it presents a series of quadrangular cells, within which there is a vertical, cylindrical tube opening upwards; opening not entirely circular.

R. CIRCULARIS (n. s.), *fig.* 82.

Fig. 82.

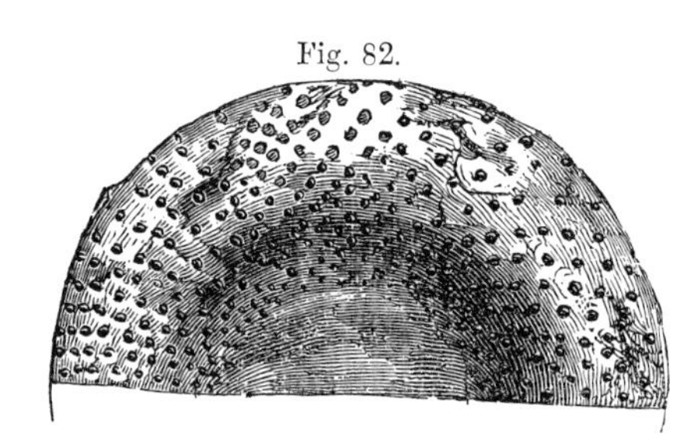

This coral is in the form of a thick, flattened ring, studded with circular cells, arranged in regular lines traversing it rather obliquely. It belongs to the Loraine shales.

AULOPORA (*Goldfuss*).

Aulopora arachnoidea, *fig.* 83.

Parasitic, fine and web-like, ramifying, branching; branches anastomosing with each other; tubes linear, narrow, enlarged at the point where the oval, slightly elevated, mouth is placed. Figure very much magnified. Trenton limestone.

Fig. 83.

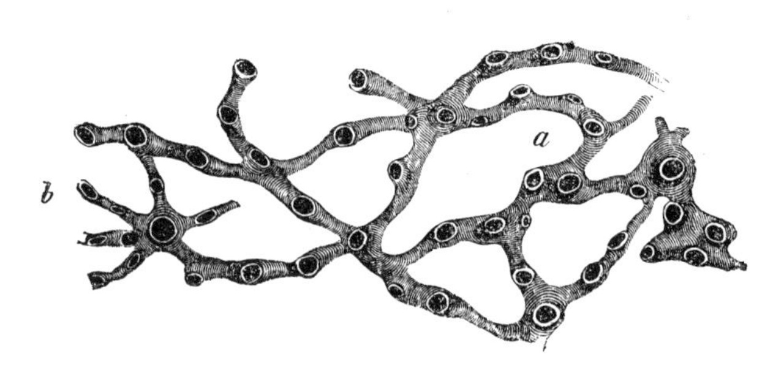

GRAPTOLITES.

Diplograpsus pristis.

Graptolites dentatus (*Vanuxem's Report*).

Pl. 12, *fig.* 3.

Stem narrow; serrations acute; numerous varieties occur in which the serrations are more obtuse; the stem widens near its distal extremity, and is subject to variations from causes which affect the integrity of the rock in which it occurs. Utica slate.

PLANTS OF THE LOWER SILURIAN ROCKS.

The plants which have hitherto been discovered in the lower Silurian rocks, are without doubt marine. They are mostly in the condition of casts; the interior parts being replaced by amorphous matter We have not as yet found any in which the tissues are preserved, so as to afford us any clue to their structure. This is not all, it is the stem only which is preserved, its form being more or less distorted, while the softer and more perishable parts, the leafy appendages, are rarely met with.

The stems are more or less round, but somewhat flattened. These casts usually exist in an entangled thicket, so as to obscure still more their most common characteristics.

In the lower part of the calciferous sandstone we meet with the first beds of these obscure plants, which occur in what have been called fucoidal layers. The presence of these beds proves that a marine vegetation must have been luxuriant in certain localities. The fact itself, the existence of beds in the early palæozoic period, is that which most interests us. This, taken in connection with the probable fact that no land plant existed in the lower Silurian period, is one of the most decisive proofs that there has been a progression in the order of creation and that that progression may be observed as well in the vegetable as in the animal kingdom. It is admitted by our most acute and learned zoologists that marine animals and plants have a lower grade of organization than the terrestrial, and hence that they take a lower rank in the scale of being.

FUCOIDES DEMISSUS (*Conrad*).

Phytopsis tubulosum (*Hall*).

In the Birdseye limestone the characteristic fossil has been referred to the vegetable kingdom. I allude to the Fucoides demissus of Conrad, the Phytopsis tubulosum of Hall.

Fig. 84.

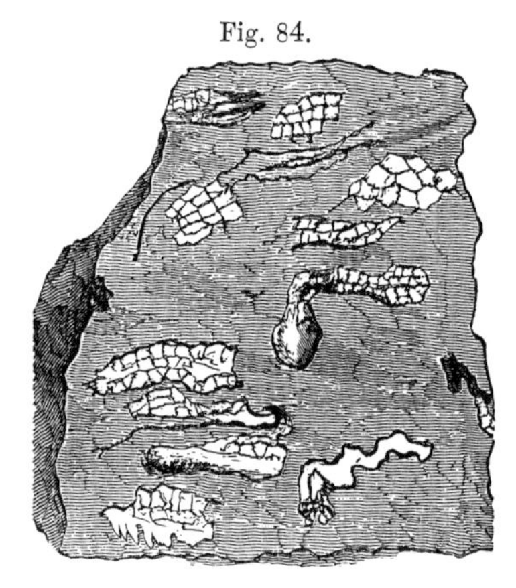

In my Geological Report of the second district of New York, I expressed the opinion that this fossil belonged rather to the animal than the vegetable kingdom. It certainly has few characters in common with the vegetables, whether we regard its external form, its growth, anastomosis or its internal structure. In connection with these peculiarities, I observed upon its surface at many places open circular cells which, under the microscope appeared to be connected with its internal cellular structure. These cells, it is true, may be incrusting ones; and foreign to the fossil, still, they appear connected at their bases and sides with the structure of the fossil. If so, it appears that the Phytopsis may be a Bryozoon, inasmuch as those cells which appear attached to its external surface belong to this division of the Molusca. It is clear, however, that it is not a coral.

Pl. 4, fig. 12, 13, exhibits its mode of growth traversing a bed vertically, anatomosing and diverging without showing that it has a main axis of development like a plant. Fig. 84 exhibits its structure obscurely, as it appears when upon the weathered surface of the Birdseye limestone.

APPENDIX,

Containing additional Descriptions of Fossils of the Taconic and Lower Silurian Systems.

GASTEROPODA.

STRAPAROLLUS PLANISTRIA (*Hall*).

Pl. 4, *fig.* 16, 17.

Shell small, depressed; volutions three or four; outer volution sharply angular; striæ flat and plain.

BELLEROPHON PROFUNDUS.

Pl. 17, *fig.* 7.

Volutions about three, contiguous, marginated, direct, gradually increasing in size; umbilicus deep; aperture somewhat lunate. Watertown, in the lower part of the Trenton limestone.

PLEUROTOMARIA PERCARINATA.

Pl. 5, *fig.* 7.

This fossil is found in the Trenton limestone at Middleville, associated with Atrypa hemiplicata, Cyrtolites compressus, and C. trentonensis, fig. 22. The gasteropoda of this locality though numerous, are rarely smooth and perfect, and hence there is some difficulty in identifying them.

P. SUBTILISTRIATA.

Pl. 6, *fig.* 11, 12.

Small, conical; spire depressed; volutions four or five; smooth, or the striæ extremely fine; aperture transverse and subtriangular. Trenton limestone.

DISCOLITES (n. g).

Shell minute, discoid; volutions one or two; texture corneo-calcareous; section circular, or apparently so; aperture round; no dilatation.

D. MINUTUS (n. s).

Shell microscopic, convolute, very thin, smooth, glossy, brown and fragile; volutions direct, scarcely contiguous; umbilicus distinct, smooth. It occurs rather abundantly in the Blue limestone of Cincinnati, associated with Berichia.

ACEPHALA.

CARDIOMORPHA SUBANGULATUS.

Pl. 12 *fig*, 7.

Edmondia subangulatus (*Hall*).

Shell large, wide; ventral margin wide, rounded; umbones rather prominent and thick or obtuse; no incurvation, prominence of the umbones extending in a low ridge towards the ventral margin. Surface marked with concentric lines of growth.

LYONSIA VETUSTA.

Cardiomorpha vetusta (*Hall*).

Pl. 13, *fig*. 8.

Form subrhomboidal; umbones prominent, obtuse; the concentric surface lines rather strong, distant, with finer intermediate ones.

L. MYTILOIDEA (*d'Orb.*).

Pl. 13, *fig*. 10, 11.

Shell elongated; anterior extremity rather narrow and gradually widening towards the margin, subcylindrical; umbones rather prominent and narrow; surface lines fine, but none left upon the cast.

L. SUBLATA (*d'Orb.*).

Pl. 13, *fig*. 1.

Mr. Hall describes it as subrhomboidal, gibbous and broadly rounded; umbones high, obtuse, not incurved, and with a shallow sinus in the margin; surface marked with imbricating lamina.

BRACHIOPODA.

ORTHIS COSTALIS (*Hall*).

Pl. 3, *fig.* 26 *a b and* 21 *c b.*

Valves strongly plicated; simple, valves unequally convex; ventral quite convex, the other rather flat; plaits about thirty-two, rounded; area rather large, triangular. This fossil occurs in thin beds in the Chazy limestone, particularly near the village of Chazy, Clinton county, N. Y.

STROPHOMENA INSCULPTA (*Hall*).

Pl. 3, *fig.* 22.

Small, valves semicircular; margin thickened; striæ sharp, elevated. Appear in the upper part of the calciferous sandstone, but more abundant in the Chazy limestone.

S. FASCIATA (*Hall*).

Pl. 3, *fig.* 24.

Semielliptical, somewhat punctated; striæ bifurcate near the margin. Chazy limestone.

S. LAEVIS.

Pl. 3, *fig.* 8.

Rather small, semielliptical, smooth. Birdseye limestone.

SPIRIFER TRENTONENSIS.

Pl. 15, *fig.* 20.

The spirifer occurs in the upper or gray portion of the Trenton limestone, at Watertown.

CRUSTACEA.

ILLAENUS ARCTURUS (*Hall*).

Pl. 3, *fig.* 12.

The distinctive characters appear in the width of the cephalic lobe at its junction with the thorax, the lateral extension of the cheek pieces, and in the more distinct development of the lobes of the cephalic shield. It belongs to the upper part of the calciferous sandstone. Chazy.

ASAPHUS MARGINALIS.

Pl. 3, *fig.* 16.

Axis with seven or eight distinct articulations; lateral lobes furrowed or with false articulations; margins entire.

ISOTELUS CANALIS.

Pl. 3, *fig.* 17, 18, 19.

The margin of the caudal shield is traversed with a rather deep furrow or channel; figs. 17 and 19 were found by myself at Chazy, in the calciferous sandstone, and were regarded by Mr. Conrad as new. In fig. 17 the margin only was preserved, the middle part having been worn away.

ASAPHUS OBTUSUS.

Pl. 3, *fig.* 14.

This fragment has received the name Asaphus obtusus, though too imperfect for determination.

CALYMENE CONRADI (n. s.).

Small, wide across the cheeks; cheek angles obtuse or rounded; posterior lobes of the glabella comparatively large and globular; thoracic lobes very convex, with a row of tubercles in the furrow or between the axis and lateral lobes. Loraine shales.

GRAPTOLITHINA.

DIPLOGRAPSUS LACINIATA (n. s.).

Pl. 1, *fig.* 24.

Stem narrow, cell mouths arranged; laciniate, or gashed serrations; acute, and rather elongated.

GLOSSOGRAPSUS SETACEUS (n. s.).

Pl. 1, *fig.* 20.

Stem rounded at the extremities, cell mouths or serrations bearing seta; serrations unequal, and in parts obsolete, or it may be obliterated by atmospheric influences. This species is evidently injured, and probably has lost its outer integument.

DIPLOGRAPSUS.

Pl. 1, *fig.* 3.

It is thin, olive green and foliaceous, but its characters too obscure to be determined with certainty.

DIPLOGRAPSUS AMPLEXICAULE.

Pl. 7, *fig.* 11 *a b*.

Cell mouths situated with or upon the sheathing scales or folioles, which are subacute.

DESCRIPTION OF PLATES.

PLATE I.

[NOTES RESPECTING PLATE I.—Fig. 4, which is referred to Orbicula is an outside cast in a very soft slate; it has been impossible to determine the texture of the shell. It resembles an Ancylus in shape and direction of the apex, but there are no certain marks by which I can determine whether the apex is directed forwards or backwards. It will probably prove to be a Helcion.

Fig. 9. I have allowed to stand as in the text a lingula, but the state of the valve at the umbo and the delicacy of the fossil, and the softness of the rock has raised a doubt in my mind respecting its perfection.]

Fig. 1. Cypricardia.
2. Lateral view of the mouths of cells of an indeterminate species.
3. Diplograpsus.
4. Orbicula excentrica.
5. Diplograpsus dissimilaris.
6. Nemagrapsus elegans.
7. do capilaris.
8. Microdiscus quadricostatus.
9. Lingula ?
10. Obolus ?
11. Diplograpsus secalinus.
12. Cladograpsus sp. indet.
13. Diplograpsus foliosus.
15. Cladograpsus dissimilaris.
16. Atops trilineatus.
17. Lingula striatus.
18. Eliptocephalus asaphoides.
19. Diplograpsus ciliatus.
20. Glossograpsus setaceus.
21. Staurograpsus dichotomous.

22. Diplograpsus obliquus.
23. Undetermined.
24. Diplograpsus laciniatus.
25. Glossograpsus ciliatus.
26. Diplograpsus rugosus.
27. Monograpsus elegans.
28. Monograpsus rectus.

PLATE II.

Fig. 1. Buthotrephis rigida.
2. Nereograpsus Jacksoni.
3. Nereograpsus Deweyi.
4. Nereograpsus lanceolata.
5. Nereograpsus Loomisi.
6. Nereograpsus gracilis.
7. Nereograpsus robustus.
8. Palæochorda marina.

PLATE III.

Fig. 1. Enallopora ?
3. *a*, E. aspera.
4. *a*, 5 *a*, Sulcopora fenestrata.
6. Straparollus sordidus (weathered).
7. *a*, 7 *l*, 6, Streptoplasma expansa.
8. Strophomena lævis.
9. Crinoid joints and fragments.
10. Actinocrinus.
11. Asterias plate.
12. Illaenus arcturus.
13. I. Crassicauda.
14. Asaphus obtusus.
15. Pygidium, asaphus ?
16. Asaphus marginalis.
17. Isotelus canalis.
18. 19. I. canalis.
20. Spine of the Ceraurus.
21. *b c*, Strophomena plicifera.
21, 22, 23. Strophomena insculpta.
23. Atrypa dubia
24. Strophomena fasciata.
25. Atrypa pleva.

26. Orthis costalis.
27. Atrypa acutirostra.
28. *a b*, Atrypa plena.
29. Atrypa plicifera.
30. Atrypa altilis.

PLATE IV.

Fig. 1. 1 *a*, and 14, Ptilodictya ramosa.
2. Straparallus labiatus.
3. and 14, Ptilodictya labyrinthica; 3, enlarged.
4. Bellerophon sulcatinus.
5. Cast of the Orthis, testudinaria.
6. Orbiculoidea.
7. Lingula prima.
8. Orthocera multicameratus.
9. Lingula acuminata.
10. Columnaria alveolata.
11. Murchinsonia abbreviata.
12. Fucoides demissus enlarged cells.
13. do do
15. Straparollus magnus.
16, 17. Scalites.
18, 22. Ptilodictya labyrinthica, pores enlarged.
21. Scalites angulatus.
20. Scalites striatus.

PLATE V.

Fig. 1. Straparollus levatus.
2. S. complanatus.
3. Crinoidean plate.
4. Orthoceras primigenius; 4, *a b*, Pleurotomaria umbilicata.
5. Pleurotomaria indenta, left corner; and P. ambigua, right corner.
6, and 11. Murchinsonia bicincta.
7. Pleurotomaria percarinata.
8. 18 *a* 18 *b*, Turbo obliquus.
12. 1 *a b*, Murchinsonia bellicincta.
13. M. abbreviata.
14. 14 *a*, Lituites undatus.
17. Turbo symmetricus.
22. Cyrtolites trentonensis?

PLATE VI.

Fig. 1. Helcion subrugosa.
2, 3, 22, 23, 24. Cyrtolites bilobatus.
4, 5. Cyrtoltes acutus.
7. Bellerophon expansus.
8, 9, 27. Bellerophon bidorsatus.
10. Pleurotomaria rotuloides.
13. P. lenticularis.
11, 12. Var. subtilistriata ?
19. *a b*, Turbo ventricosus.
20. T. Americanus.
21. Loxonema subelongata.
25, 26. Cyrtolites subcarinatus.

PLATE VII.

Fig. 1. 1 *a*, *b*, *d*, Ptilodictya recta.
3. *a*, *b*, *c*, Ptilodictya acuta.
4. *a* to *g*, Ptilodictya elegantula.
5. Enallopora perantiqua.
6. Aulopora arachnoides.
7. 7 *a*, Ptilodictya ramosa.
8. *a*, *b*, *c*, Subretopora reticulata.
9. Retopora foliacea.
10. Stellipora antheloidea.
11. Diplograpsus amplexicaule.

PLATE VIII.

Fig. 1. Lingula attenuata.
2. L. riciniformis.
3. L. equalis.
4. L. quadrata.
5. L. elongata.
6. L. curta.
7. L. obtusa.
8. L. crassa.
9. Orbiculoidea filosa.
11. O. lamellosa.
11. Trematis terminalis.

PLATE IX.

Fig. 1. Orthis testudinaria, 1 *a* to *l*.
2. O. subequata.
3. O. bellarugosa.
4. O. disparilis.
5. O. perverta.
6. O. æqualis.
7. O. fissicosta.
8. O. tricenaria.
9. O. plicatella.
10, 11. O. pectinella.
12. O. insculpta.
13. O. dichotoma.

PLATE X.

Fig. 1. Atrypa extans.
2. A. nucleus.
3. A. bisulcata.
4. A. deflecta.
5. A. recurvirostra.
6. A. exigua.
7. A. circulus.
8. 9. A. ambigua.
10. A. hemiplicata.
11. A. extans.
12. A. subtrigonalis.
13. A. increbescens.
14. A. dentata.
15. A. modesta.
16. A. sordida.

PLATE XI.

Fig. 1, 5 *b*, 3, 7. Strophomena alternata.
2. S. planumbona.
4. *a*, *b*, *c*, *d*, S. tenuistriata.
6. *a*, *b*, *c*, *d*, *e*, *f*, S. sericea.
8. *a*, *b*, 9 *c*, *e*, S. filitexta.
7, 10. S. alternistriata.
12. S. planoconvexa.

PLATE XII.

Fig. 1. Endoceras proteiforme.
2. Oncoceras constrictum.
3. Diplograpsus pristis.
4. Orthoceras arcuoliratum. 4. Cyrtolites filosum. O. tereteforme.
5. Endoceras proteiforme.
7. *a*, *b*, Bellerophon exhaustus.
10 *a*, *b*, 5 and 11. Cyrtolites compressus.
13, 14 *a*, *c*, *d*, 15 *a*, *b*, *c*. Trocholites ammonius.
12, 11, 5. Bellerophon punctifrons.

PLATE XIII.

Fig. 1. Lyonsia sublata.
2. Cypricardia subtruncata.
3. Lyonsia subaviculoides.
5, 6. Posidonomya bellistriata.
7. Cardiomorpha subangulata.
8. Lyonsia vetusta.
13, 15. Cardiumorpha ventricosa.
14. Orthonota parallela.
16, 17. Cardiomorpha subtruncata.
18, 19. Ambonychia orbicularis.
20, 21. Posidonomya amygdalina.
22. Posidonomya orbiculata.
23, 25. Ambonychia subundata.
27. Avicula elliptica.
28, 29, 30. A. trentonensis.
31. A. subarcuata (not referred to).

PLATE XIV.

Fig. 1 *a*, *b*, 1. Receptaculites neptuni.
2. Lyonsia sanguinolaroidea.
3. Lyonsia gibbosa.
4. Lysonia trentonensis.
5, 6. Cardiomorpha ventricosa.
7, 8. Lysonia dubia.
14, 15. Lysonia faba.
10. Leda levata.
11. Cypricardia americana.
9 *a* to *p*. Orthis lynx.

PLATE XV.

Fig. 1. *a* to 1 *k*, Ceraurus pleurexanthemus, and its parts.
2. *a* to 2 *e*, Ceraurus vigilans.
2, 5, 18, parts of the Lichas trentonensis.
3. *c d e* and 7 7 *a b c*, Phacops calicephalus.
4. *a*, head of the Trinucleus concentricus; 4 *b*, showing the length of the spines as it usually occurs in the Trenton limestone.
6. Acidaspis Spiniger.
7. *a*, head; and *c*, eye of the P. calicephalus.
8. and 11. Asaphus extans.
12. Triarthrus beckii.
13. Illaenus trentonensis.
15. Illaenus crassicauda.
16. Calymene senaria.
18. Pygidium.
19. Head and parts of the Senaria, showing the direction of the great suture, and the form of the rings.
20. Spirifer trentonensis.
22. Strophomena nasuta.

PLATE XVI.

Fig. 1. 1 *a b* 3, varieties of the Endoceras proteiforme showing the surface markings.
4. *a b c d e f*, Conularia trentonensis.
5. *a b*, C. granulata.
6. C. papilata.
7. *a*, C. gracile.
7. Head of the Trinucleus concentricus.
8. Part of the head of the Ceraurus pleurexanthemus.
9. Part of the lobe of the Isotelus gigas.
10. Middle lobe of the head of the Isotelus.
11. Epistoma of the Isotelus.
12. Isotelus gigas.

PLATE XVII.

1. Trinucleus concentricus of the Loraine shales.
2. Strophomena alternata.
3. S. nasuta.
4. Lyonsia subtruncata.
5. Orthis crispata.

6. Tentaculites flexuosa.
7. Bellerophon or Cyrtolites profundus.
8. Pleurotomaria indenta.
8, *a* 8. Lyonsia submodiolaris.
9, and 16. Murchinsonia gracilis, cast.
9, *a b*, Pleurotomaria subconica.
10, *c*, Cyrtolites bilobatus.
10. Avicula demissa.
12. Orthis testudinaria of the Loraine shales.
13. Triarthrus beckii. Utica slate.
13. *a*, Pleurotomaria lenticularis.
14. Lyonsia anadontoides.
15. Avicula insueta.
16. Murchinsonia gracilis.
17. Periploma planulata.
18. Lingula quadrata.
19. Orthoceras.
20. Heterocrinus heterodactylus, joint.
21. Porcelia ornata.
22. Cardiomorpha poststriata.

PLATE XVIII.

Fig. 1, and 2. Encrinal heads disputed, exhibiting the mode of distinguishing their different elements. B, is applied to the immediate part to which the pelvis is attached; E, pelvis; F, costal plate; S, intercostal plate; H, scapular plate; I, interscapular plate.

2. K, arms Lecanacrinus macropetalus; L, cuneiform, joint; M, hand; N, fingers.

3. Elementary parts of the trilobite from Burmeister; under side of the Asaphus cornigerus; 4, upper side of the same; *a*, clypeus; *b b*, antennae bearing lobes; *c c*, lateral lobes; *e e*, mandibulae; *f f*, indentations into which the lateral lobes are placed when the animal rolls itself up; *h*, rectum; *d*, Labrum.

5. Illaenus crassicauda.

6, 7, 8. Views of the Asaphus, when rolled up.

9. Front view of the Illaenus rolled up.

INDEX.

PART I.

PART II.

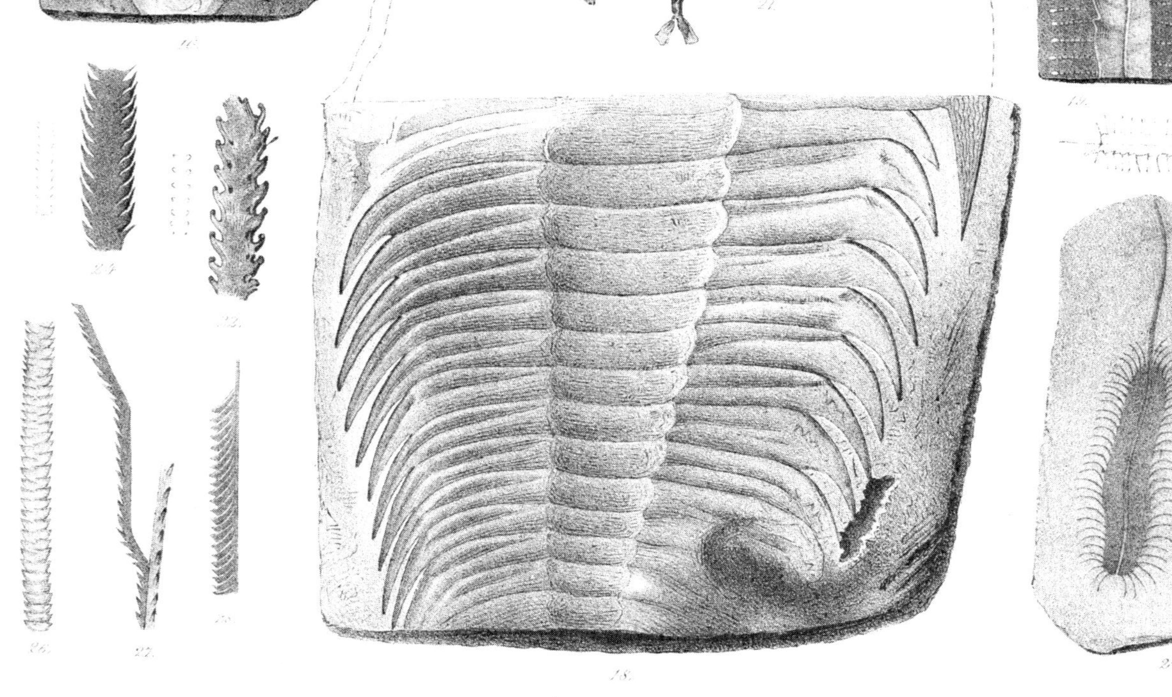

E. Emmons Jr. Del. On Stone by F. J. Swinton. Lith. of J. Murray, Albany.

Plate 1.

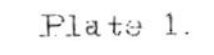

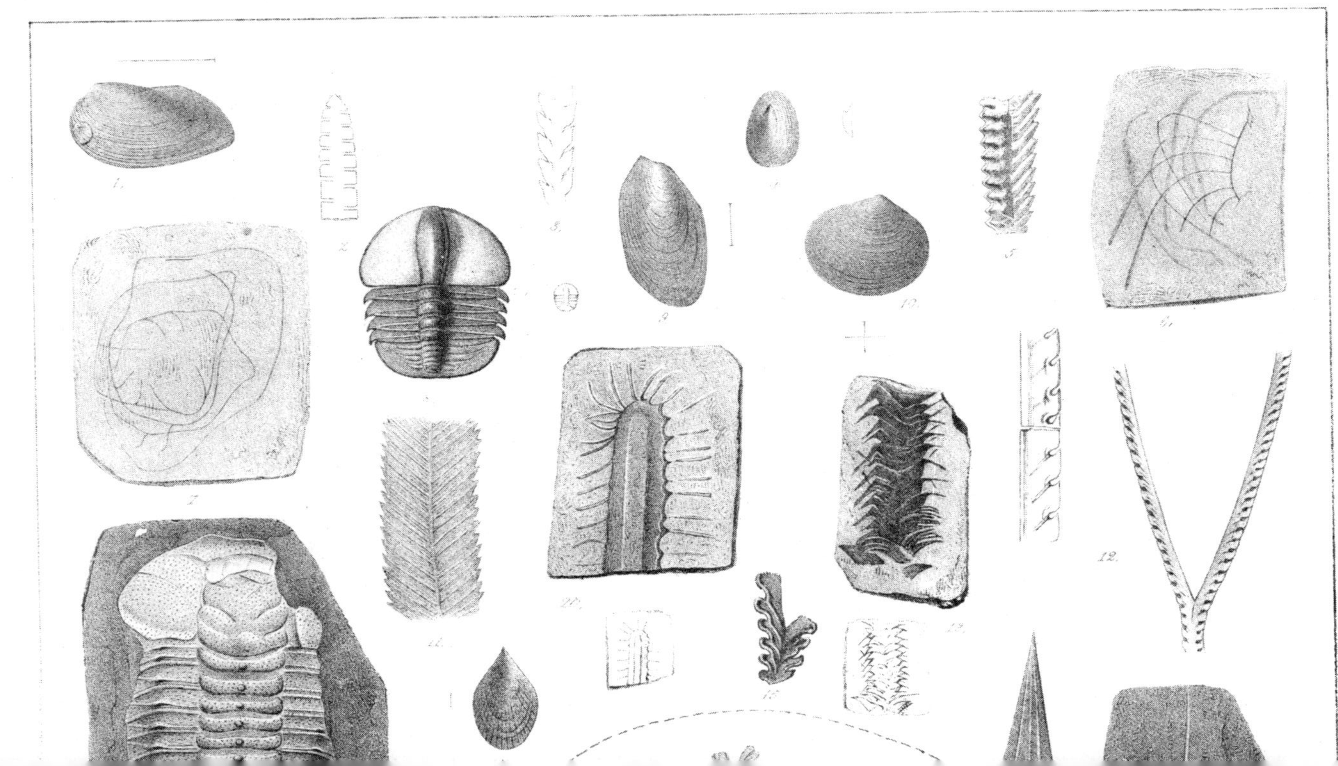

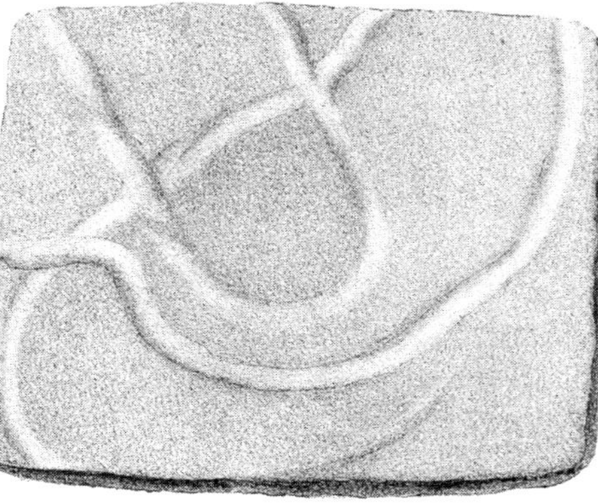

E. Emmons Jr. Del. | On Stone by F. J. Swinton. | Lith of J. Murray, Albany.

Plate 2.

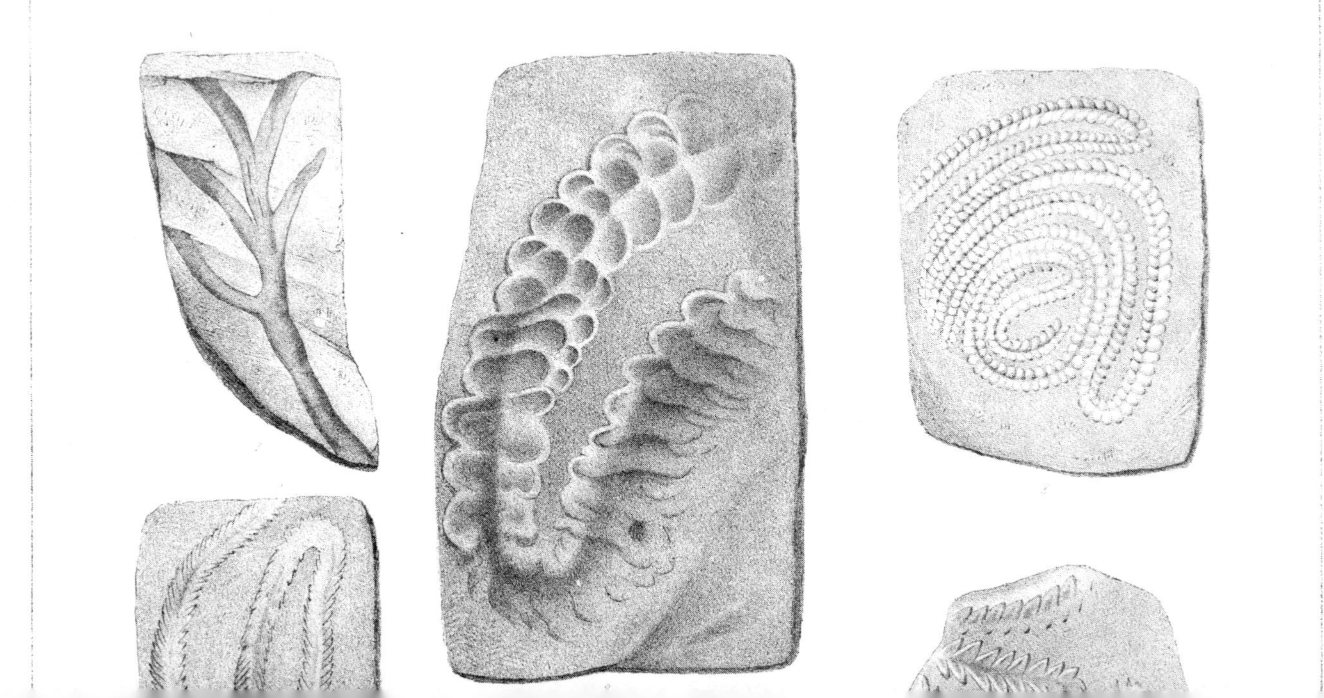

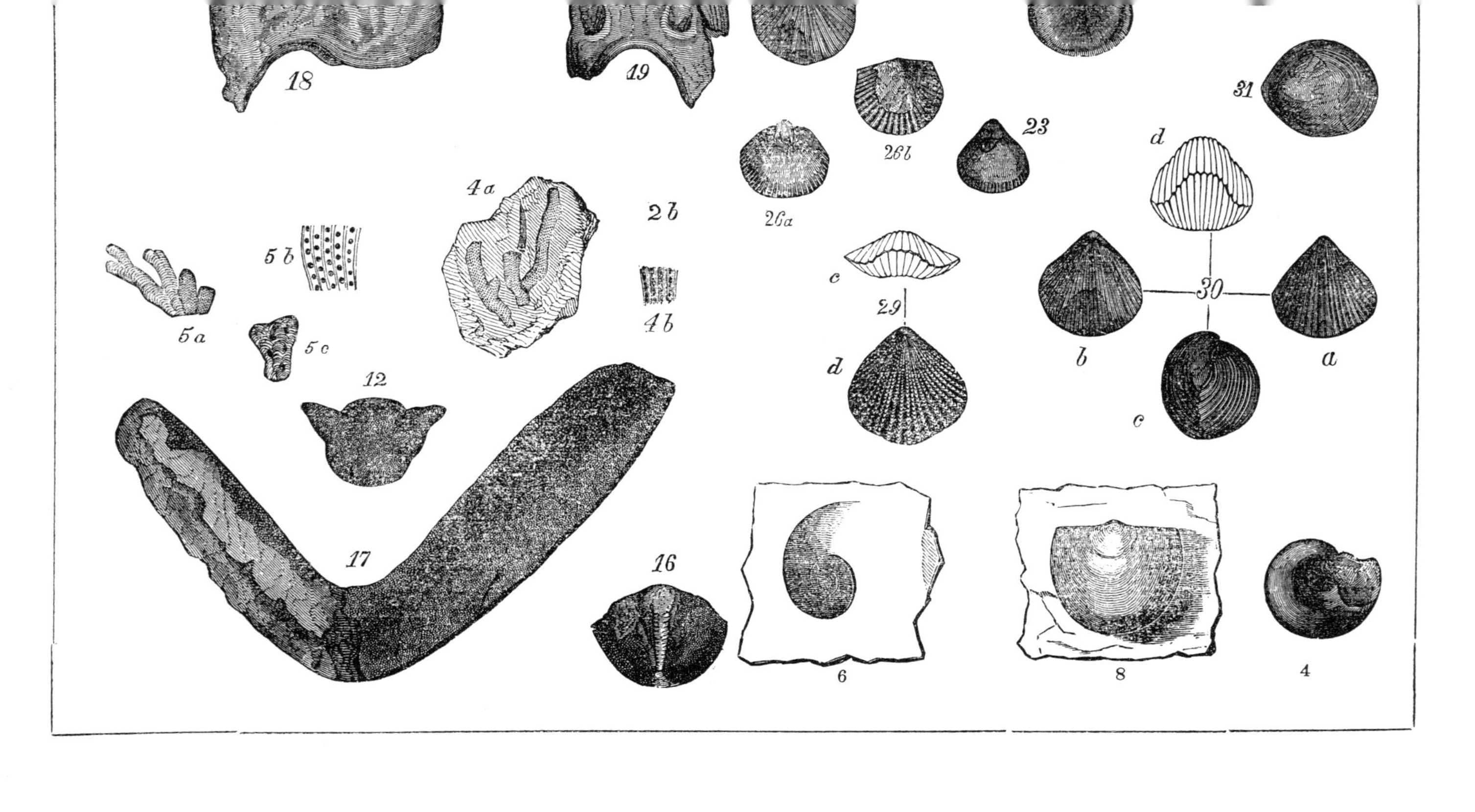
18
19
31
23
26b
d
4a
2b
26a
5b
c
30
29
4b
5a
5c
b
a
d
12
c
17
16
6
8
4

PLATE III.

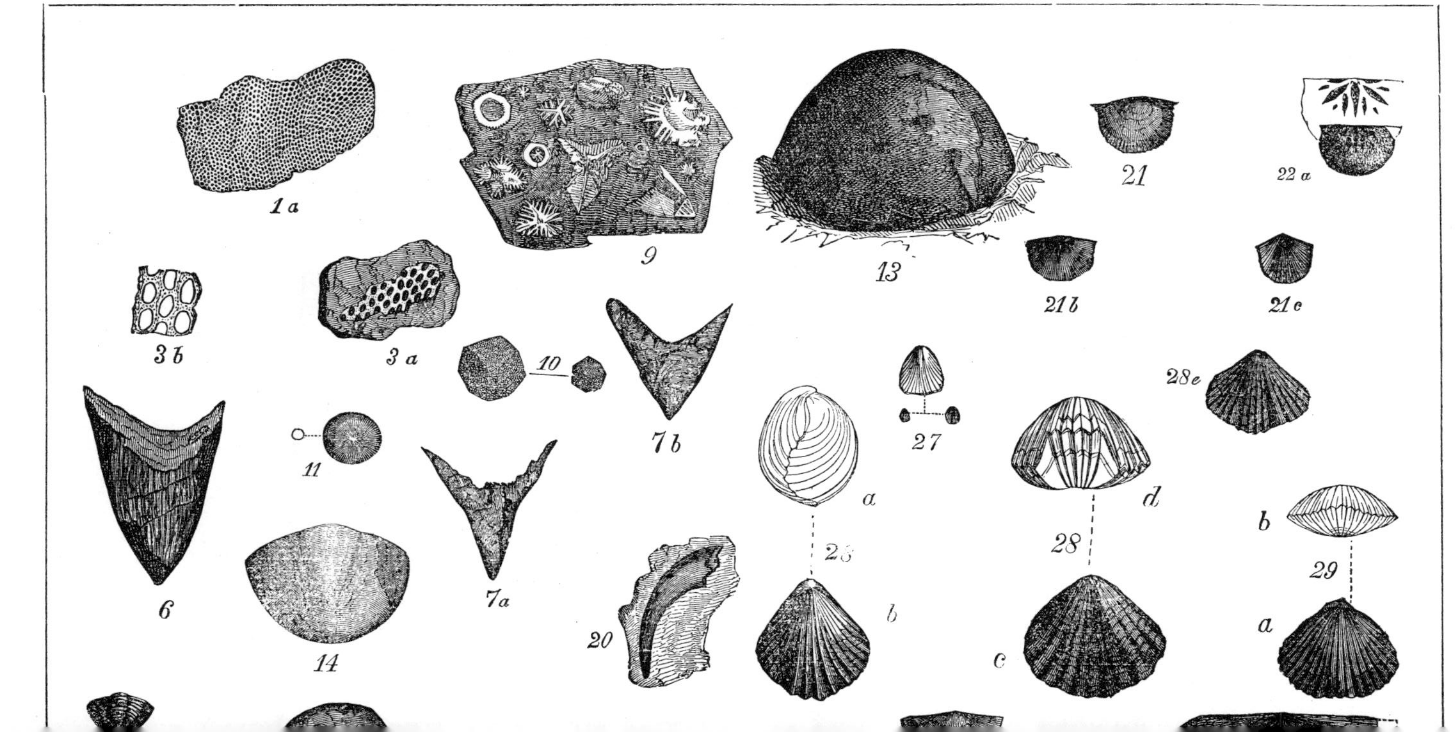

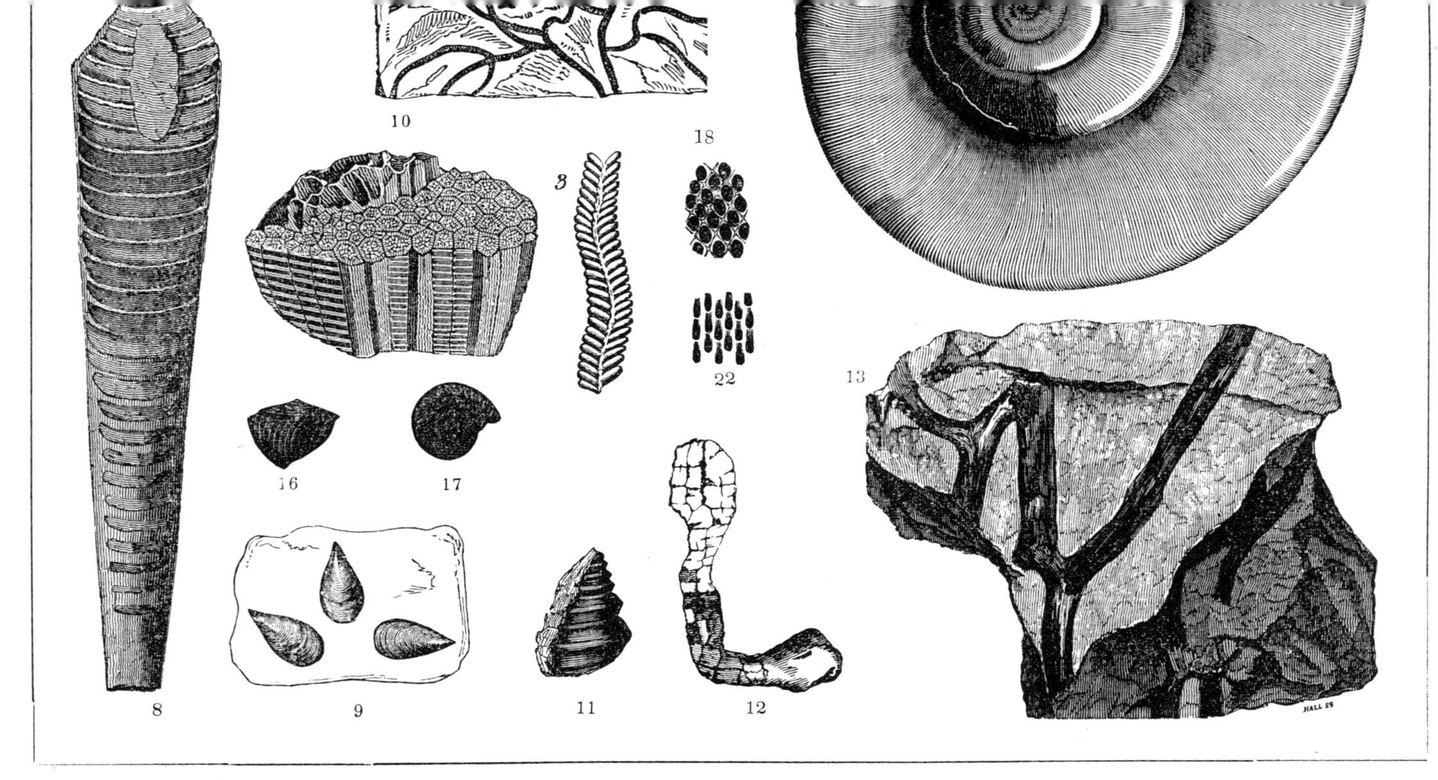

10
18
3
13
22
16
17
8
9
11
12
HALL SC

PLATE IV.

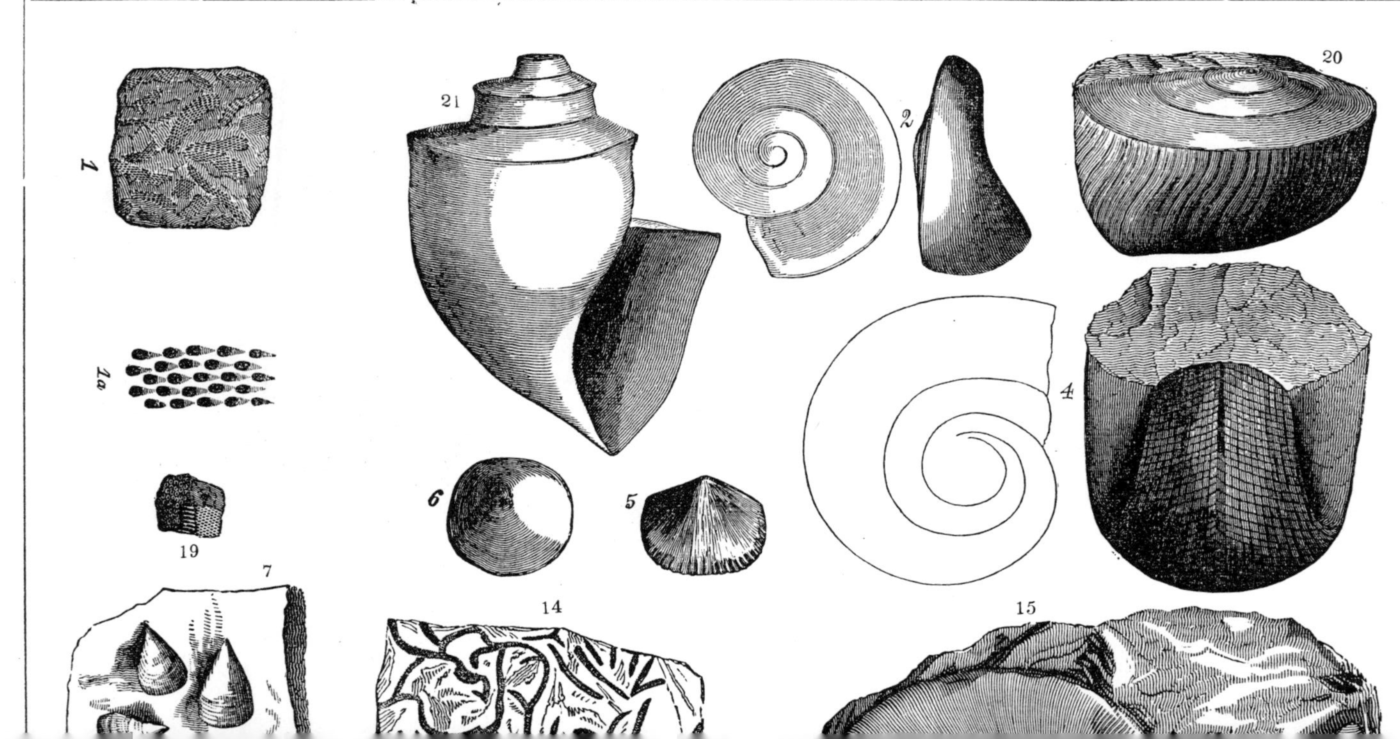

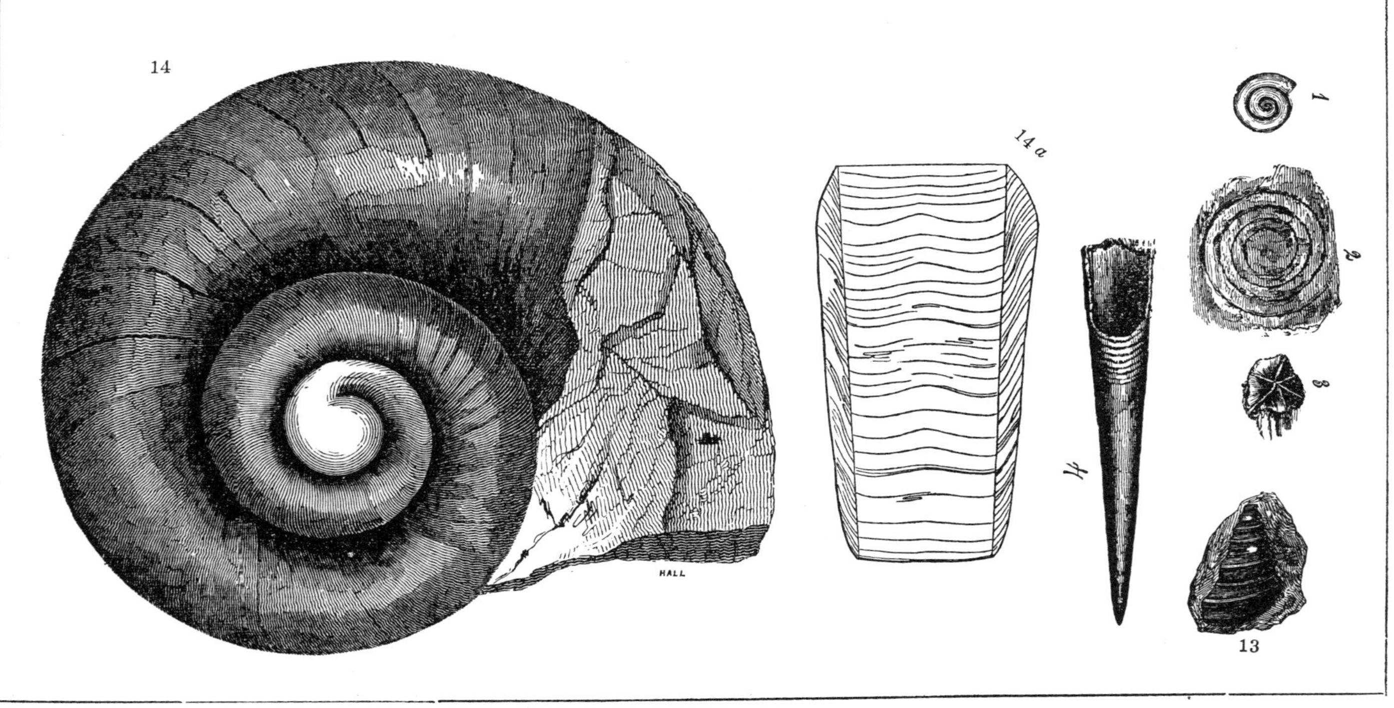

11
14
14a
1
2
3
4
13
HALL

PLATE V

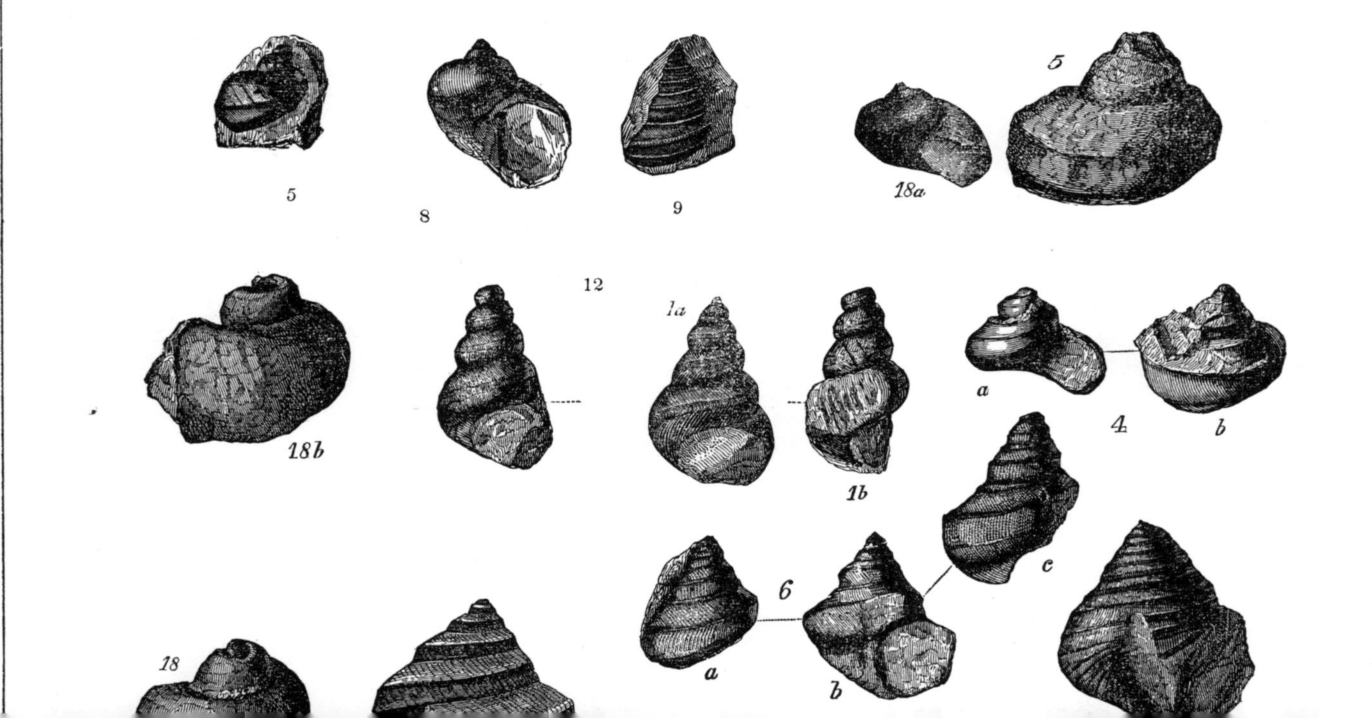

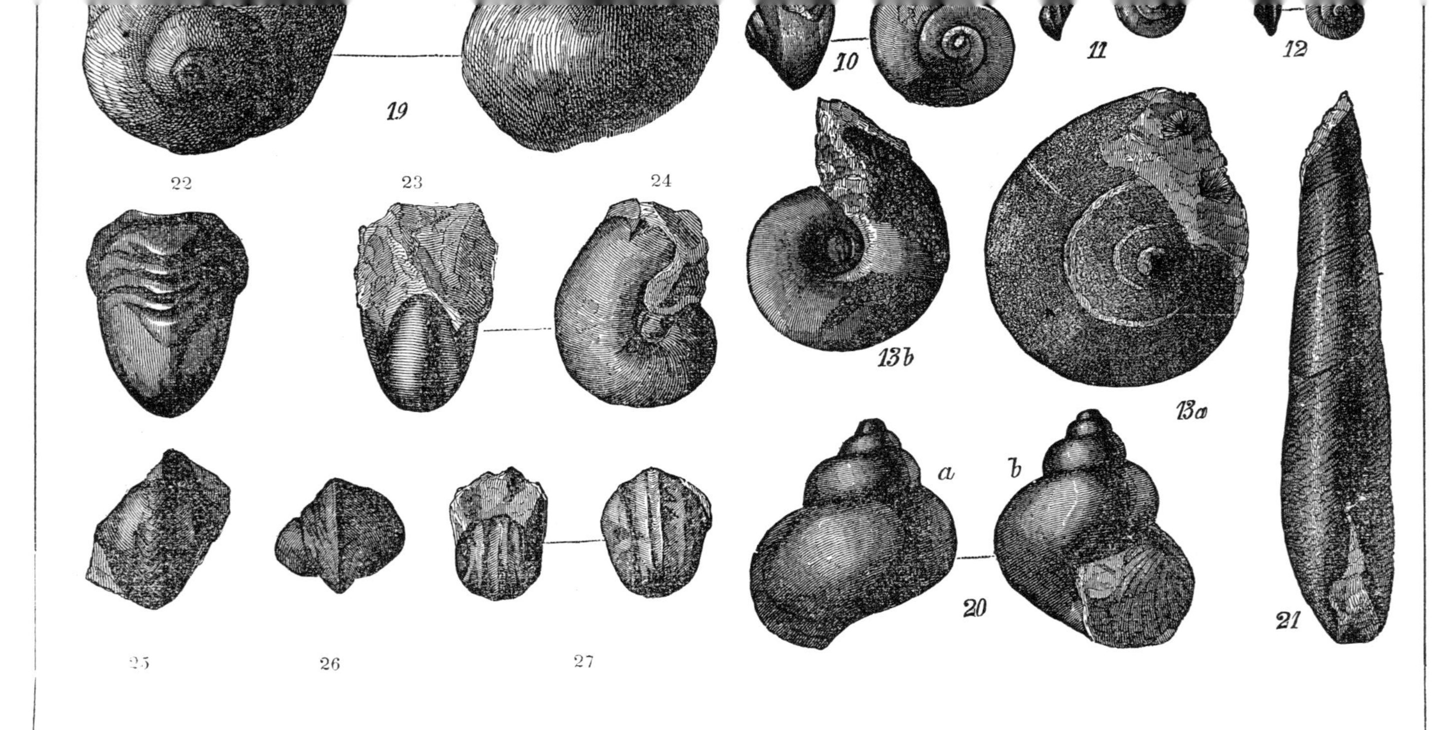
10
11
12
19
22
23
24
13b
13a
a
b
20
21
25
26
27

PLATE VI.

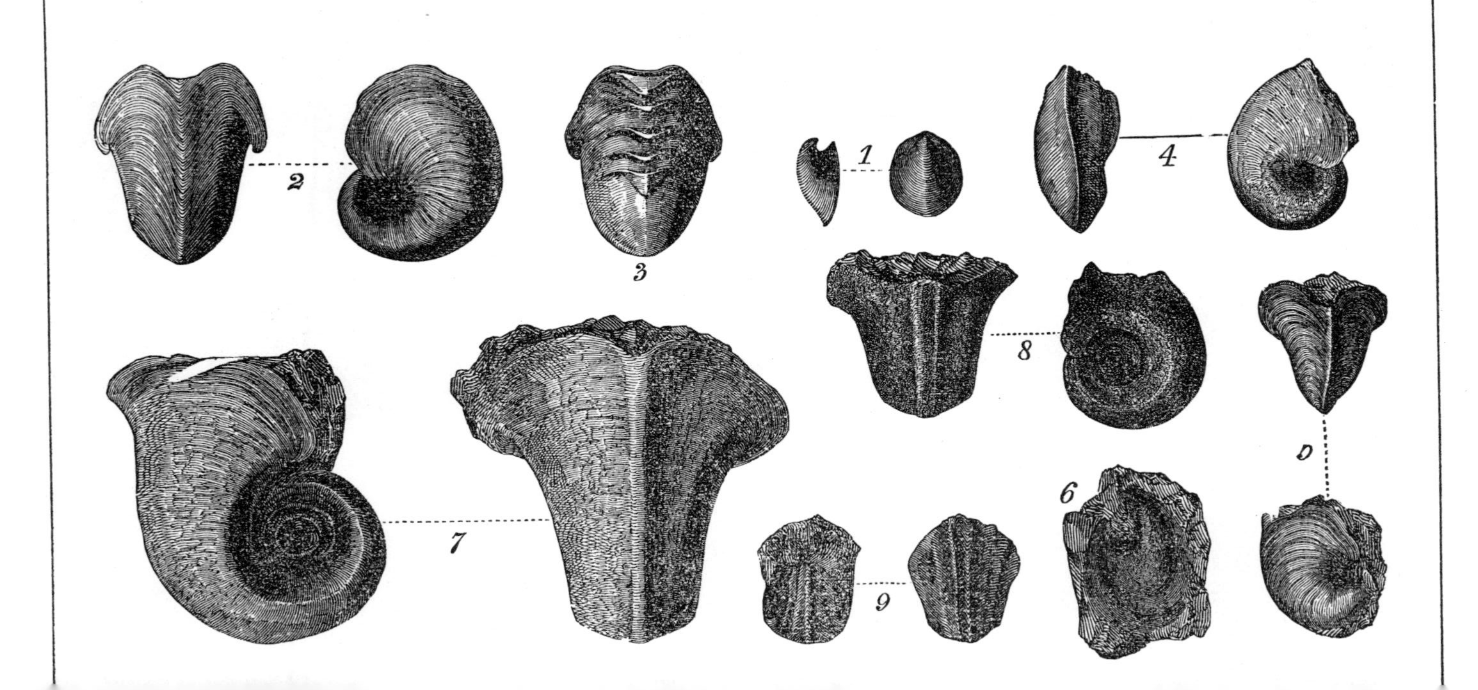

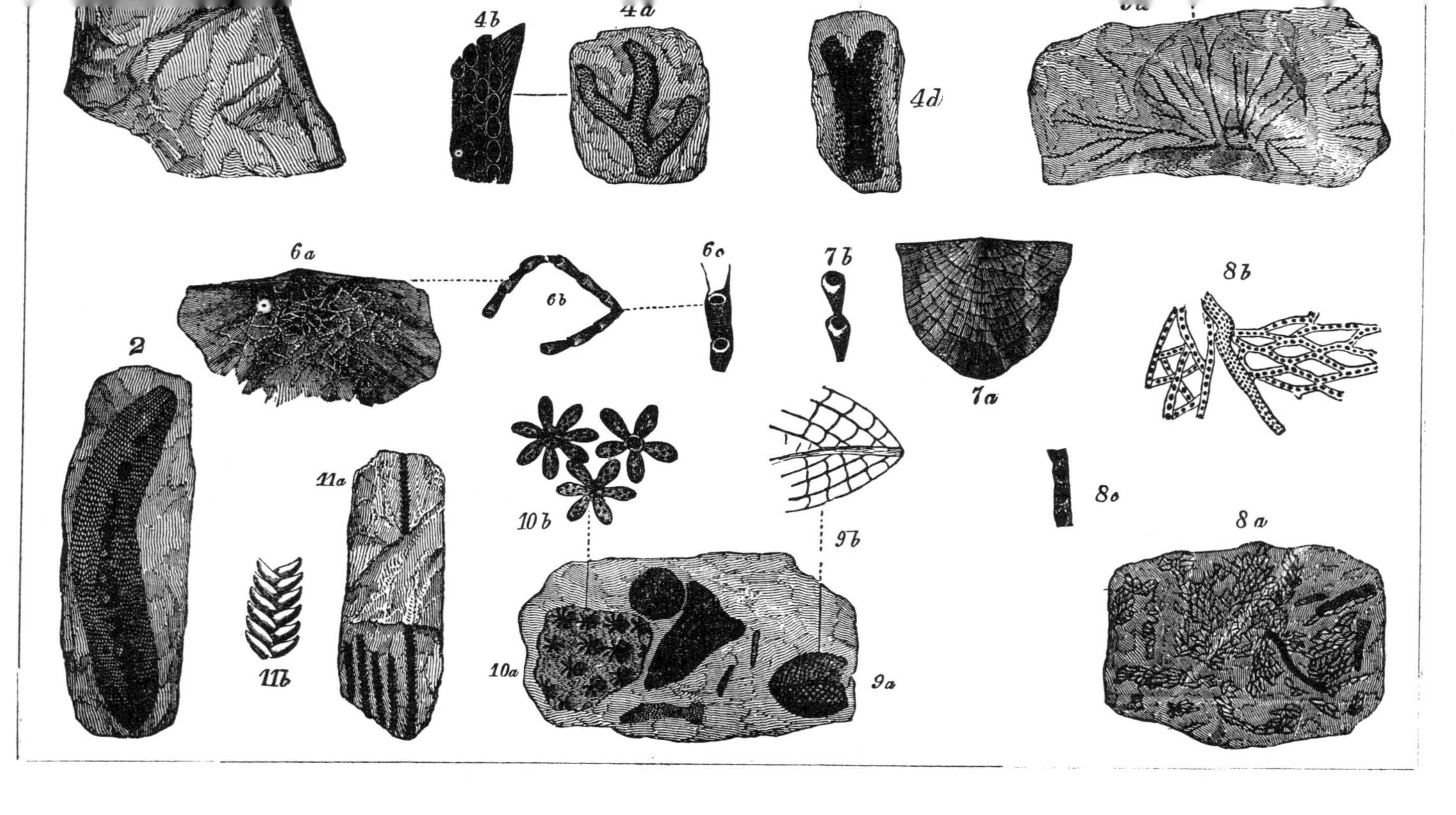
4b
4a
4d
6a
6b
6c
7b
7a
8b
2
11a
11b
10b
10a
9b
9a
8c
8a

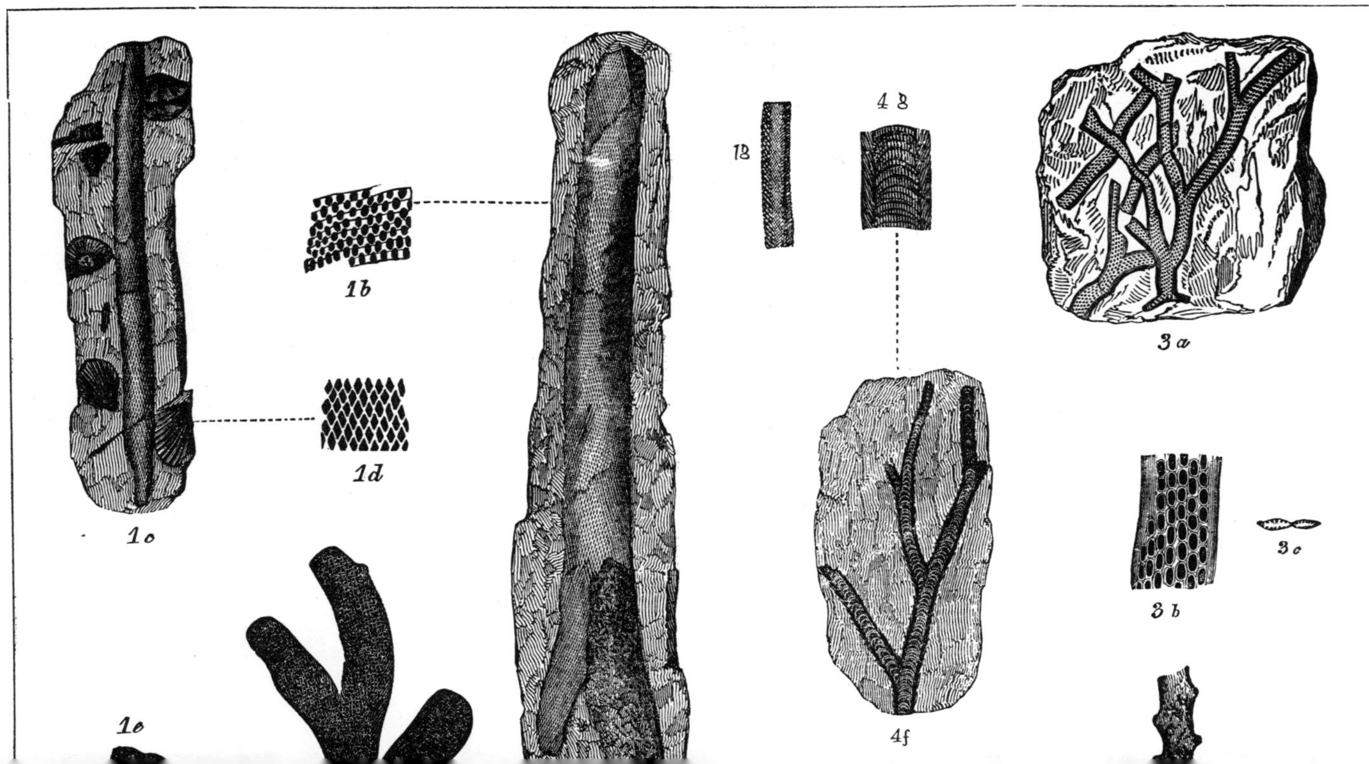
1c
1b
1d
1e
18
4 8
4f
3a
3b
3c

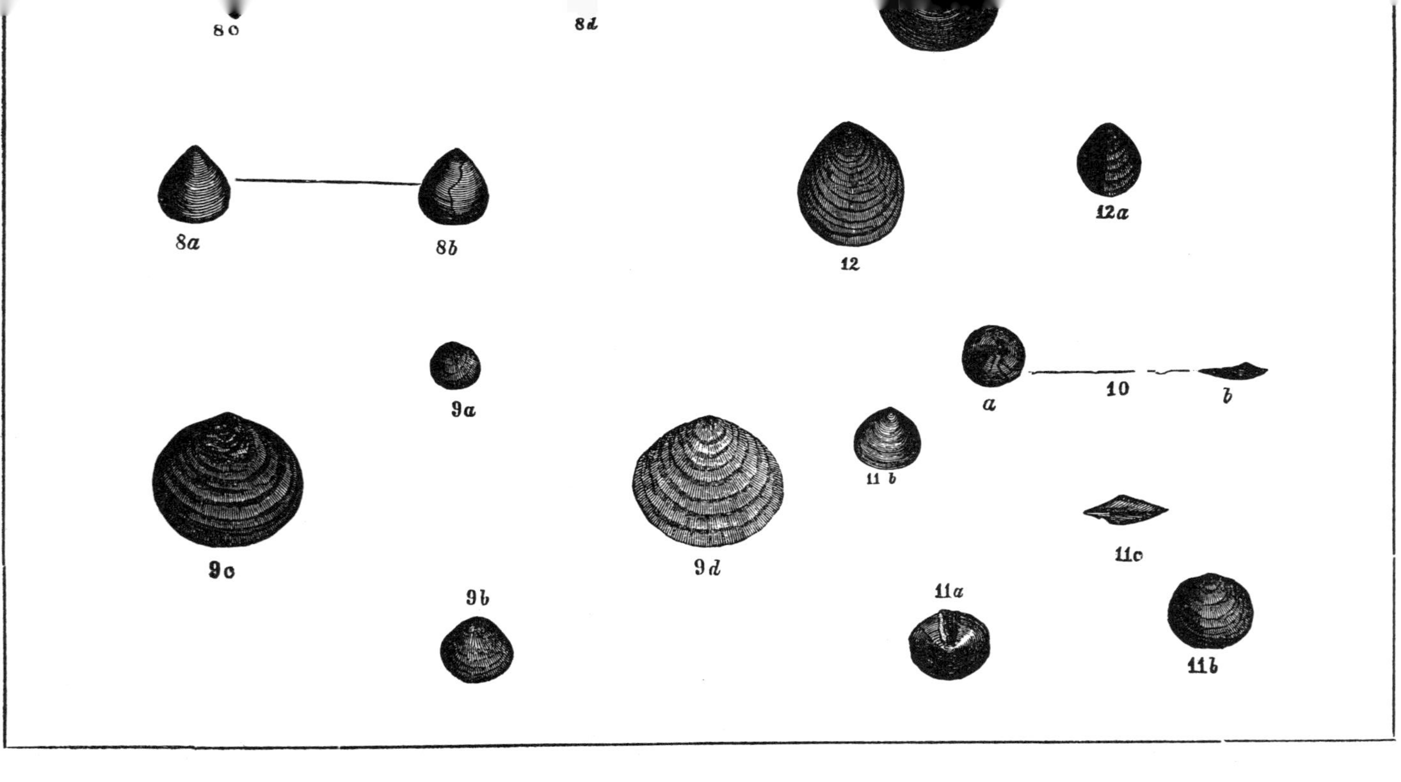
8c
8d
8a
8b
12
12a
9a
a
10
b
11 b
9c
9d
11c
9b
11a
11b

PLATE VIII.

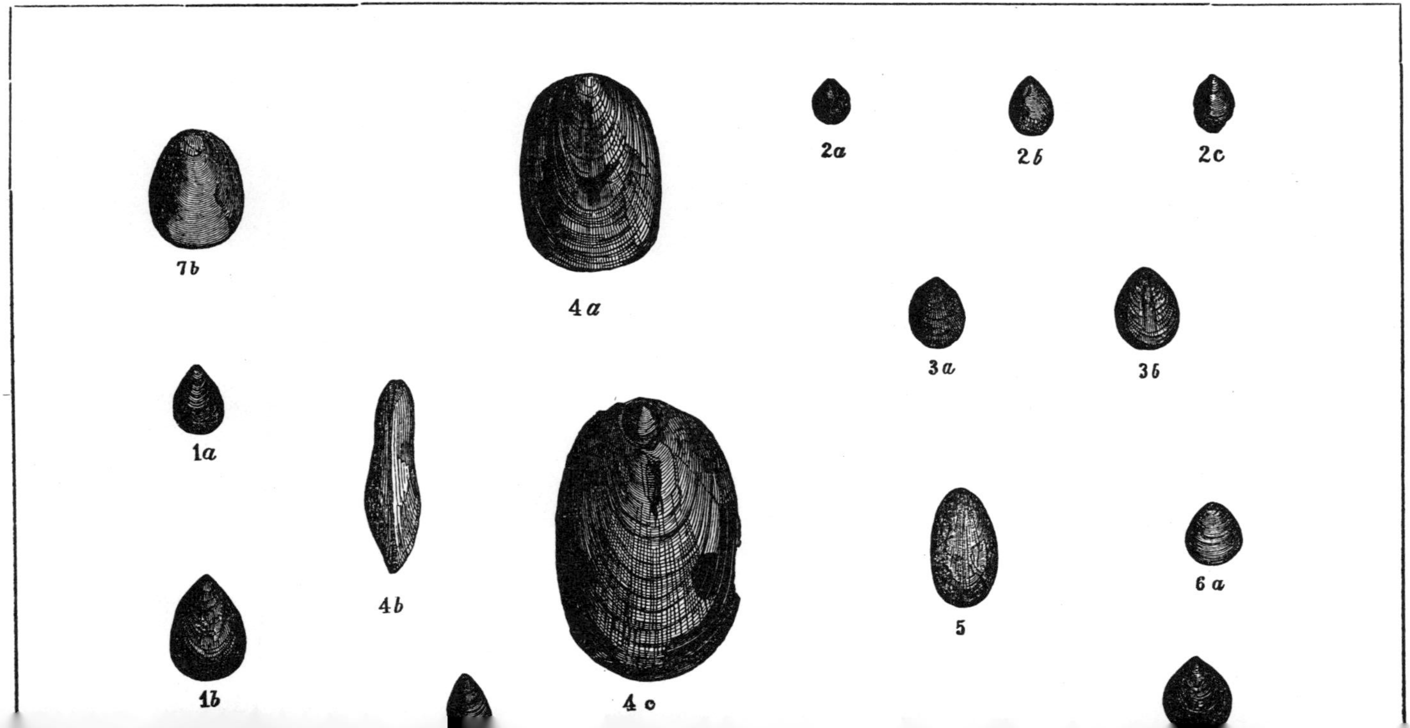

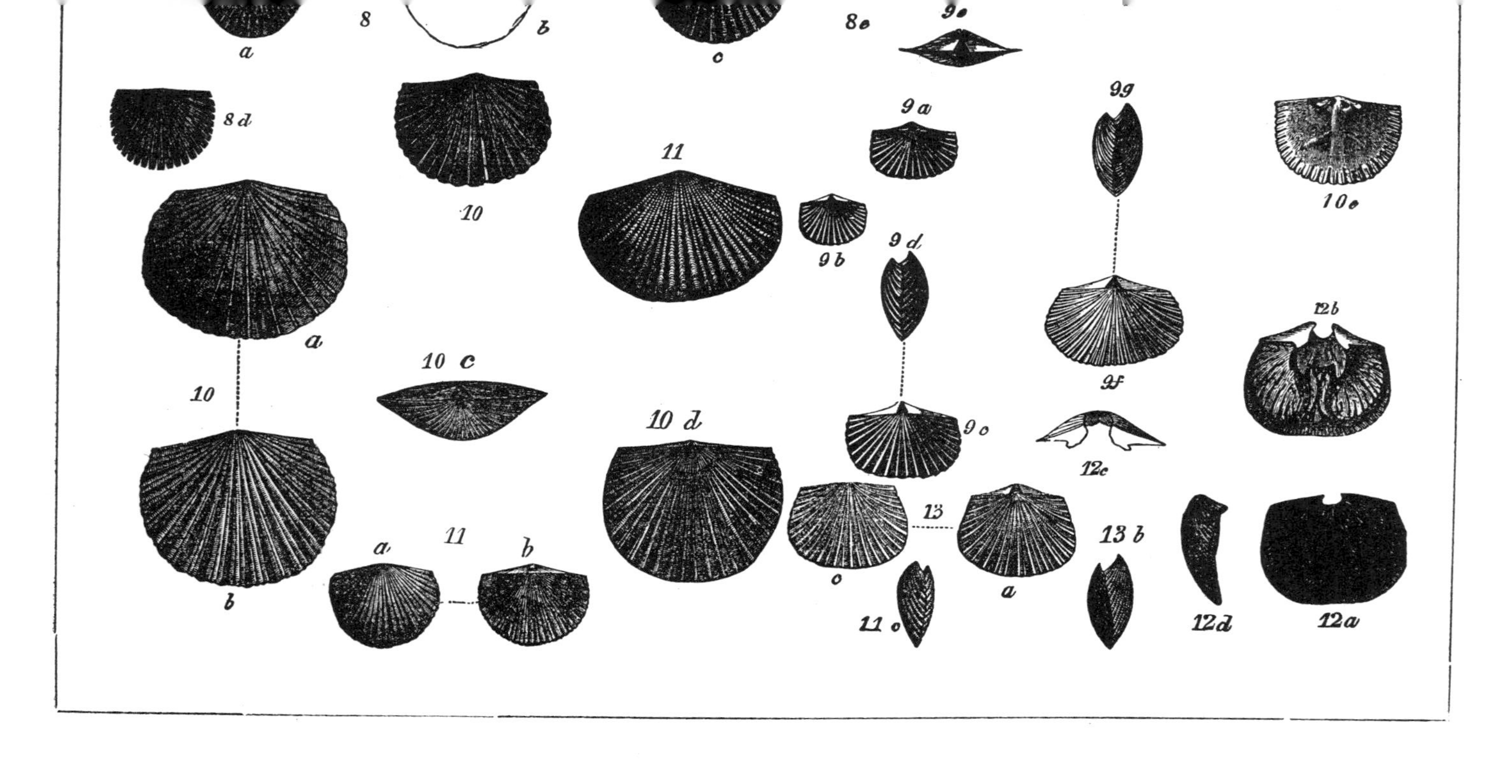
8
a
b
c
8e
9e
8d
10
11
9a
9g
10e
9b
9d
a
10
10 c
10 d
9c
9f
12b
12c
11
a
b
b
13
c
a
13 b
11 c
12d
12a

PLATE IX.

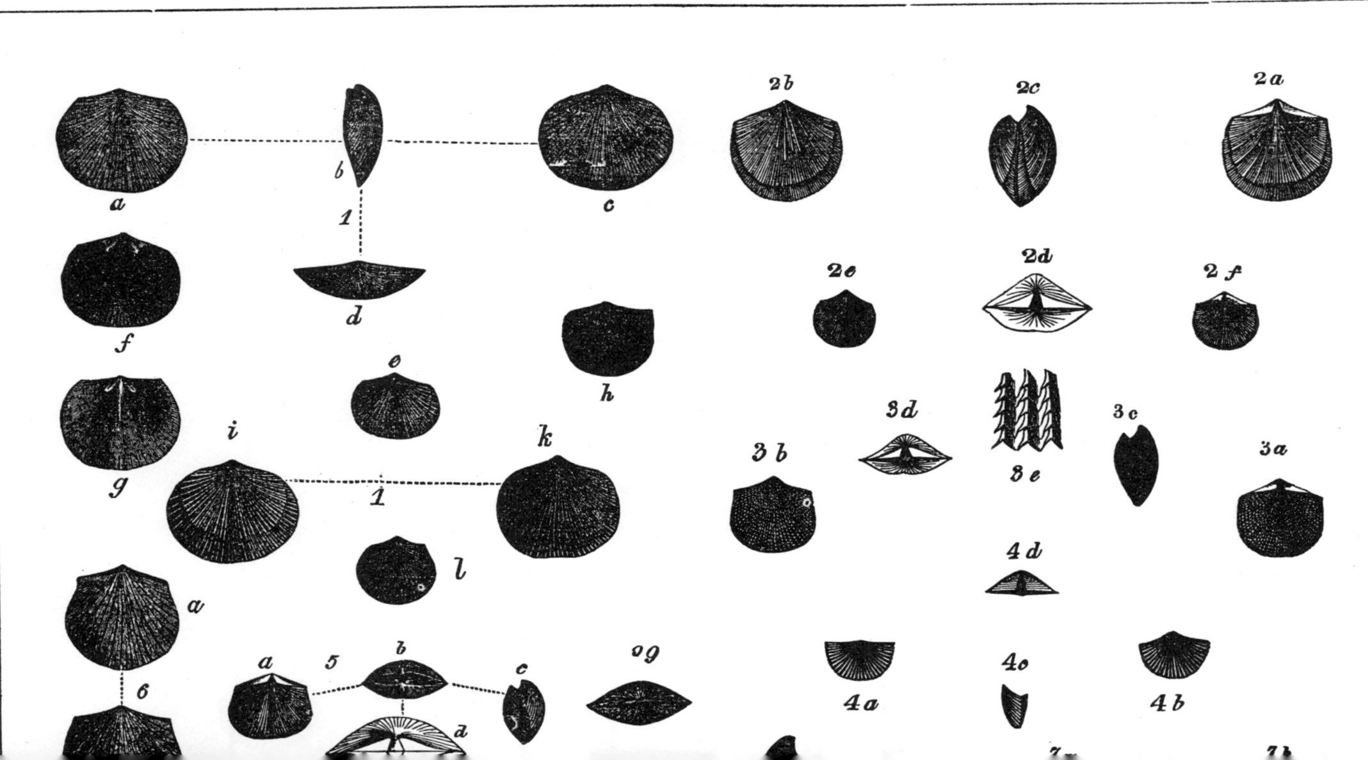

a
b
1
c
2b
2c
2a
f
d
2e
2d
2 f
e
h
3d
3c
g
i
k
3 b
3 e
3a
1
l
4 d
a
a
5
b
c
og
4o
6
d
4a
4b

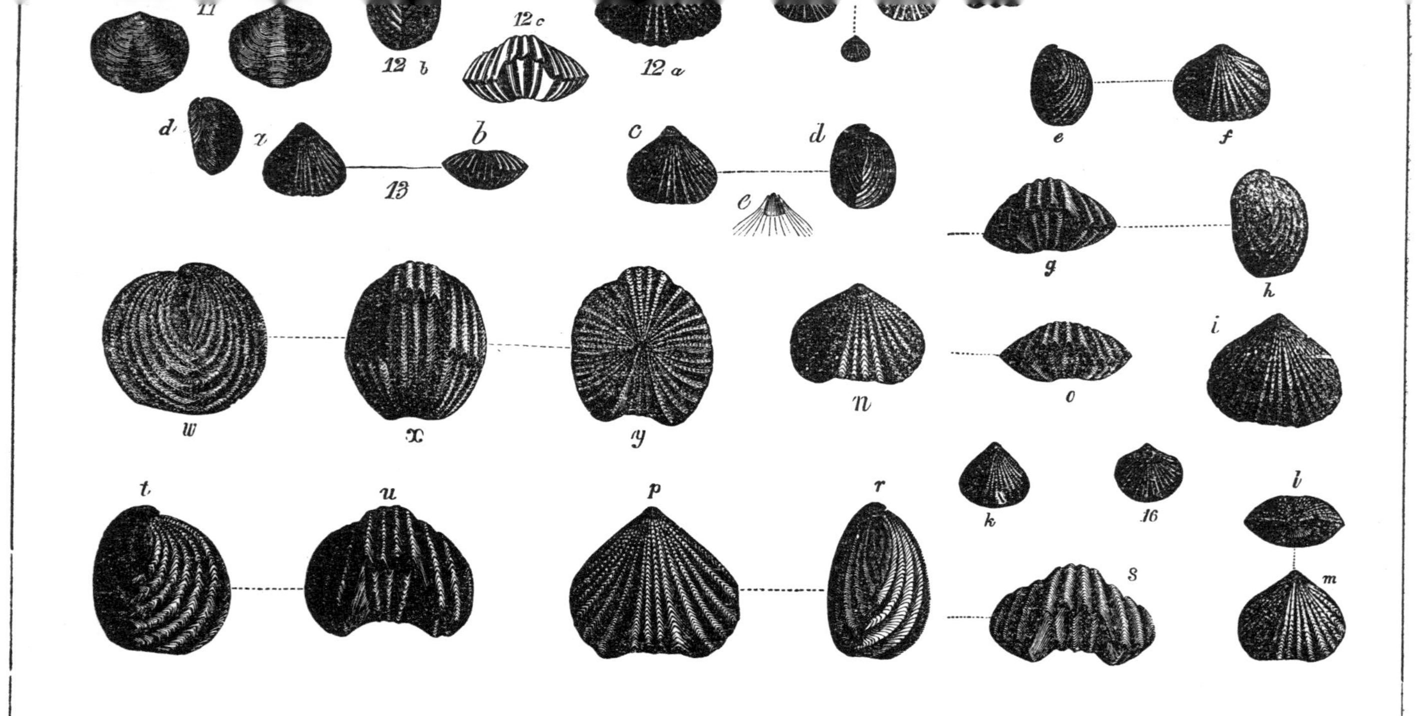
11
12 b
12 c
12 a
e
f
d
x
b
13
c
d
c
g
h
w
x
y
n
o
i
t
u
p
r
k
16
l
s
m

PLATE X.

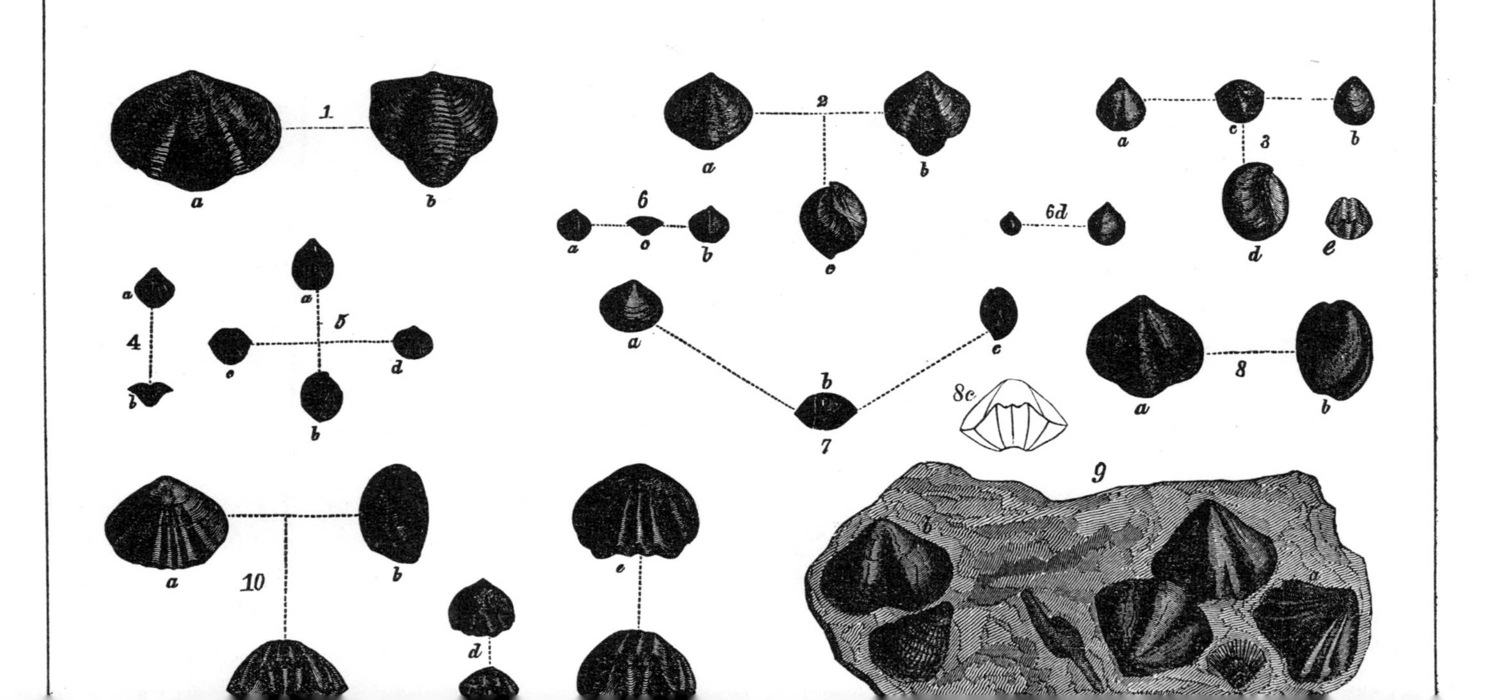

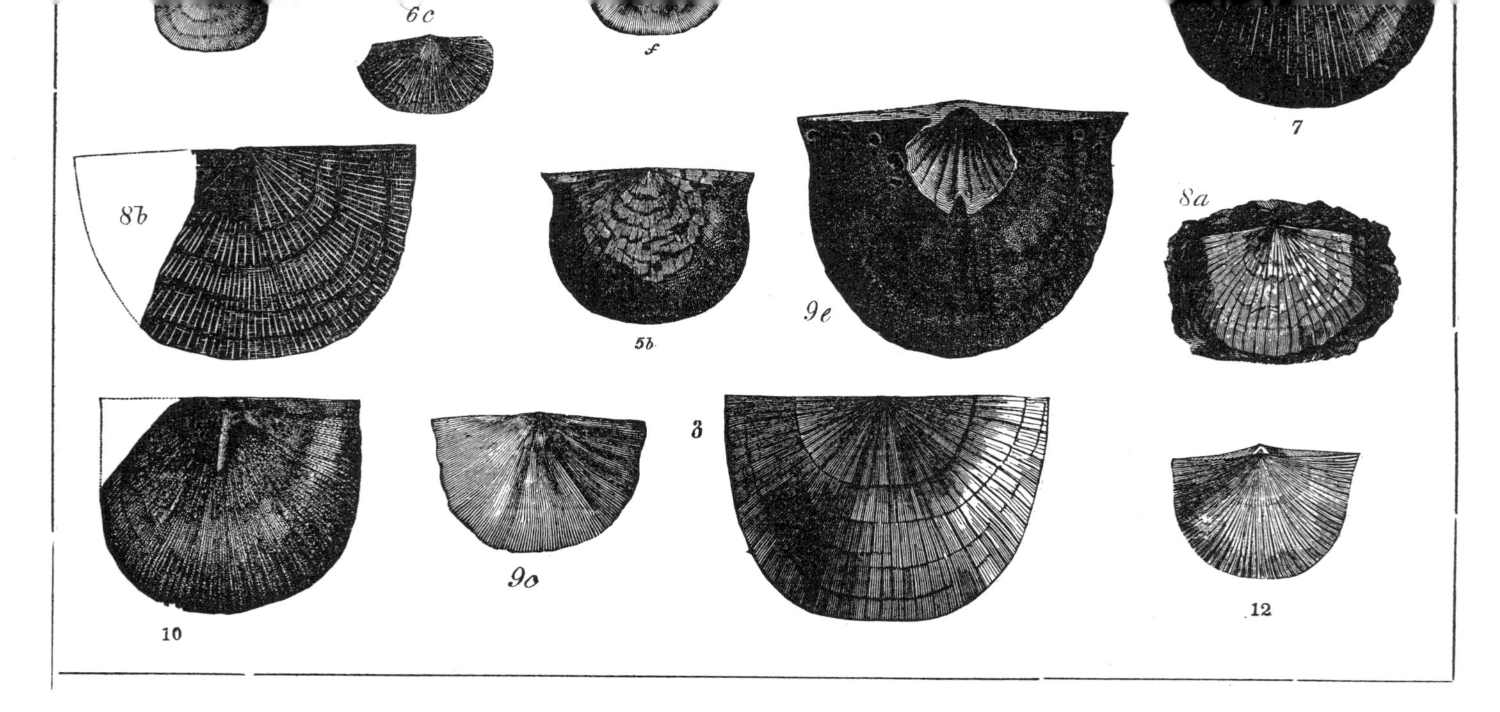
6c
f
7
8b
5b
9e
8a
10
9a
8
12

PLATE XI

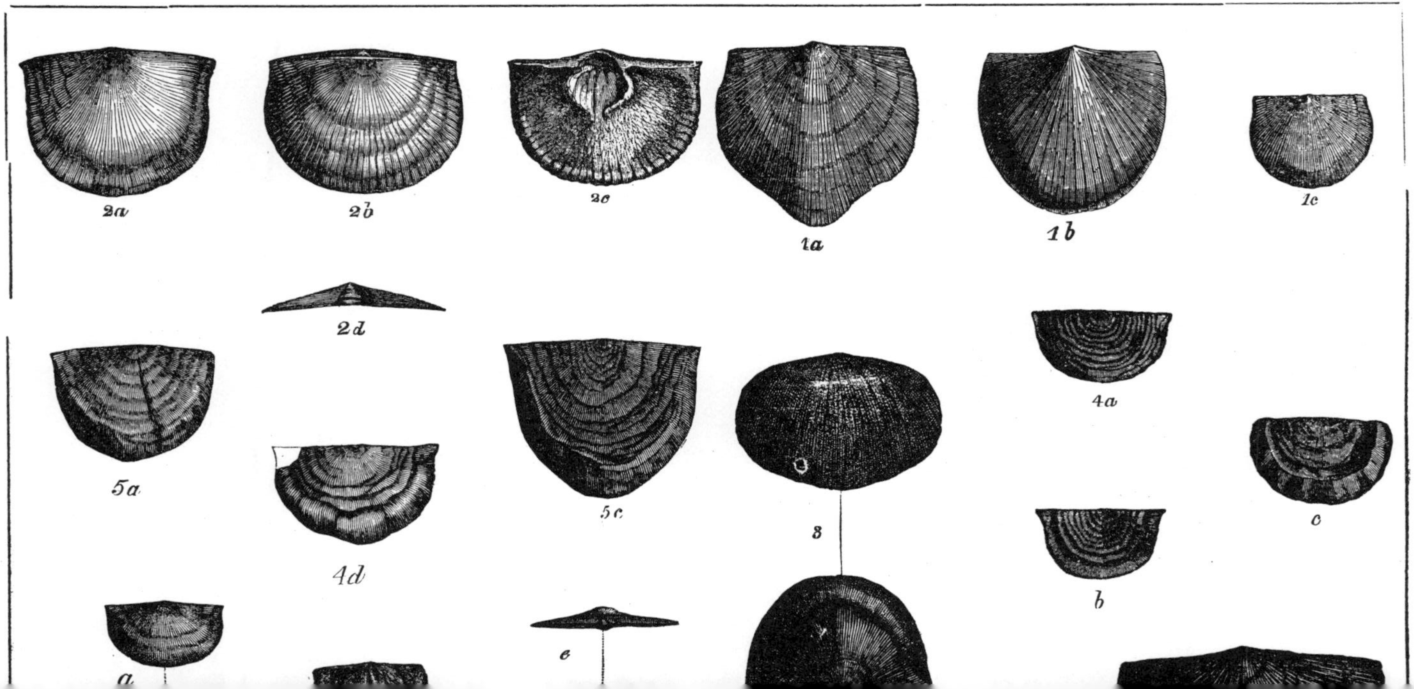

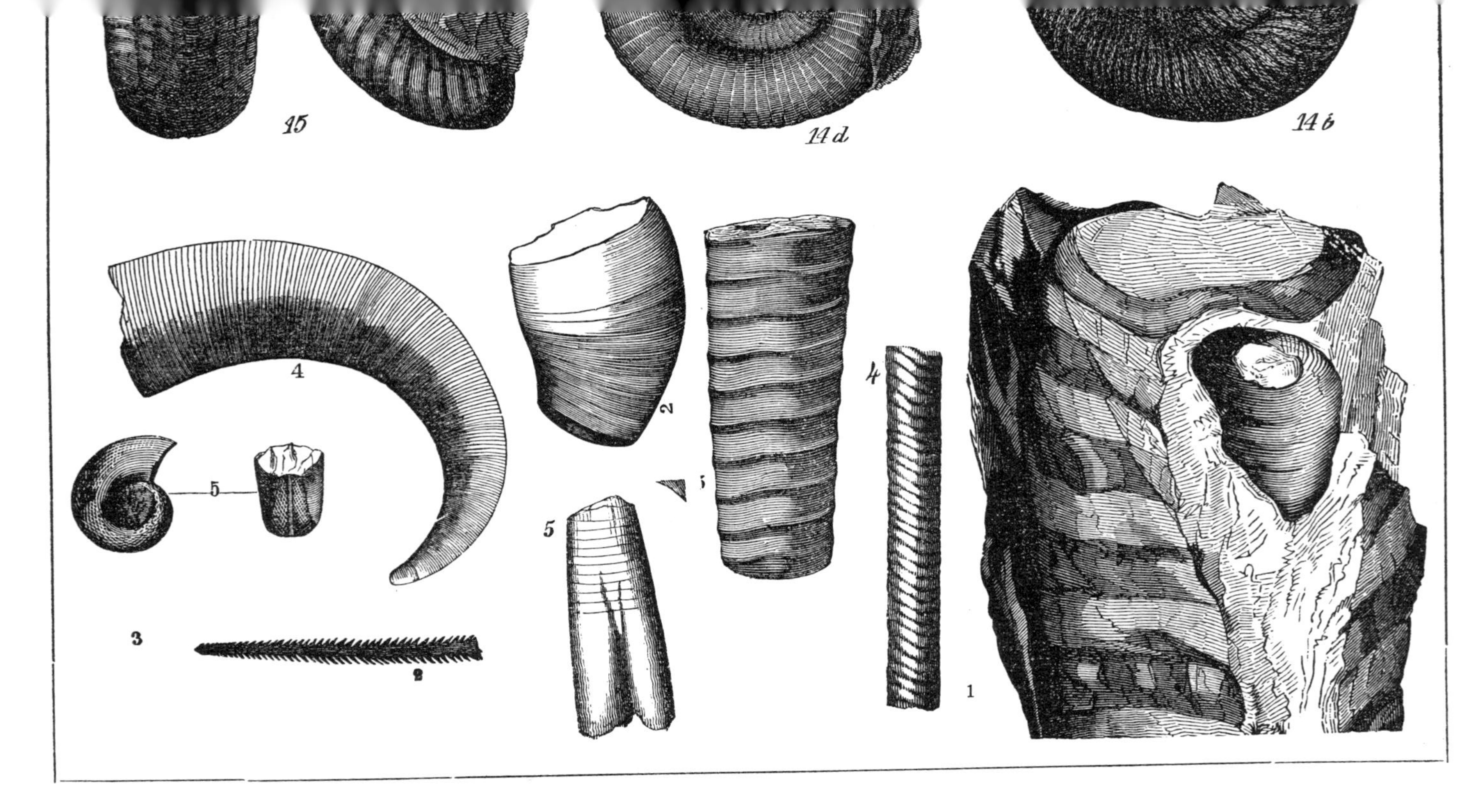
15
14 d
14 b
4
2
5
5
4
1
3
2

PLATE XII.

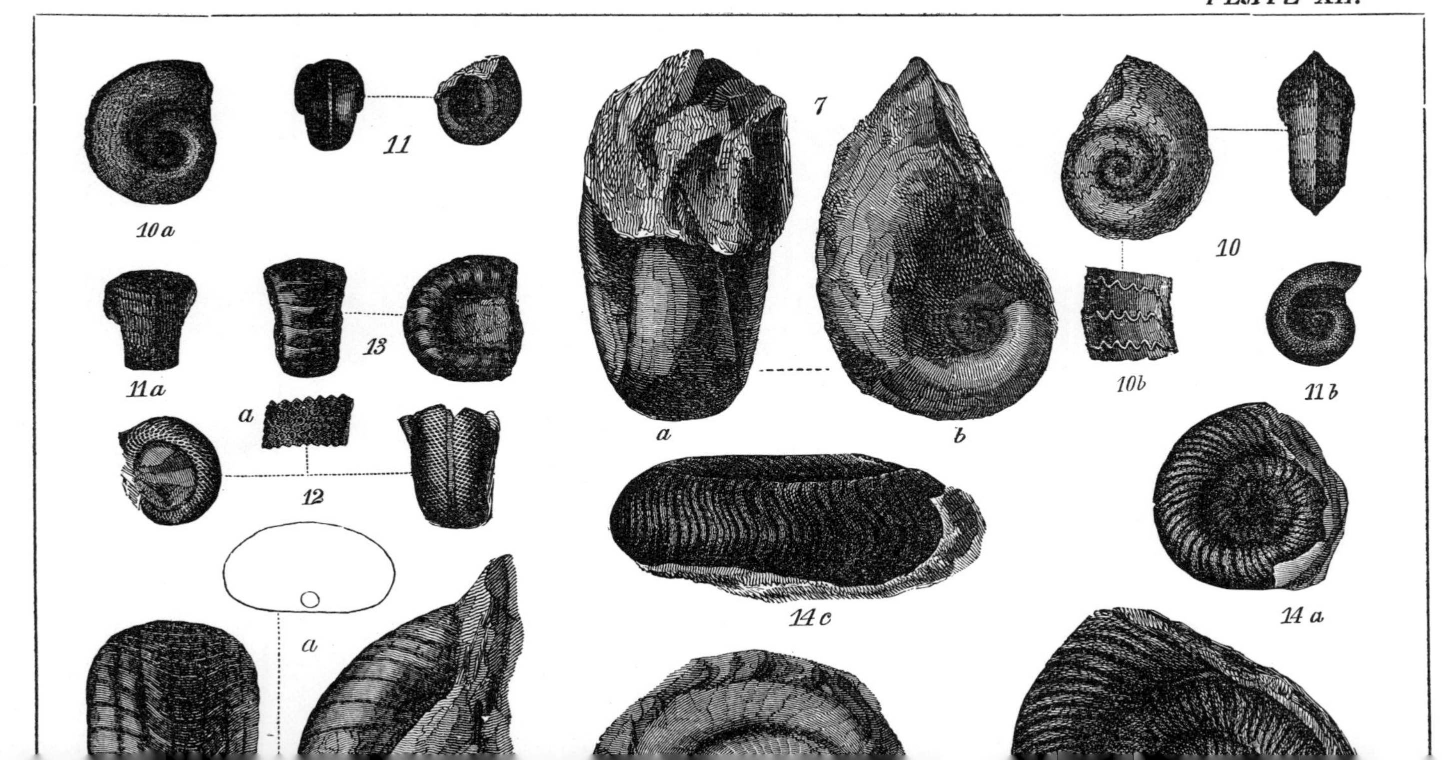

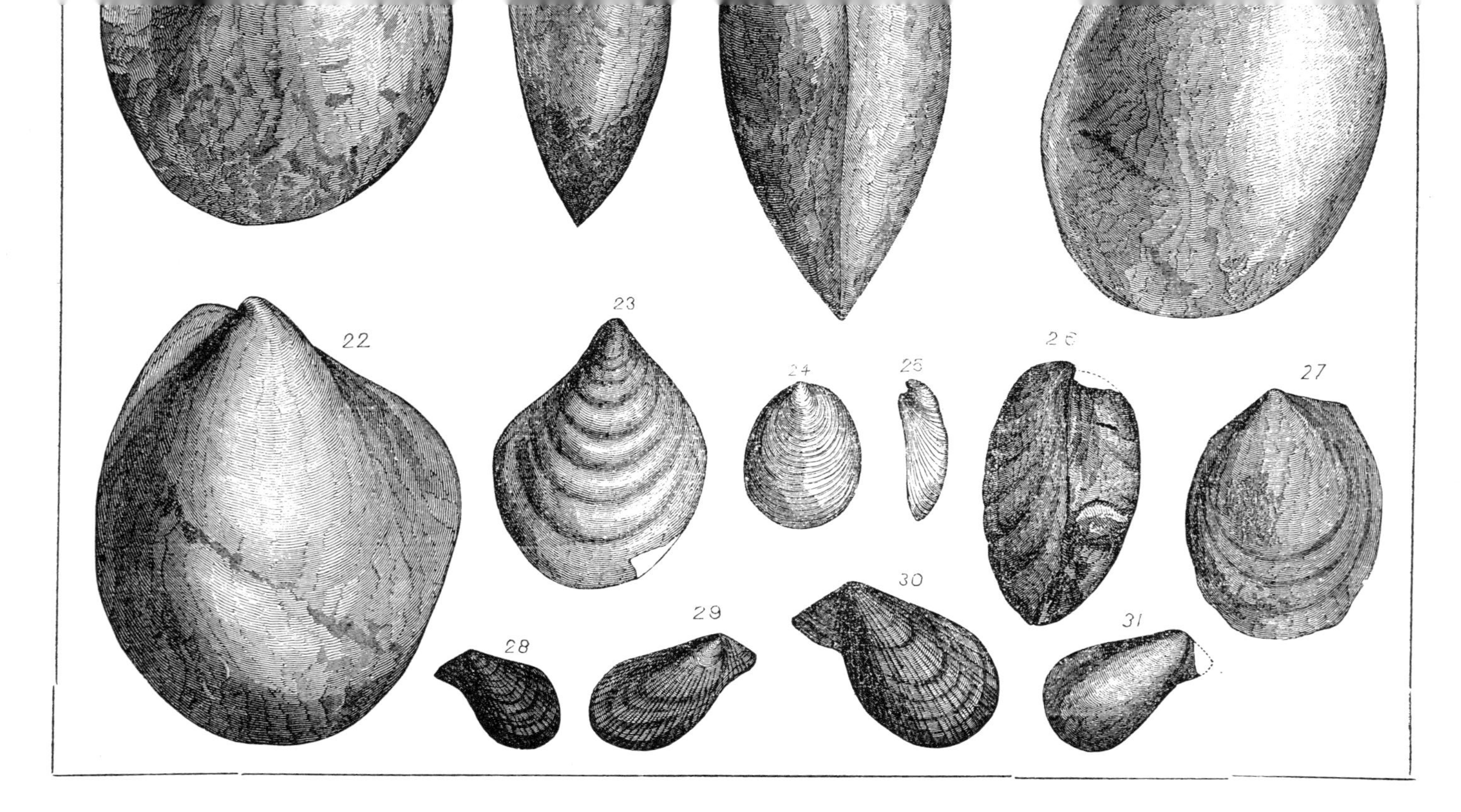
22
23
24
25
26
27
28
29
30
31

PLATE XIII.

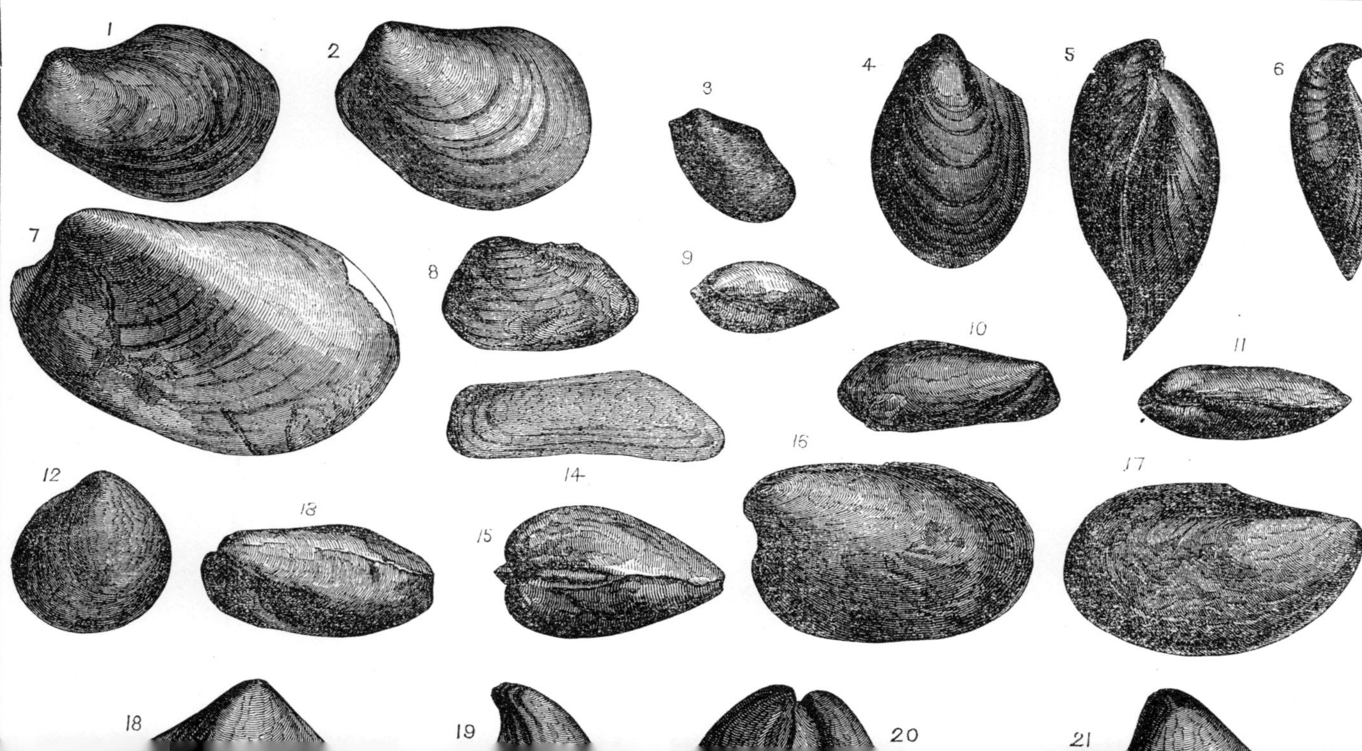

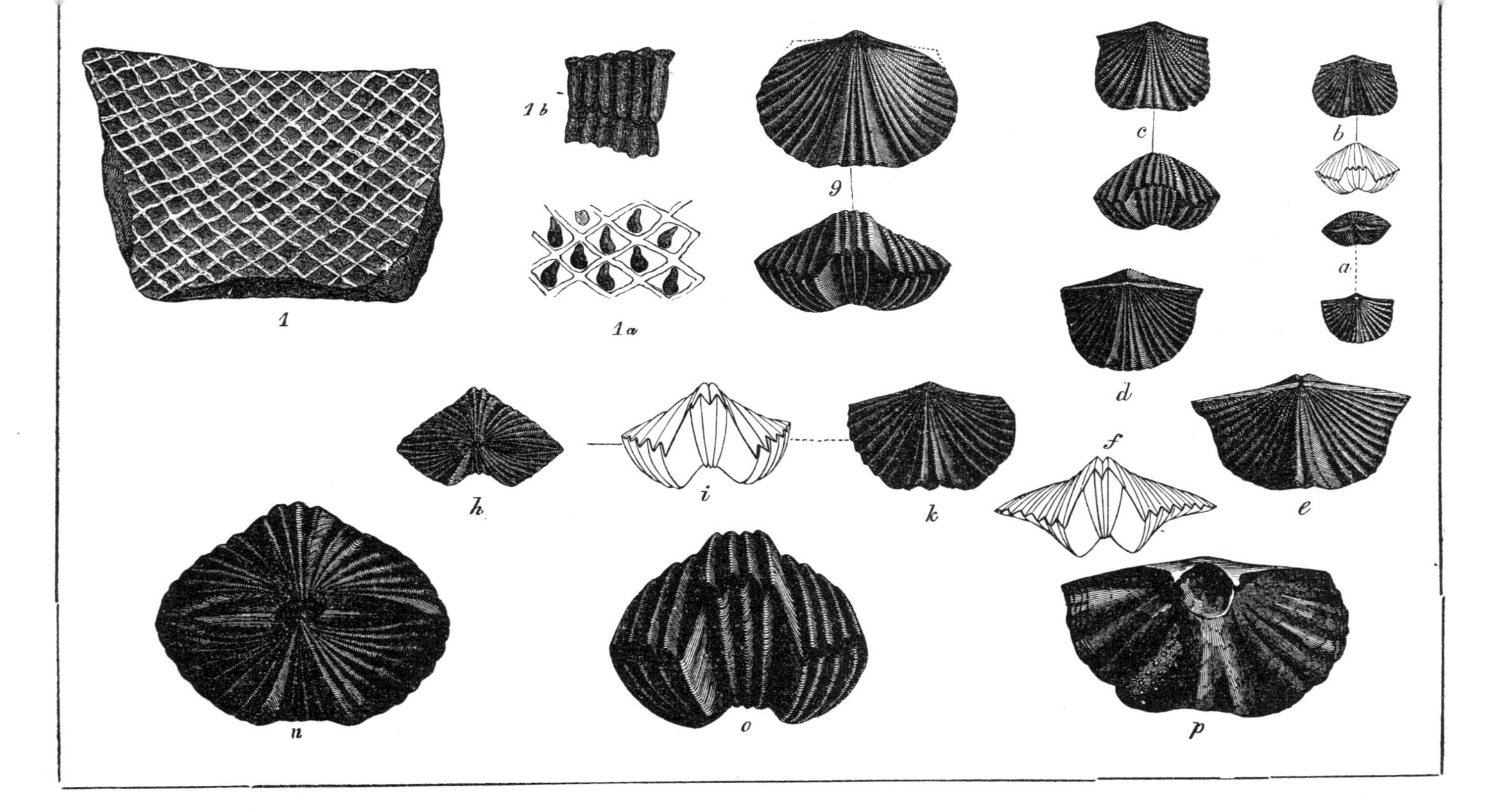
1
1b
1a
9
c
b
a
d
h
i
k
f
e
n
o
p

PLATE XIV.

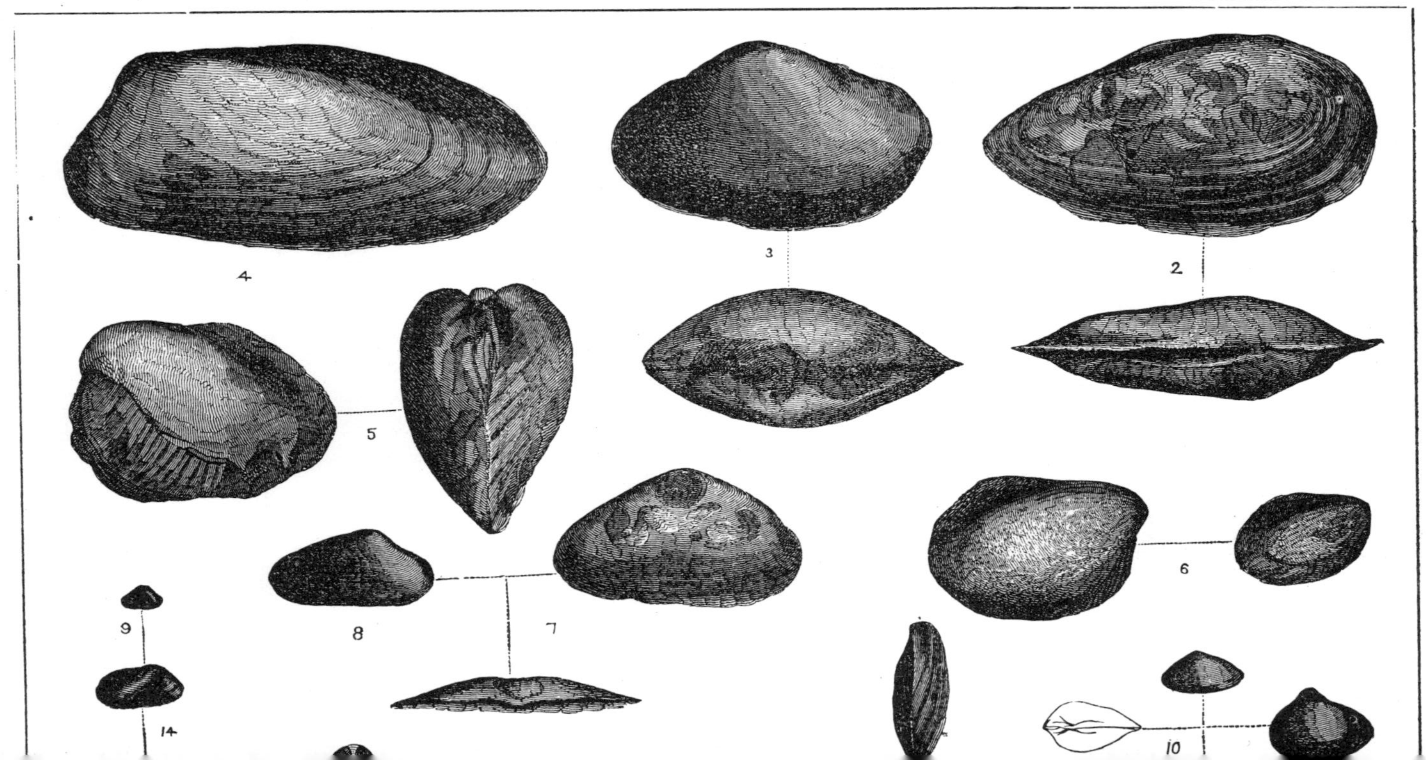

PLATE XV.

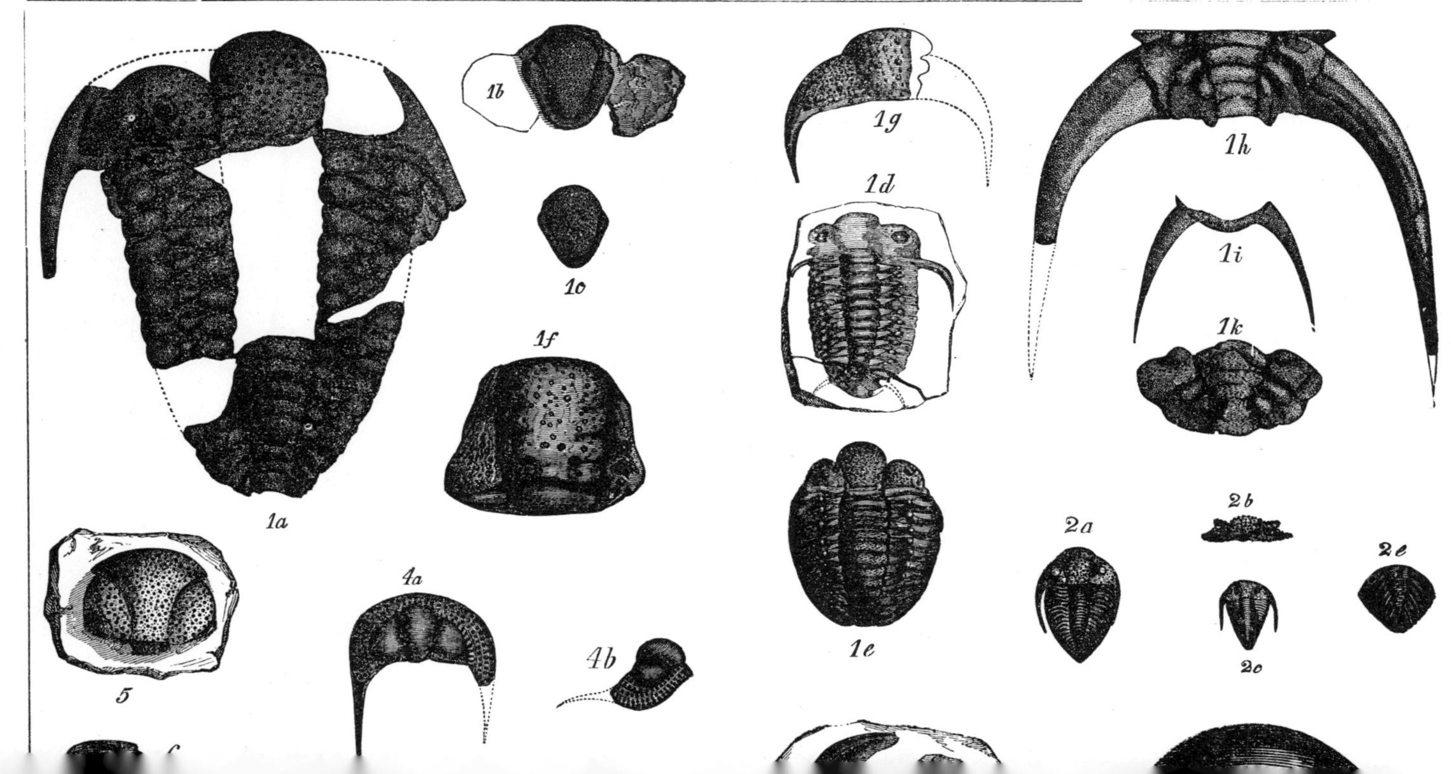

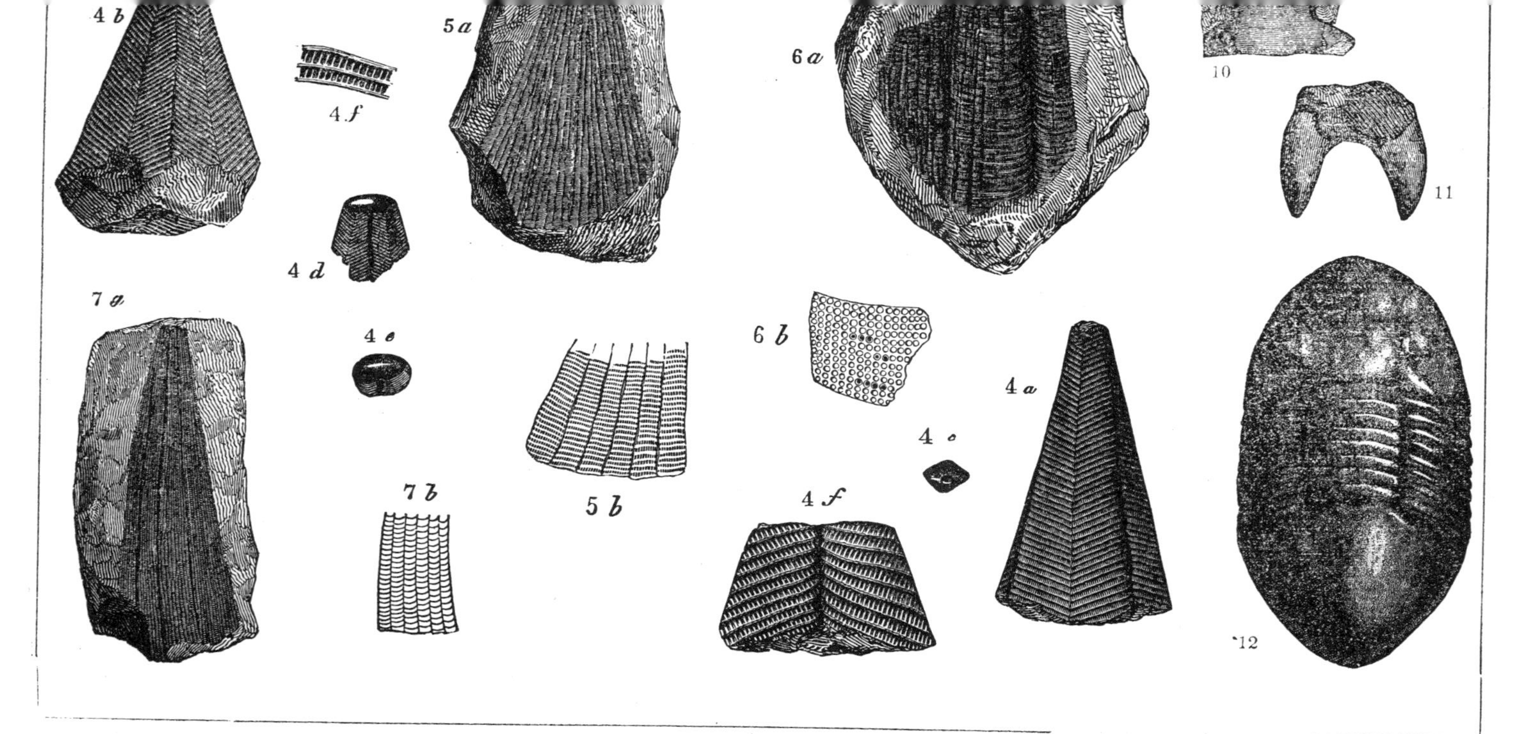

4 b
4 f
5 a
6 a
10
11
4 d
7 a
4 e
6 b
5 b
7 b
4 a
4 e
4 f
12

PLATE XVI.

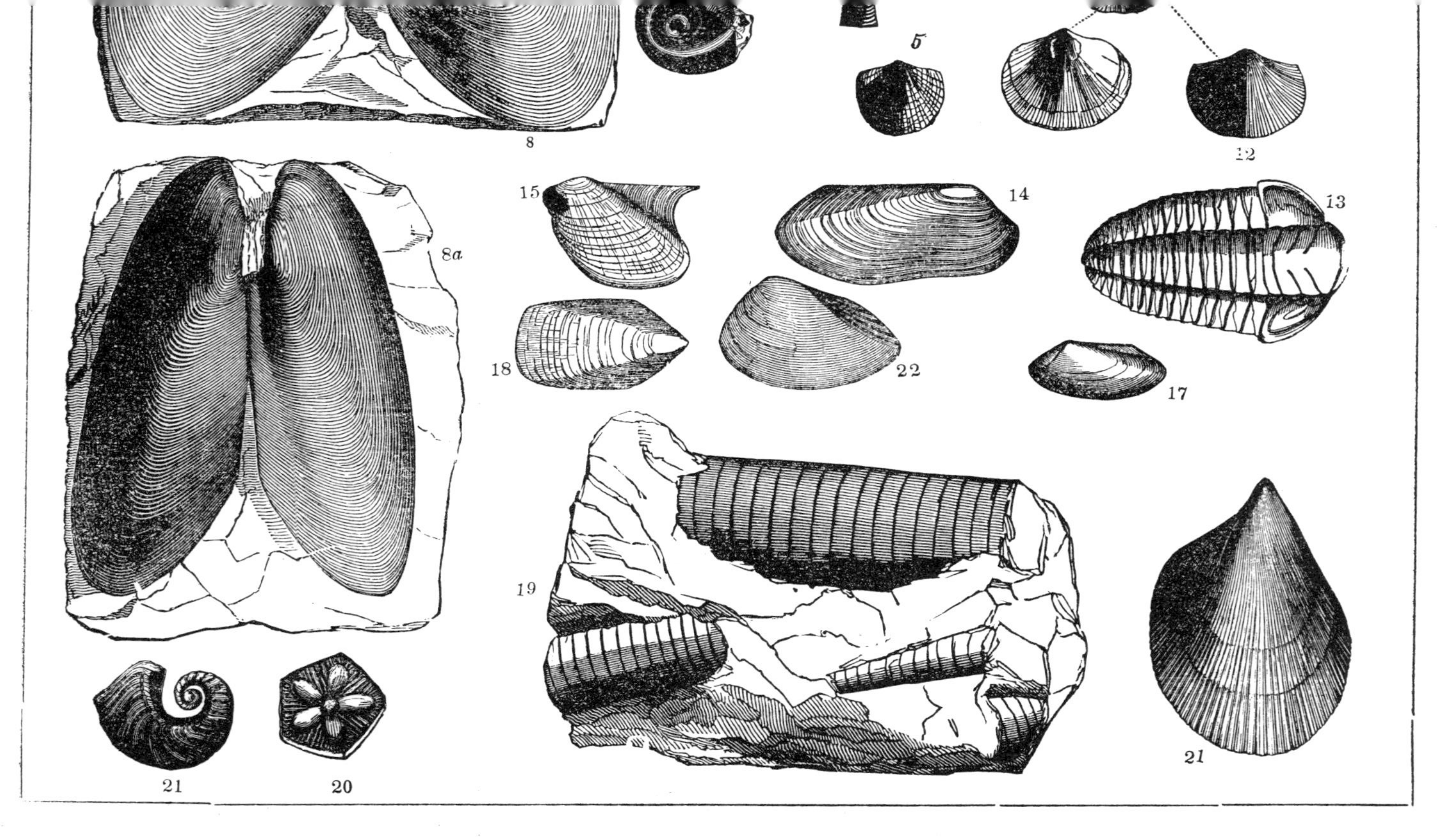

5
12
8
15
14
13
8a
18
22
17
19
21
20
21

PLATE XVII.

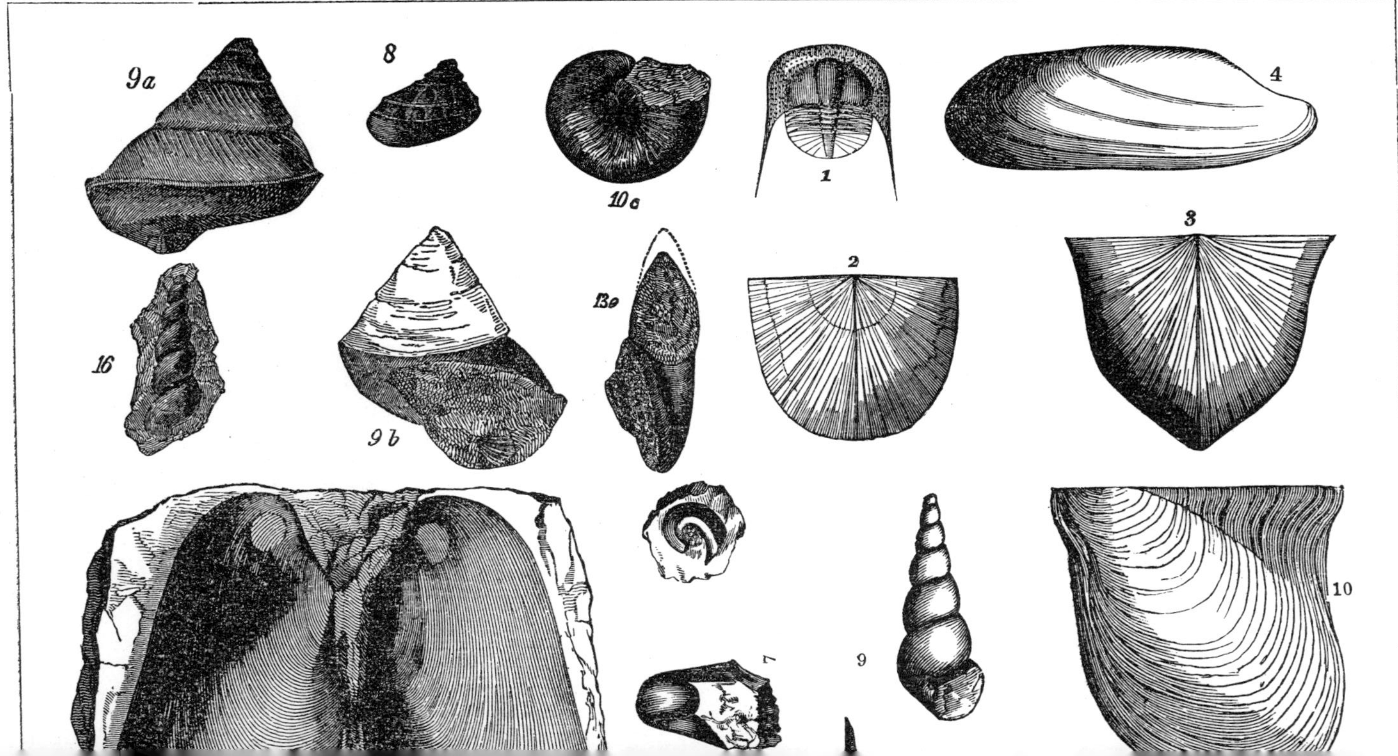

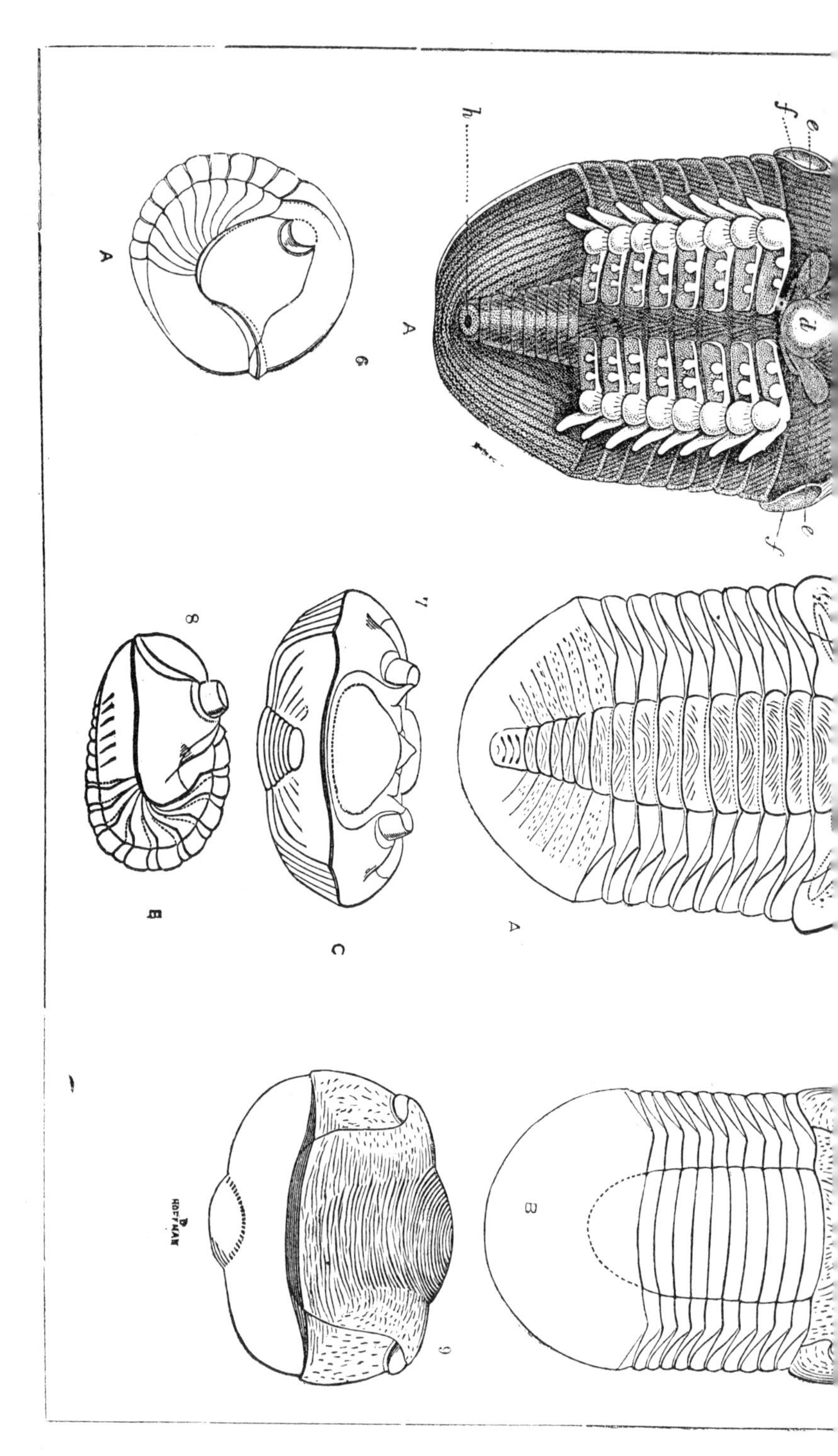

PLATE XVIII.

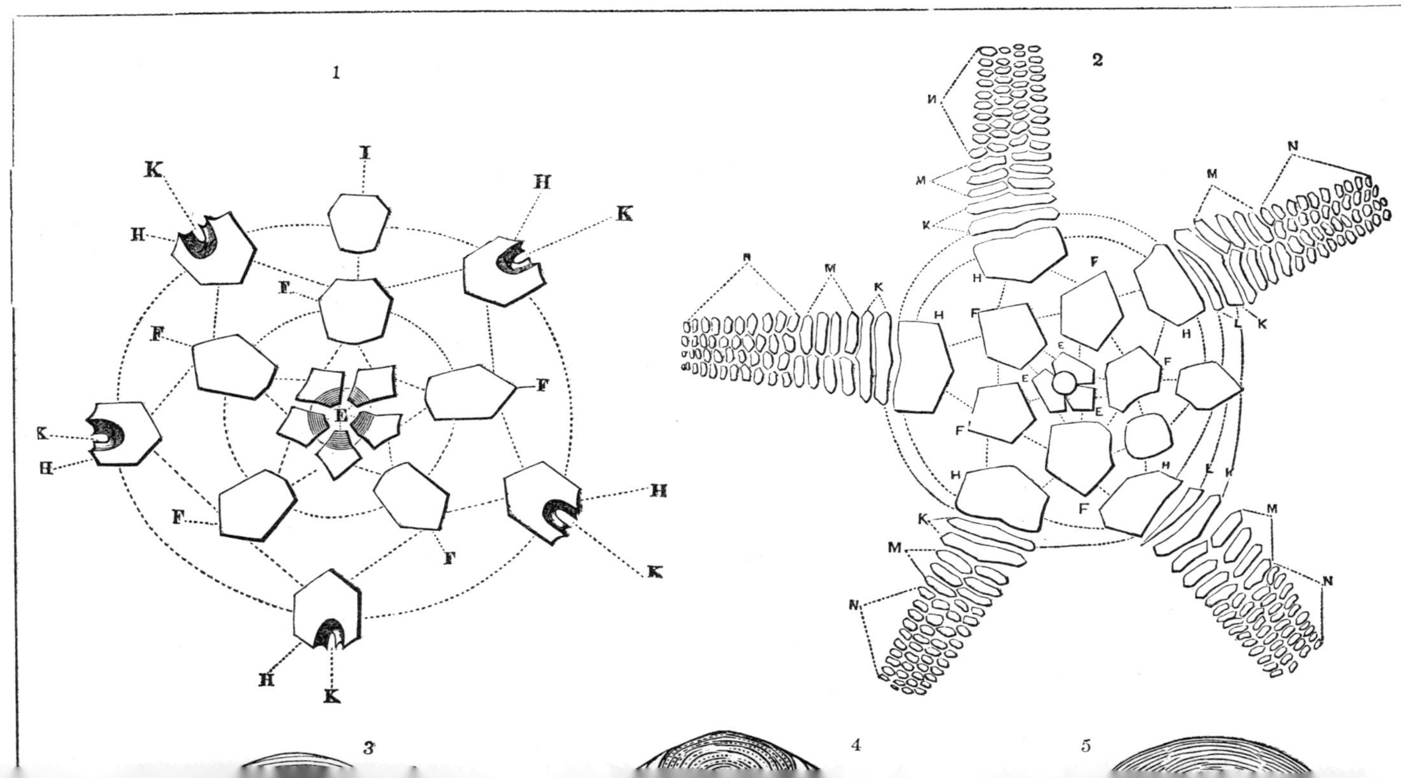

NATURAL SCIENCES IN AMERICA

An Arno Press Collection

Allen, J[oel] A[saph]. **The American Bisons,** Living and Extinct. 1876

Allen, Joel Asaph. **History of the North American Pinnipeds:** A Monograph of the Walruses, Sea-Lions, Sea-Bears and Seals of North America. 1880

American Natural History Studies: The Bairdian Period. 1974

American Ornithological Bibliography. 1974

Anker, Jean. **Bird Books and Bird Art.** 1938

Audubon, John James and John Bachman. **The Quadrupeds of North America.** Three vols. 1854

Baird, Spencer F[ullerton]. **Mammals of North America.** 1859

Baird, S[pencer] F[ullerton], T[homas] M. Brewer and R[obert] Ridgway. **A History of North American Birds:** Land Birds. Three vols., 1874

Baird, Spencer F[ullerton], John Cassin and George N. Lawrence. **The Birds of North America.** 1860. Two vols. in one.

Baird, S[pencer] F[ullerton], T[homas] M. Brewer, and R[obert] Ridgway. **The Water Birds of North America.** 1884. Two vols. in one.

Barton, Benjamin Smith. **Notes on the Animals of North America.** Edited, with an Introduction by Keir B. Sterling. 1792

Bendire, Charles [Emil]. **Life Histories of North American Birds** With Special Reference to Their Breeding Habits and Eggs. 1892/1895. Two vols. in one.

Bonaparte, Charles Lucian [Jules Laurent]. **American Ornithology:** Or The Natural History of Birds Inhabiting the United States, Not Given by Wilson. 1825/1828/1833. Four vols. in one.

Cameron, Jenks. **The Bureau of Biological Survey:** Its History, Activities, and Organization. 1929

Caton, John Dean. **The Antelope and Deer of America:** A Comprehensive Scientific Treatise Upon the Natural History, Including the Characteristics, Habits, Affinities, and Capacity for Domestication of the Antilocapra and Cervidae of North America. 1877

Contributions to American Systematics. 1974

Contributions to the Bibliographical Literature of American Mammals. 1974

Contributions to the History of American Natural History. 1974

Contributions to the History of American Ornithology. 1974

Cooper, J[ames] G[raham]. **Ornithology.** Volume I, Land Birds. 1870

Cope, E[dward] D[rinker]. **The Origin of the Fittest:** Essays on Evolution and **The Primary Factors of Organic Evolution.** 1887/1896. Two vols. in one.

Coues, Elliott. **Birds of the Colorado Valley.** 1878

Coues, Elliott. **Birds of the Northwest.** 1874

Coues, Elliott. **Key To North American Birds.** Two vols. 1903

Early Nineteenth-Century Studies and Surveys. 1974

Emmons, Ebenezer. **American Geology:** Containing a Statement of the Principles of the Science. 1855. Two vols. in one.

Fauna Americana. 1825-1826

Fisher, A[lbert] K[enrick]. **The Hawks and Owls of the United States in Their Relation to Agriculture.** 1893

Godman, John D. **American Natural History:** Part I — Mastology and **Rambles of a Naturalist.** 1826-28/1833. Three vols. in one.

Gregory, William King. **Evolution Emerging:** A Survey of Changing Patterns from Primeval Life to Man. Two vols. 1951

Hay, Oliver Perry. **Bibliography and Catalogue of the Fossil Vertebrata of North America.** 1902

Heilprin, Angelo. **The Geographical and Geological Distribution of Animals.** 1887

Hitchcock, Edward. **A Report on the Sandstone of the Connecticut Valley,** Especially Its Fossil Footmarks. 1858

Hubbs, Carl L., editor. **Zoogeography.** 1958

[Kessel, Edward L., editor]. **A Century of Progress in the Natural Sciences: 1853-1953.** 1955

Leidy, Joseph. **The Extinct Mammalian Fauna of Dakota and Nebraska,** Including an Account of Some Allied Forms from Other Localities, Together with a Synopsis of the Mammalian Remains of North America. 1869

Lyon, Marcus Ward, Jr. **Mammals of Indiana.** 1936

Matthew, W[illiam] D[iller]. **Climate and Evolution.** 1915

Mayr, Ernst, editor. **The Species Problem.** 1957

Mearns, Edgar Alexander. **Mammals of the Mexican Boundary of the United States.** Part I: Families Didelphiidae to Muridae. 1907

Merriam, Clinton Hart. **The Mammals of the Adirondack Region,** Northeastern New York. 1884

Nuttall, Thomas. **A Manual of the Ornithology of the United States and of Canada.** Two vols. 1832-1834

Nuttall Ornithological Club. **Bulletin of the Nuttall Ornithological Club:** A Quarterly Journal of Ornithology. 1876-1883. Eight vols. in three.

[Pennant, Thomas]. **Arctic Zoology. 1784-1787.** Two vols. in one.

Richardson, John. **Fauna Boreali-Americana;** Or the Zoology of the Northern Parts of British America, Containing Descriptions of the Objects of Natural History Collected on the Late Northern Land Expeditions Under Command of Captain Sir John Franklin, R. N. Part I: Quadrupeds. 1829

Richardson, John and William Swainson. **Fauna Boreali-Americana:** Or the Zoology of the Northern Parts of British America, Containing Descriptions of the Objects of Natural History Collected by the Late Northern Land Expeditions Under Command of Captain Sir John Franklin, R. N. Part II: The Birds. 1831

Ridgway, Robert. **Ornithology.** 1877

Selected Works By Eighteenth-Century Naturalists and Travellers. 1974

Selected Works in Nineteenth-Century North American Paleontology. 1974

Selected Works of Clinton Hart Merriam. 1974

Selected Works of Joel Asaph Allen. 1974

Selections From the Literature of American Biogeography. 1974

Seton, Ernest Thompson. **Life-Histories of Northern Animals: An Account of the Mammals of Manitoba.** Two vols. 1909

Sterling, Keir Brooks. **Last of the Naturalists:** The Career of C. Hart Merriam. 1974

Vieillot, L. P. **Histoire Naturelle Des Oiseaux de L'Amerique Septentrionale,** Contenant Un Grand Nombre D'Especes Decrites ou Figurees Pour La Premiere Fois. 1807. Two vols. in one.

Wilson, Scott B., assisted by A. H. Evans. **Aves Hawaiienses:** The Birds of the Sandwich Islands. 1890-99

Wood, Casey A., editor. **An Introduction to the Literature of Vertebrate Zoology.** 1931

Zimmer, John Todd. **Catalogue of the Edward E. Ayer Ornithological Library.** 1926